Weiliang Xu and John E. Bronlund

Mastication Robots

Studies in Computational Intelligence, Volume 290

Editor-in-Chief
Prof. Janusz Kacprzyk
Systems Research Institute
Polish Academy of Sciences
ul. Newelska 6
01-447 Warsaw
Poland
E-mail: kacprzyk@ibspan.waw.pl

Further volumes of this series can be found on our homepage: springer.com

Vol. 269. Francisco Fernández de de Vega and Erick Cantú-Paz (Eds.)
Parallel and Distributed Computational Intelligence, 2009
ISBN 978-3-642-10674-3

Vol. 270. Zong Woo Geem
Recent Advances In Harmony Search Algorithm, 2009
ISBN 978-3-642-04316-1

Vol. 271. Janusz Kacprzyk, Frederick E. Petry, and Adnan Yazici (Eds.)
Uncertainty Approaches for Spatial Data Modeling and Processing, 2009
ISBN 978-3-642-10662-0

Vol. 272. Carlos A. Coello Coello, Clarisse Dhaenens, and Laetitia Jourdan (Eds.)
Advances in Multi-Objective Nature Inspired Computing, 2009
ISBN 978-3-642-11217-1

Vol. 273. Fatos Xhafa, Santi Caballé, Ajith Abraham, Thanasis Daradoumis, and Angel Alejandro Juan Perez (Eds.)
Computational Intelligence for Technology Enhanced Learning, 2010
ISBN 978-3-642-11223-2

Vol. 274. Zbigniew W. Raś and Alicja Wieczorkowska (Eds.)
Advances in Music Information Retrieval, 2010
ISBN 978-3-642-11673-5

Vol. 275. Dilip Kumar Pratihar and Lakhmi C. Jain (Eds.)
Intelligent Autonomous Systems, 2010
ISBN 978-3-642-11675-9

Vol. 276. Jacek Mańdziuk
Knowledge-Free and Learning-Based Methods in Intelligent Game Playing, 2010
ISBN 978-3-642-11677-3

Vol. 277. Filippo Spagnolo and Benedetto Di Paola (Eds.)
European and Chinese Cognitive Styles and their Impact on Teaching Mathematics, 2010
ISBN 978-3-642-11679-7

Vol. 278. Radomir S. Stankovic and Jaakko Astola
From Boolean Logic to Switching Circuits and Automata, 2010
ISBN 978-3-642-11681-0

Vol. 279. Manolis Wallace, Ioannis E. Anagnostopoulos, Phivos Mylonas, and Maria Bielikova (Eds.)
Semantics in Adaptive and Personalized Services, 2010
ISBN 978-3-642-11683-4

Vol. 280. Chang Wen Chen, Zhu Li, and Shiguo Lian (Eds.)
Intelligent Multimedia Communication: Techniques and Applications, 2010
ISBN 978-3-642-11685-8

Vol. 281. Robert Babuska and Frans C.A. Groen (Eds.)
Interactive Collaborative Information Systems, 2010
ISBN 978-3-642-11687-2

Vol. 282. Husrev Taha Sencar, Sergio Velastin, Nikolaos Nikolaidis, and Shiguo Lian (Eds.)
Intelligent Multimedia Analysis for Security Applications, 2010
ISBN 978-3-642-11754-1

Vol. 283. Ngoc Thanh Nguyen, Radoslaw Katarzyniak, and Shyi-Ming Chen (Eds.)
Advances in Intelligent Information and Database Systems, 2010
ISBN 978-3-642-12089-3

Vol. 284. Juan R. González, David Alejandro Pelta, Carlos Cruz, Germán Terrazas, and Natalio Krasnogor (Eds.)
Nature Inspired Cooperative Strategies for Optimization (NICSO 2010), 2010
ISBN 978-3-642-12537-9

Vol. 285. Roberto Cipolla, Sebastiano Battiato, and Giovanni Maria Farinella (Eds.)
Computer Vision, 2010
ISBN 978-3-642-12847-9

Vol. 286. Alexander Bolshoy, Zeev (Vladimir) Volkovich, Valery Kirzhner, and Zeev Barzily
Genome Clustering, 2010
ISBN 978-3-642-12951-3

Vol. 287. Dan Schonfeld, Caifeng Shan, Dacheng Tao, and Liang Wang (Eds.)
Video Search and Mining, 2010
ISBN 978-3-642-12899-8

Vol. 288. I-Hsien Ting, Hui-Ju Wu, Tien-Hwa Ho (Eds.)
Mining and Analyzing Social Networks, 2010
ISBN 978-3-642-13421-0

Vol. 289. Anne Håkansson, Ronald Hartung, and Ngoc Thanh Nguyen (Eds.)
Agent and Multi-agent Technology for Internet and Enterprise Systems, 2010
ISBN "Pending"

Vol. 290. Weiliang Xu and John E. Bronlund
Mastication Robots, 2010
ISBN 978-3-540-93902-3

Weiliang Xu and John E. Bronlund

Mastication Robots

Biological Inspiration to Implementation

Springer

Prof. Weiliang Xu
Massey University
School of Engineering & Advanced Technology
Massey University
Private Bag 102 904
Albany Campus, Auckland
New Zealand
E-mail: w.l.xu@massey.ac.nz

Prof. John E. Bronlund
School of Engineering and Advanced Technology
Massey University
Private Bag 11 222
Turitea Campus, Palmerston North
New Zealand
E-mail: j.e.bronlund@massey.ac.nz

ISBN 978-3-540-93902-3 e-ISBN 978-3-540-93903-0

DOI 10.1007/978-3-540-93903-0

Studies in Computational Intelligence ISSN 1860-949X

Library of Congress Control Number: 2010928433

Typeset & Cover Design: Scientific Publishing Services Pvt. Ltd., Chennai, India.

Printed on acid-free paper

9 8 7 6 5 4 3 2 1

springer.com

Preface

The biological systems we see around us as birds, trees and animals have evolved over millennia. They are complex systems adapted for a multitude of functions including flight, running, reproduction, and digestion. There has been an increasing awareness of these systems by engineers who are now looking to biology as inspiration for design. Mechanical and robotic designs have been developed based on the biomechanics of humans and animals. Similarly computer algorithms, artificial intelligence and control systems have been developed from genetic or neurological principles.

In the study of food science, there has been an increasing trend for engineers and technologists to look further and further down the food product life cycle. In the past the focus has been on food 'from farm-gate to the plate'. In recent times research has looked past the plate to study what happens to the food in the mouth, during swallowing and throughout the digestion process. Food structures are now being designed from an understanding how the food properties interact with the physiological processes occurring in the mouth and the gut. In this way foods aimed at prevention of diet related disorders can be developed.

In this book we look to merge these two perspectives by developing robots to simulate human mastication. The development of these robots can be applied to physical simulation of human chewing and provide insights into the relationships between food structure and perceived texture during mastication. By developing chewing robots, mechanical design, artificial intelligence and control strategies might be developed for use in other completely unrelated applications.

The robotic simulation of the human mastication system offers a chance to study the biomechanics and control of a complex system. The human mandible is compact and yet it can apply large forces. It can adapt to effectively breakdown disparate materials ranging from fibrous structures like bran to hard brittle foods such as nuts or tough sticky materials like toffee. The jaws trajectory can be controlled to achieve rapid movement without biting the surrounding oral structures such as the cheeks or tongue. Biting is abruptly stopped if even a small unexpected hard object (such as a fragment from an olive pit) is detected between the occlusal surfaces of the teeth. The jaw muscles are strong enough to enable Malaysian strongman "King Tooth" to pull a seven coach train over 4 metres with his teeth and yet we control this force well enough to move a marshmallow with our teeth and barely leave a mark. Understanding how nature achieves this amazingly broad functionality can potentially lead to a next generation of mechatronic devices or methodologies for their control.

The use of robotic systems to simulate chewing of foods by humans offers tremendous advantages over existing food texture characterisation methods. A robotic jaw can be fully instrumented to allow real time measurement of the magnitude and direction of the forces applied to the food. The true trajectory used to fracture foods can be studied in order to understand if or how we choose optimal strategies to convert the food into a swallow safe bolus in the mouth. From this understanding it may be possible to design food structures that can evade breakdown during chewing and therefore alter the stomach emptying rate or the rate of nutrient release during digestion. Alternatively foods could be structured in order to provide the flavours and tastes we desire in the food at lower concentrations in the food. After all we can't sense flavours that are released from foods after they are swallowed.

Our work on chewing robots began with an impromptu visit in 2002 to the Dentistry School at Université d'Auvergne, Clermont Ferrand, France. There we visited Prof Alain Woda and Dr Marie- Agnès Peyron who were engaged in characterising how food texture affected the way in which it was chewed by human subjects. From this initial meeting the intriguing idea of mechanically reproducing human trajectories to provide textural information of foods was sparked. From this point we formed the unusual combination of food engineering and robotics researchers working together in collaboration to design chewing robots.

To develop biologically inspired masticatory robots we have needed to carry out research in a broad multidisciplinary team. Over the years we have collaborated with food scientists, physiologists, dentists, biomechanics engineers, mechatronics engineers and particle technologists. Without this combination of different perspectives, development of useful systems would be much more difficult.

Much of the hands on work has been done by postgraduates at Massey University. Short term projects have been completed by Ben Daumas, Stefan Seitz, Mark Greenbrook, Aimery Charbonneau, Gabriel Feraudet, Luc Getto, Sébastien Martin, Thomas Surrel, Thibaud Roquier, Adrien Lauranceau-Moineau, Pierre-Henri Bouhet, Pierre Cambonie, Cyril Nicolotto, Vincent Parinet, and Lars Kuhnert. Masters and PhD students Sebastian Pap, Darren Lewis, Charles Chen, Otmar Nitsche, Jonathan Torrance, Richard Sun and Gauss Li have all worked on varying aspects of robot development and control. PhD students Christine Flynn, Scott Hutchings, Hongyan Yao and Jean Ne Cheong have also contributed through development of the knowledge of human chewing behaviour and its relationship to texture. Many 4th year undergraduates at Albany campus were also involved in simulating our robots for their kinematics and dynamics.

We would like to acknowledge the significant inputs of our collaborators at Massey University. We specifically thank Dr Loulin Huang and Dr Johan Potgieter for joint supervision of robotics focused postgraduate students and Prof Jim Jones, Prof Roger Lentle and Dr John Grigor for joint supervision of human food mastication focussed research projects. Dr Dong (Walter) Xie also contributed to this work with respect to developing knowledge systems for food mastication.

Special thanks go to Dr Kylie Foster who has had a very active role in human chewing behaviour based research.

We have also had guidance and collaboration from the wider research community. The advice and encouragement from the members of the Biomouth research group (http://www.biomouth.org) has been appreciated, particularly the valuable contributions on occlusion and dental materials from Prof Jules Kieser and his associates at the School of Dentistry, University of Otago, New Zealand. Robotic design has been assisted by the insights on jaw biomechanics provided by Prof Andrew Pullan from the Bioengineering Institute, University of Auckland, New Zealand, and Dr Oliver Röhrle from the Institute of Applied Mechanics, University of Stuttgart, Germany.

We have enjoyed a significant and ongoing collaboration with Marco Morgenstern and his colleagues from The New Zealand Institute of Plant and Food Research. This research has been very productive in terms of both robotic design and in understanding food texture. A similar collaboration has been established with the members of the Riddet Institute based at Massey University, New Zealand, linking understanding of foods in the mouth to the understanding of foods in the gut.

Through these collaborations we have benefited from good research funding. In particular we would like to acknowledge the funding received from Massey University and the New Zealand Foundation for Research, Science and Technology through our collaborations with Plant and Food Research, The Bioengineering Institute, and the Riddet Institute.

We also appreciate the willingness of researchers from around the world to engage in discussions and the sharing of ideas. The interactions with Dr Marie-Agnès Peyron, Prof Alain Woda, Dr Martin Wickham, Prof Jose Aguilera, Assoc/Prof Anuchita Moongngarm, Prof Paul Singh, Prof Osvaldo Campanella and Dr Yacine Hemar have been especially useful.

Much of the content of this book has been produced through the collaborations listed above. Specific contributions are acknowledged in each chapter. Thanks to Julie Rayner for help in editing the manuscript.

In Chapter 1, mastication in humans is reviewed together with details of mastication robots reported in the literature. Chapter 2 outlines robotic models of the mastication system as a basis for robot design. Chapters 3 and 4 outline alternative designs for robots of various parallel configurations, demonstrating the sequence of steps made in the design process. A simpler chewing robot based on a 4-bar linkage mechanism is presented in Chapter 5. This device was developed for use by food engineers interested in understanding the dynamics of food texture.

The later chapters in the book are aimed at application of the chewing robots to provide understanding of real food systems. In Chapter 6, the measurement of human chewing trajectories is discussed together with how they can be adapted to make them suitable for implementation on the robotic systems. Chapter 7 details experimental work carried out to validate the force measurements made using the 6RSS chewing robot for a number of simplified and real jaw trajectories on a variety of food products. In Chapter 8 attention is turned to how the results generated from the robotic devices discussed in this book and in the literature can be interpreted to understand food texture. The key functionality required for this purpose is reviewed and the current capability of each device is evaluated.

Chapter 9 outlines novel approaches to high level control of mastication robots using neural control systems inspired from the way the human chewing process is controlled using the central nervous system. Finally Chapter 10 outlines the application of knowledge frameworks as a tool to consolidate the data on food mastication collected from experiments on humans and robots. Such systems are required in order to make sense, through methods such as data mining, of the large repository of information available.

This book summarises a large body of research on masticatory robots and its content demonstrates how complex robotic design can be carried out to provide useful tools for the food industry. We expect that the human chewing inspired designs and control strategies developed here can be implemented as novel solutions to other industrial applications. While further development of masticatory robots is still required, accurate reproduction of human chewing is not far away.

Auckland, New Zealand, 2010

Weiliang Xu
John Bronlund

Contents

Chapter 1
Introduction*

Abstract. A masticatory robot refers to a robot that can perform at least some defined human masticatory function. This chapter briefly reviews the masticatory system, masticatory measurements and computational models of mastication that are relevant to masticatory robotics. Also critically reviewed, is state-of-the-art in robotics research with respect to the engineering of the human jaw system. The masticatory system has two rigid components: a fixed maxilla (upper jaw) and a mobile mandibular (lower jaw). These are joined by two temporomandibular joints (TMJ). Unique features of the TMJ are described. The muscles of mastication are explained with regards to their role in the rhythmic opening and closing of the mandible in three-dimensional (3D) space. Because the breakdown of foods is performed directly by the teeth; the functionalities of the incisors, premolars and rear molars are presented. Two computational models of the masticatory system are presented in which Hill-type muscle models are used. We also describe masticatory robots developed for dental training, jaw simulation, food texture and breakdown analysis, speech therapy with regard to muscle modelling, TMJ models, masticatory biomechanics and controls of actuation. Finally, we discuss the major accomplishments and challenges in masticatory modelling and robotics; and we compare a number of such robots in light of relevant biomechanical aspects of the mastication system. The application of artificial intelligence in robotic mastication is also described briefly.

1.1 Why Mastication Robots

Mastication is a complex process whereby food taken into the mouth is processed together with saliva into a form suitable for swallowing (bolus). The bolus is brought to approximate body temperature before transfer to the stomach, where digestion, absorption and utilisation begin [1]. The masticatory sequence can be broken into three phases Firstly **ingestion**, or the transfer of food to the teeth by the tongue, secondly the **main sequence**, during which food is formed into a bolus through rhythmic chewing, and thirdly **clearance and swallowing** [2].

Traditionally, masticatory robots are concerned with the main sequence, features of which can be measured quantitatively.

---

* Reprinted with modification from Xu WL, Bronlund JE, Potgieter J, Foster KD, Röhrle O, Pullan A and Kieser J (2008) Review of the human masticatory system and masticatory robotics. Mech. Machine Theory **43:**1353-1375, with permission from Elsevier.

W. Xu and J.E. Bronlund: Mastication Robots, SCI 290, pp. 1–33.
springerlink.com

Masticatory performance is associated with the quantitative movement parameters of duration (rhythm), velocity and displacement of the mandible together with bite force, in relation to the chewing cycle. Trajectories of reference points such as the incisor, working condyle and/or balancing condyle in the frontal, sagittal and/or horizontal planes, have been used for the analysis of masticatory performance [3-5]. Improvement of such performance is associated with a reduced duration or an increased rhythm of the chewing cycle, and an increased mandibular velocity [3]. At the occlusal phase in the horizontal plane, the incisal movement exhibits distinct closure paths for foods of different textures, even though the overall paths and chewing patterns are different across individual subjects [4].

A number of studies have examined both the range of the occluding phase at the lower incisal point during chewing movements, and how chewing cycle excursions and velocities change with varying food hardness [6-10]. Hardness has been shown to directly affect the number of chewing cycles, time (sequence and cycle duration) and the muscular activity required to produce a bolus ready for swallowing [9,10]. It has been found that the opening amplitude during chewing is only slightly affected by food hardness [9, 10] and is more affected by the rheological properties of the food [10]. Although the jaw velocities have been shown to be influenced by food hardness, they are more affected by the rheological behaviour of the food [10]. The maximum bite force can be obtained by electromyography (EMG) which measures both muscular activity and jaw movement. It has been found to be reproducible within a single subject when measured on different occasions, although maximum bite force is highly variable between subjects [11].

To analytically characterise masticatory efficiency, the measurements must be made continuously over the mastication sequence. These measurements may include the frequency, length of chewing, tracing of jaw movement, force distribution, application of compression and shear forces on the food plus particle size and structure of the bolus just prior to swallowing. As expected, these vary between subjects (due to differences in jaw geometry, teeth shape and sensitivity to pain) together with food texture (rheological behaviour, hardness and adhesion especially to dentures etc.). Because of the complex nature of the chewing process, there is a real need for the development of quantitative methods for evaluating the capability of a person to effectively chew foods.

The biomechanical behaviour of the masticatory system has been modeled mathematically by a number of researchers to test hypotheses concerning masticatory functions and in order to analyse the effects of surgical or orthodontic interventions [12]. Possibly the first attempt at 3D modeling of the human masticatory system for static biting forces, was that of Osborn et al. [13,14] who theorised that the masticatory system was mechanically redundant. Hence, different muscle activation patterns could be applied to produce a given bite force.

More recently, Koolstra et al. considered physiological constraints in modeling and patterns of muscle activation [15, 16]. These authors then developed a 3D dynamic model of jaw motion by applying Newton's laws to the masticatory system where a geometrically simplified joint model was used for the TMJ and a linear displacement model applied for all masticatory muscles. However, the simplified muscle recruitment and material properties severely limited the model. Furthermore the reliability of the model was computationally demanding, often reaching the limits of high-performance desktop PCs. A six-degree-of-freedom model for jaw dynamics was later developed. This could simulate the changes in lengths and contraction velocity of the sarcomeres of the human jaw-opening and jaw-closing muscles, as well as the consequences of force production during jaw opening and closing movements [17-19].

In the jaw dynamics models [20, 21], the muscles were modeled as Hill-type flexible, single-line actuators and the jaw motion was simulated using ADAMS, a software package for multi-body mechanical systems. Recently, a full 3D model of the muscles of mastication has been built [22]. The geometry used in this model was based on anatomical data with biomechanical behaviour derived from key physiological parameters of active and passive muscle properties.

Since the early 1990s, there have been a number of attempts at developing masticatory robots for the purpose of providing dental patient training [23, 24], jaw simulation [25], food texture assessment [26, 27] and speech therapy [28, 29]. The heuristic value of this approach lies in the fact that it is able to perform actual mastication thereby enabling one to understand different scenarios, explore different ideas, develop novel hypotheses, and to gain insight into the consequences of variations in masticatory function between and within individuals. To further advance this new and significant field of robotics, it is essential to have a thorough understanding of the human masticatory system and to critically review the state-of the-art in masticatory robotics.

The rest of this chapter is organized as follows. The masticatory system, from a biomechanics viewpoint, is briefly described in Section 1.2. The quantitative measurements of the mastication process are presented in Section 1.3. Section 1.4 reviews the computational models for masticatory kinematics and dynamics. In Section 1.5, a comprehensive review of mastication robots is made, followed by a short review of artificial intelligence being applied to robotic mastication in Section 1.6. Section 1.7 addresses the major accomplishments and challenges in masticatory robotics and Section 1.8 concludes the chapter.

1.2 The Masticatory System

1.2.1 Jaw and Temporomandibular Joint

The masticatory apparatus consists of an upper and lower jaw, as shown in Fig. 1.1. The upper jaw is called the maxilla and is attached to other bones that make up the skull. The lower jaw is referred to as the mandible and is attached to the skull by muscle. The mandible is pivoted at the condyle via the temporomandibular joint

(TMJ) at each side of the jaw. The mandible moves in relation to the skull under control of the central nervous system (CNS). Both the maxilla and the mandible have teeth attached for use in the cutting and grinding of foods [30].

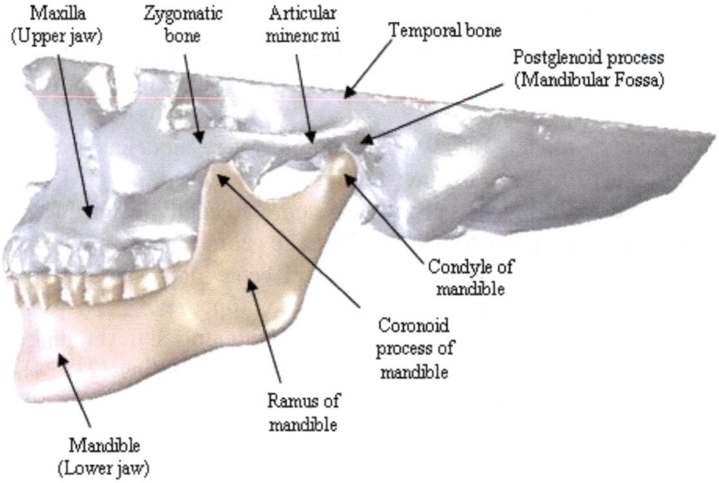

Fig. 1.1 Bones of skull and mandible in lateral view

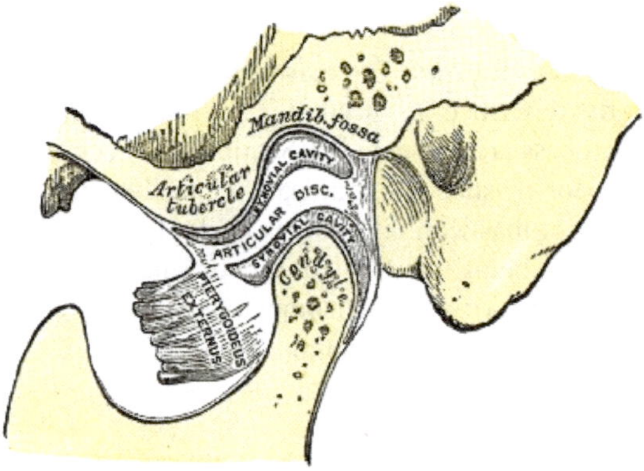

Fig. 1.2 Sagittal section of the temporomandibular joint (©1918 Lea & Febiger, Reprinted from [30], with permission)

The TMJ is literally the joint between the temporal bone of the skull and the condyle of the mandible. As illustrated in Fig. 1.2, an articular disc (soft tissue) separates the condyle and the temporal bone and enables the jaw to move along the mandibular fossa. The articular disc absorbs the shocks to the TMJ from chewing and other movements. In contrast to other joints of the human body, for example the hip or knee joint which rotate around a more or less fixed joint axis, the TMJ is guided by two articular surfaces linked across the midline by a rigid mandible. As a consequence, the TMJ movement is not restricted to a fixed trajectory; rather its movement occurs within an envelope of motion in a three-dimensional space [31, 32]. Fig. 1.3 shows the trajectories of the TMJ while chewing hard foods [33].

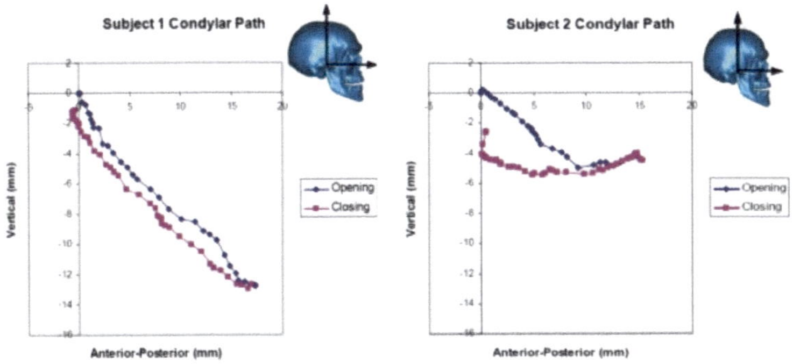

Fig. 1.3 TMJ movement of two healthy subjects (© 2005 Lehman-Grimes SK, Reprinted from [33], with permission)

1.2.2 Muscles of Mastication

The masticatory system is driven by a complex assembly of contracting muscle groups working as an ensemble. This includes muscle groups on both sides of the midline [19]. The main muscles of mastication include the masseter, temporalis, medial pterygoid, lateral pterygoid and digastric muscles[20], referring to Fig. 2.1. The digastric muscle is attached to the chin and the bottom of the skull. The large temporalis muscles are attached to the side of the skull and the top of the lower jaw behind the teeth and consist of vertical and horizontal muscle fibres. The masseter muscles are attached to the cheek on the skull and to the lower-rear section of the lower jaw. The medial pterygoid muscles are attached to the inside of the skull and the lower jaw. The lateral pterygoid muscles are attached to the skull and the lower jaw in a horizontal fashion.

Using these muscle groups in different ways allows the lower jaw to move in six degrees of freedom [34]. From a classical anatomical point of view, the muscles of mastication can be divided into mouth-opening and mouth-closing muscle groups. Mouth-opening muscle groups, such as the geniohyoid, the mylohyoid, and the digastric, are muscle groups with relatively small physiological cross-sectional areas (PCSA). Supported by the force of gravity, they are able to abduct (open) enabling the mandible to move away from the maxilla at a high velocity under application of small forces. Mouth-closing muscle groups, such as the temporalis, the masseter and the medial pterygoid have large PCSA, in comparison to the mouth-opening muscle groups, that allow movement of the mandibular teeth against those of the maxilla with high forces. The lateral pterygoid muscle group completes the muscular system, and governs side to side movements of the jaw.

1.2.3 Teeth

The maxilla and the mandible have teeth, the primary function of which is to perform mastication [35]. The front teeth are referred to as incisors and are used to take an initial bite of a piece of food. The food is then broken up further by the pre-molars which are located between the front incisors and the posterior molars. When the food particles are sufficiently broken up by each type of tooth, the tongue and cheeks then move the food across the tooth row for further breakdown.

Different classes of teeth are characterised by different shapes. Those at the front of the jaw have sharp edge-like blades, while teeth at the back have peaks called cusps. The cusps on the post canine teeth are dull, as sharp cusps are more likely to plastically deform food particles rather than fracturing them [35]. As shown in Fig. 1.4, the cusps can thus be thought of as spherical, resulting in an indentation when pressed into a food particle [36]. The fracture force resulting from this indentation is proportional to the radius of the spherical cusp [37].

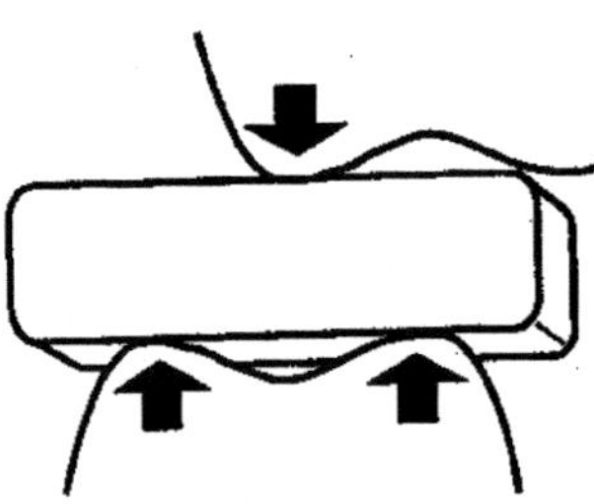

Fig. 1.4 Three-point bending principle adapted to mastication for fracturing food (© 1996 Elsevier, Reprinted from [36] with permission)

The dental layout of the human jaw can be seen in Fig. 1.5. The teeth that are located on the mandible and the maxilla make a curve in both the frontal and sagittal planes. While the frontal curve is referred to as the 'Curve of Wilson', the curve in the sagittal plane is called the 'Curve of Spee' [39].

The curve of Spee is thought to protect the pre-molars and molars by allowing the incisors and canines to contact before the molars and pre-molars. By doing so, the strong forces applied by the muscles of mastication are removed from the molars and pre-molars [41]. This is required to stop excessive wear on the molars and pre-molars thereby enabling the ability to chew food to be retained over an individual's entire lifetime. An ideal curve of Spee would extend though the condyle with a radius of approximately four inches [39].

The curve of Wilson allows the teeth to move without any interference at occlusion [41]. This can be thought of as ball and socket joint in which the ball is able to move freely inside the socket without excessive interference. The teeth are therefore able to move with minimal interference thereby reducing the chance of damage.

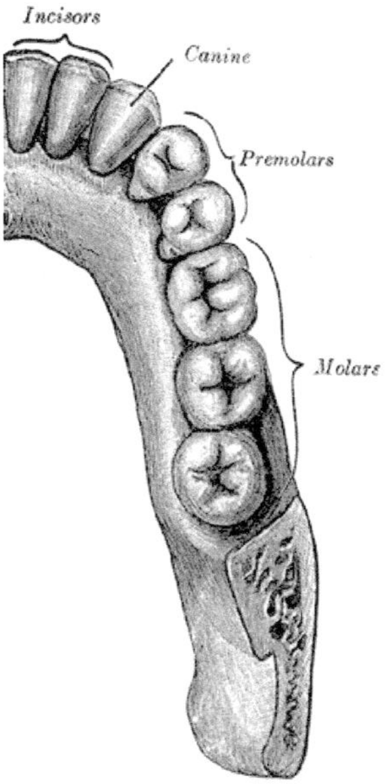

Fig. 1.5 Teeth in the human mouth (©1918 Lea & Febiger, Reprinted from [30] with permission)

1.3 Measurement of Mastication Movement

Mastication movement is characterised by a rhythmic jaw opening and closing movement involving mandibular displacement, the velocity of opening and closing plus masticatory frequency. Muscular activity causing mastication movement, can be indirectly measured using EMG (electromyography) and is thought to be related to those forces required to break down a given food particle. Obviously, such forces vary between subjects, due to differences in jaw geometry, teeth shape, muscles of mastication, sensitivity to pain, and oral status. They also vary with changes in food texture, e.g. elasticity, plasticity, hardness together with the degree of adhesiveness of the food.

1.3.1 Jaw Movement

1.3.1.1 Posselt Envelope

Posselt [42] described an envelope of the maximum mandibular movement. This may be represented in a three-dimensional figure which traces the path of the lower incisor teeth during guided jaw movements. Fig. 1.6 illustrates the Posselt envelope in the sagittal, frontal, and horizontal planes.

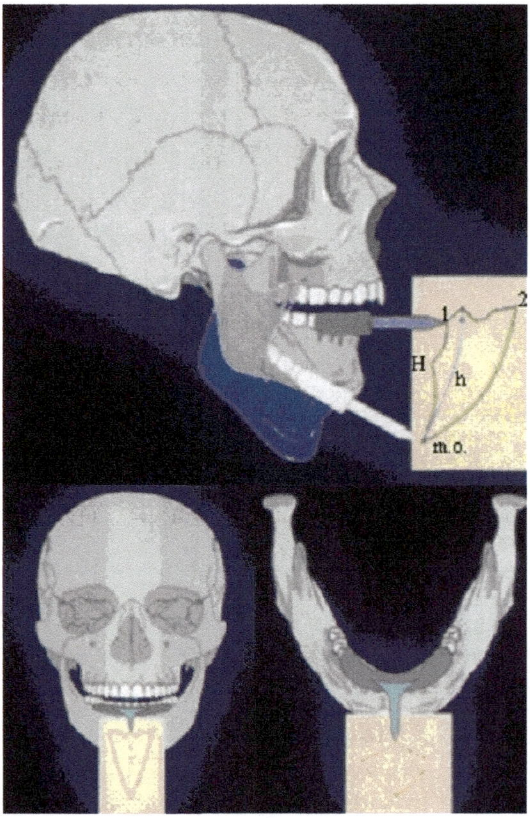

Fig. 1.6 Posselt envelope in the sagittal plane, the frontal plane, and the horizontal plane (© 1997 Quintessence, Reprinted from [43] with permission)

The envelope is characterised by three extreme movement paths; an extreme posterior opening, an extreme anterior closing and a maximum upper glide (Fig. 1.6). Extreme posterior opening (indicated by line 1-m.o.) is divided into two parts. The first part describes the so-called hinge movement (1-H), which is a rotation of about 10 degrees around the hinge axis. Further jaw-opening requires the hinge movement to be overlaid with a forward movement, which is indicated by the second part of the curved line (H-m.o.). Anterior extreme opening (illustrated by line 2-m.o.) is also an overlaid hinge movement. The forward and backward glide between points 1 and 2 is irregular due to incisal and other tooth guidance. It is worth noting that a specific masticatory movement will always fall within the maximum movement space defined by the Posselt envelope.

1.3.1.2 A Cycle of Chewing

The movement of the incisor and molar teeth whilst chewing can be measured in three-dimensional coordinates for various purposes. These include the determination of pathology related to the mandibular movement, the assessment of chewing efficiency and the characterization of food texture [8-10, 44-46].

A normal chewing cycle may be divided into three phases: opening, closing, and occlusion (Fig. 1.7). The speed of the mandible in the opening phase is initially slow and increases as the mouth opens. When the mouth starts to close, the mandible moves laterally outward. It closes quickly towards the teeth and then slows for occlusion. Opening velocities are faster than closing velocities during the chewing of food [10]. The occlusal phase ends 0.5 mm from the maximum intercuspation position (MI) [44].

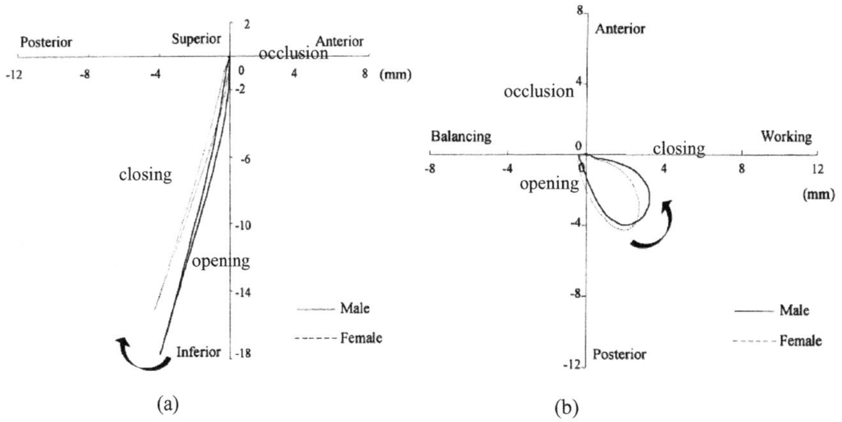

Fig. 1.7 A chewing cycle of the incisor point in (a) the sagittal plane and (b) the frontal plane (© 2000 ELSEVIER, Reprinted from [46] with permission)

1.3.1.3 A Sequence of Chewing Cycles

Peyron and co-workers described a complete sequence of chewing cycles for a model food as recorded using a Carstens 2D Articulograph [9]. Two-dimensional coordinates of the lateral and supero-inferior movements of the incisor point were determined, as plotted in Fig. 1.8. It can be seen that the subject chewed on the left side of the jaw except for three peaks of the lateral excursion movement which were performed on the right side. This can be traced back to the action of the tongue, which interacts with the chewing cycle by collecting or accurately placing food in position. Furthermore, it can be seen in the supero-inferior movement (z-direction in Figure 1.8) that the chewing amplitudes vary between 30 mm and 20 mm for the beginning and the end of the chewing cycle, respectively.

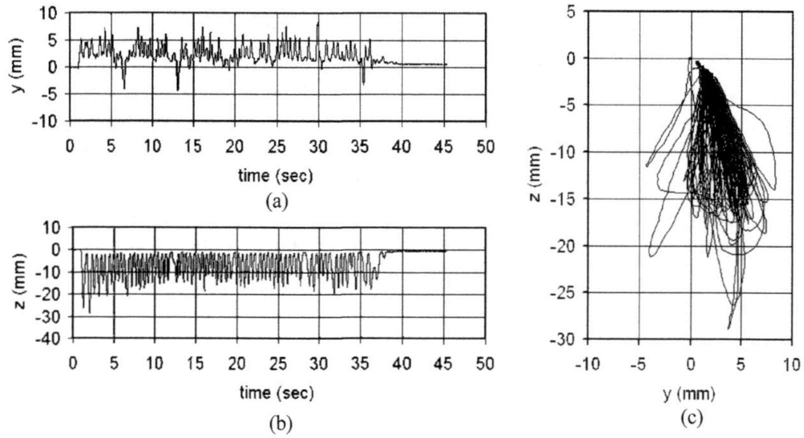

Fig. 1.8 Recorded chewing trajectory of the incisor point plotted as (a) lateral movement, (b) supero-inferior movement, and in (c) frontal plane (© 2002 Springer, Reprinted from [9] with permission)

Using a custom made brace and a motion capturing system known as VICON MX, the three-dimensional coordinates of the incisor point and the three angular movements of the mandible about its x, y and z axes were measured [45].

The trajectories used to chew different food particles differ depending on both the shape and the texture of the food particles. This is due to the fact that the teeth operate differently depending on the chewing trajectory. If a vertical chewing motion is used, the cusps of the teeth are used to fracture the food particles, whereas if a more lateral chewing motion is used, the sharp edges of the teeth function as blades to cut up food particles [34]. The chewing trajectory used is therefore adjusted to ensure that the teeth are utilised correctly to chew the desired food particles.

1.3.2 Chewing Forces

The forces applied on the teeth vary with the type of food that is being chewed. The force applied to a single tooth is also different from the total force between all the contacting teeth during chewing. On foods such as biscuits, carrots and cooked meats, forces range between 70 and 150N on any single tooth [48]. Forces on all the contacting teeth whilst chewing these foods however, range between

190 and 260N [49]. This is much smaller than the maximum bite force that can be applied to the molars, with measurements ranging from 500 to 700N [50]. It is thought that these high forces are important in the process of thegosis or grinding teeth to sharpen the cusps [51].

Helkimo and Ingervall [52] measured the biting and chewing forces of a representative group of people. The results showed that the average bite force on the incisors was 40% of the force on the molars, whereas the chewing force on the incisors was about 47% of that on the molars. Furthermore, the authors determined that the average chewing force was 52% of the average bite force on the molars and 60% on the incisors. These measurements have subsequently been confirmed by various researchers. For example, the maximum bite forces between the upper and the lower molars are usually in the range of 500 - 700N, whereas the maximum mouth opening forces generally do not exceed 150N [50, 53].

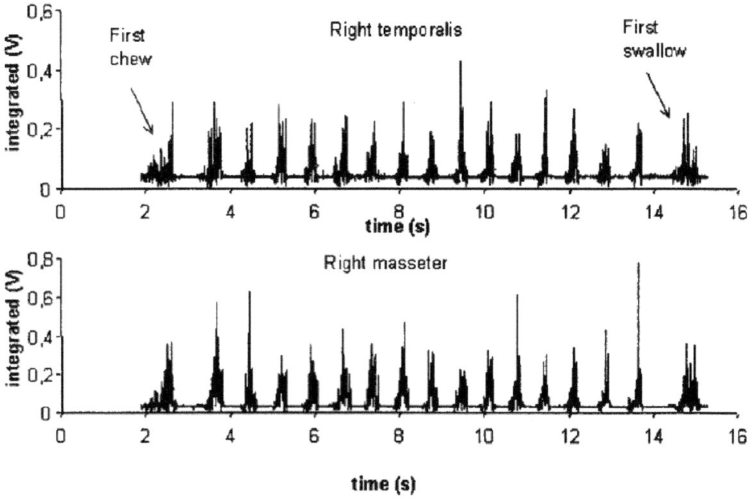

Fig. 1.9 A sample EMG recording in relation to the mastication forces [58] (© 2001 PAGE PUBLICATIONS, Reprinted from [58] with permission)

EMG is an indirect measure of the electrical activity of main masticatory muscles recorded non-invasively with surface electrodes or invasively with needle/wire electrodes. Non-invasively obtained measurements, known as surface EMGs, are becoming more widely used and are easier to obtain than invasive EMGs [54].

Such experimentally obtained data is often used in conjunction with the physiological cross-sectional area (PCSA) of the muscle in order to calculate estimates of instantaneous muscle forces [15, 16, 55-57] or to differentiate different food texture characteristics [58]. Fig. 1.9 shows, for example, a processed EMG recording of the right masseter and temporalis when a subject was chewing an almond. The

EMG is a function of time and involves a series of bursts with the measured voltage being related to the force developed by the muscle. Each burst corresponds to the jaw-closing portion for each chew.

1.4 Computational Models of the Masticatory System

Software packages that provide sufficient functionality to develop models for biomechanical applications are becoming more and more accessible. Computational models of the masticatory system can be categorised into static, dynamic and kinematic. Static models investigate the deformation of the mandible under varying loads [59-61] or muscle forces [62], together with the significance of functional aspects [64-66]. Dynamic models typically use muscle forces and directions as inputs to study various aspects of the human masticatory system [15, 18, 19, 66]. Kinematical models are more focused on the mandibular movement itself [46, 67, 68]. The dynamic and kinematic models are the two most commonly used approaches and are described in more detail in Sections 1.4.1 and 1.4.2 below.

Most models, in particular the dynamic models, rely on calculated or estimated muscle or joint forces as these cannot be measured directly. To estimate muscle forces, all studies but one (which used a full 3D description of the muscles [22]) represent the muscles of mastication, or a particular subgroup of them, as one-dimensional strings. Most studies calculate the muscle forces by linking them proportionally to the PCSA [69], combining the PCSA with EMG [55], or using single-line Hill-type actuators [19]. The lines of action are thereby obtained either by examining cadavers [70] or by determining the centreline of 3D reconstructions based on MRI [71, 72]. The muscle forces are then used, for example, to calculate the joint reaction forces [57], to predict tensions, deformations [73-75], and volumetric strain of the temporomandibular joint cartilage [76] (or to assess the loading conditions of a temporomandibular joint prosthesis [77]).

1.4.1 Dynamics Models

One example of a dynamic six-degree-of-freedom model for jaw dynamics was developed by Koolstra et al. [17-19]. Several research groups adopted this model, or parts of it, as the basis of their studies [62, 78]. The model is capable of simulating the changes in lengths and contraction velocity of the sarcomeres of the human jaw-opening and jaw-closing muscles as well as the consequences for force production during jaw open-close movements. The model shown below (Fig. 1.10) consists of a lower jaw which is accelerated by forces and accompanying torques with respect to its centre of gravity applied by muscles, joint surfaces, bite points and ligaments [17]. The biological properties of the muscle groups, such as fibre length, sacromere lengths and PCSA were obtained from the same research group [79, 80].

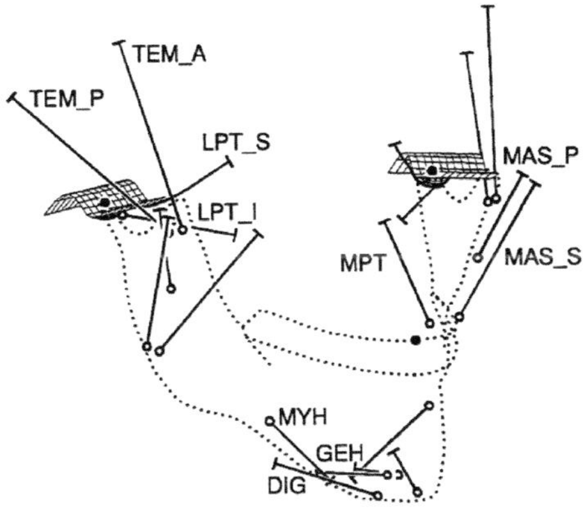

Fig. 1.10 Ventro-lateral view of the model (© 2001 Elsevier, Reprinted from [19] with permission)

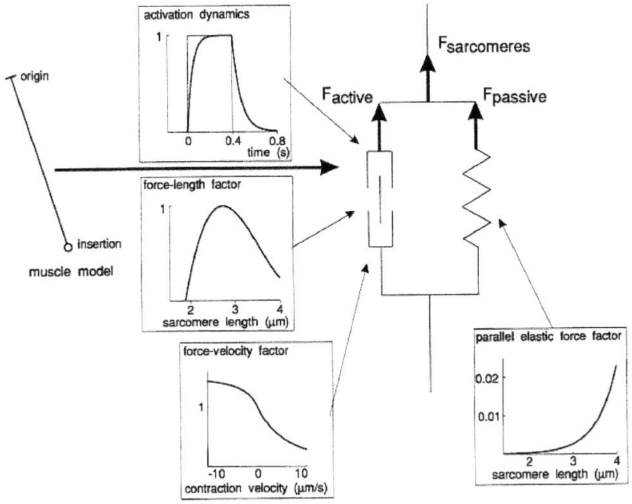

Fig. 1.11 Hill-type muscle model (© 2001 Elsevier, Reprinted from [19] with permission)

Muscles were modelled in Hill-type flexible, single-line actuators (Fig. 1.11), where the muscular (sarcomere) force generated depends on the amount of activation, sarcomere dynamics and activation dynamics [17]. The muscle is activated in the manner that the insertion point on the mandible moves towards the origin.

1.4.2 Kinematics Model

The computational model proposed in [57, 68] is based on magnetic resonance imaging to obtain an accurate anatomical representation of the movements of muscles of mastication and the mandible. Kinematics was applied to characterise the jaw movement in terms of motion screw under the influence of muscular forces. However, the determination of the muscle force direction and magnitude was estimated or set arbitrarily.

While the motion screw was characterised by a linear and angular velocity motion, the action screw represented a unique axis where all the resultant forces and torques acting on a rigid body were co-linear (Fig. 1.12a) [68]. Motion and action screws are generally skew to each other. The power expended during the motion is the sum of the scalar products of forces and velocities, the torques and angular velocities, corrected by the distance and angulation of the two screws. The generalisation of screw theory enables the quantitative assessment of the energy used to perform various types of mandibular movements. Fig. 1.12b shows a system of four forces acting on the mandible. The resultant force and movement making up the action screw, results in the mandible moving about the motion screw.

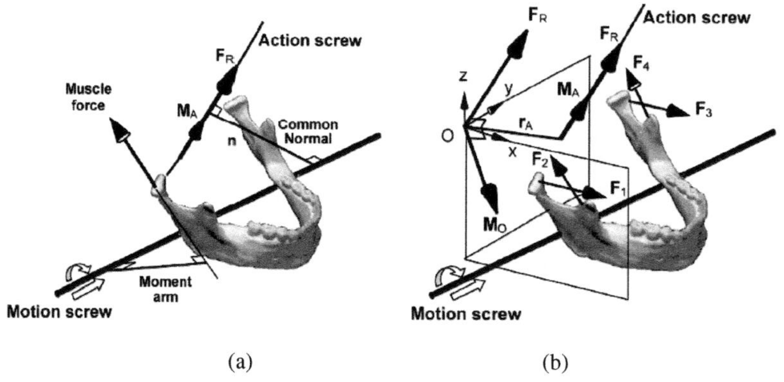

Fig. 1.12 The screw model, (a) a muscle and (b) the mandibular mechanics [68] (© 2003 Elsevier, Reprinted from [68] with permission).

In contrast, Weingartner et al. [81, 82] used a kinematical model in which muscles were represented by threads displaying the force vector of the muscles, with more than one thread being required for each muscle. These threads were described as cylindrical objects that adapted their length and orientation according to the movements of the mandible. The muscle shape was then represented by flexible surfaces (with respect to muscular deformation) and was based on computer visualisation techniques rather than biomechanical principles. Movement related changes in the shapes of the muscles of mastication were reflected in changes of a surface mesh.

In addition to the above computational models, a physical model of the masticatory system can be designed in a CAD system such as SolidWorks. Kinematics and dynamics can then be simulated in CosmosMotion [26] or SimMechanics[83]. Consequently, the physical model is instrumental to real robotic systems.

1.5 Mastication Robots

There have been a variety of machines and devices available for measuring human masticatory movements or food properties. However, these machines only have one particular function and are not able to simulate the entire suite of complex functions and movements involved during the mastication process. The challenge, therefore, has been to robotically reproduce human jaw movements in a holistic and controllable way, for use in a range of applications such as dental training, jaw simulation, food texture analysis and speech therapy.

1.5.1 Dental Training Robot

1.5.1.1 WY Series Robots

The WY (Waseda-Yamanashi) series of robots has been developed for the training of jaw disorder patients [23, 24, 84-86]. The robot is used to open and close a patient's lower jaw by mimicking the doctors' hand motion during a mouth opening training session. The most advanced version is the WY-5/WY-6 built between 1998 and 2003. This incorporates a parallel mechanism of 6 degrees-of-freedom (DOF), as shown in Fig. 1.13 [84, 85]. It is able to reproduce the same movable range and force as the human jaw.

The robot is actuated by six linear motors via ball screws. The upper mouth piece of the robot holds a patient's upper jaw and the patient's lower jaw is moved accordingly by the robot. This mechanism is more effective than the conventional technique using mouth gauges, which can only perform 1-DOF movement (opening and closing).

The most important factor during therapy is the biting force exerted on the patient via the mouth-opening gauges. The WY-5 robot measures the biting force acting on it from the patient in terms of three translational components of the force. The robot is operated remotely by a doctor robot that has two DOFs (for open/close and forward/backward movements) [84] or three DOFs (for open/close, forward/backward, and right/left movements) [85]. Clinical trials have been reported [86].

To monitor any change in the patient's jaw muscle, and hence to evaluate robotic therapy, muscular EMG measurements have been incorporated into the WY robot. Results from clinical trials found that the post-treatment EMG signal was lower than that during the treatment [87].

The WY-5 was further modified to measure food texture and a new robot named the WWT-1 (WWT for Waseda Wayo Texturobot) was developed in 2004 (www.takanishi.mech.waseda.ac.jp). The new robot is able to mimic the human mastication movement with an accuracy of 0.02 mm. With an integrated force sensor, the WTW-1 can measure the chewing force with a resolution of 0.001N.

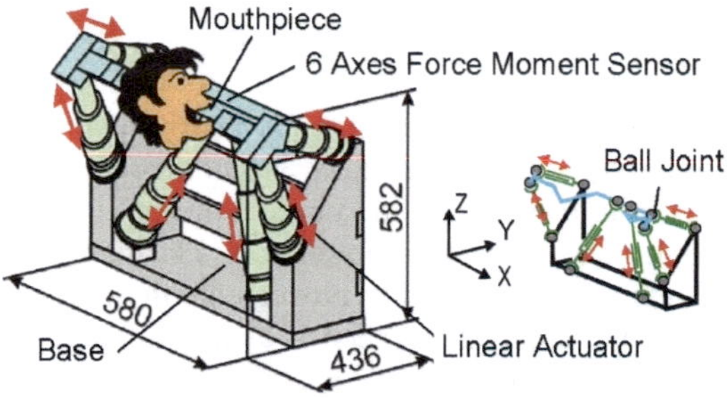

(a)

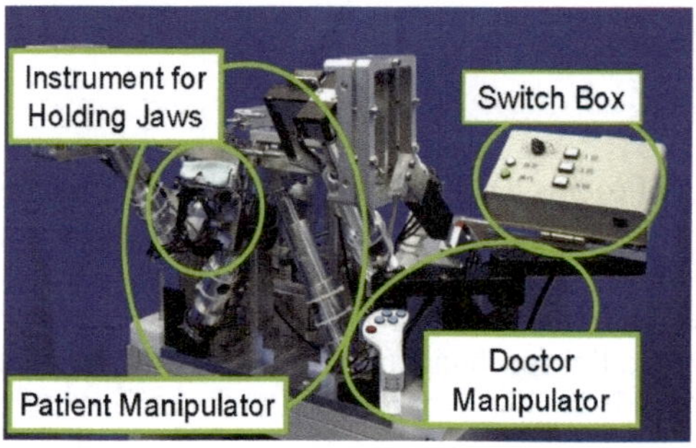

(b)

Fig. 1.13 WY-5 jaw training robot, (a) the mechanism and (b) the real robot (© 2010 TAKANISHI A, Reprinted from www.takanishi.mech.waseda.ac.jp with permission)

1.5.1.2 WJ Series Robots

The WJ (Waseda Jaw) robot series was developed in place of patients to work with a WY series robot for dental training purposes [23, 24, 87]. The WJ robot is used as a patient robot to understand patient's mastication movement and resistance forces during jaw opening and closing training. The WJ robot may also be used for evaluating and treating TMJ dysfunction. Fig. 1.14 shows a WY training robot working with a patient a) and a patient robot b).

The WJ robots were built under the simple assumption that the human jaw involves only three DOFs of movement: open/close, forward/backward and right/left [88], as shown in Fig. 1.15. The latest versions, WJ-3 and WOJ-1 (WOJ for Waseda Okina Jaws), were built in 2002 [21] (www.takanishi.mech.waseda.ac.jp) as shown in Fig. 1.16. They have three DOFs and implement artificially produced trajectories for clenching and grinding.

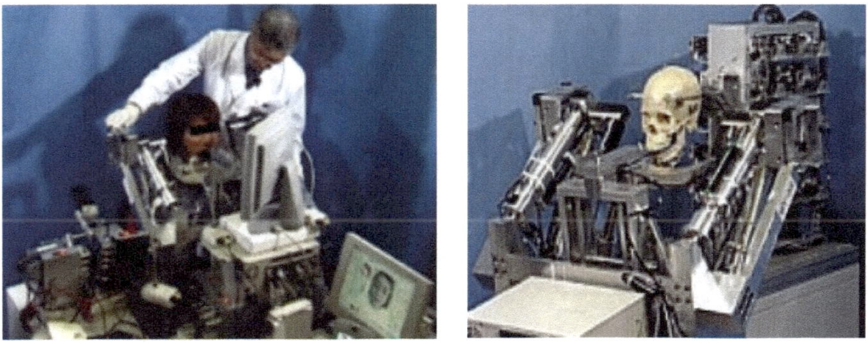

(a) (b)

Fig. 1.14 WY robot working on a) patient and b) WJ robot (© 2010 TAKANISHI A, Reprinted from www.takanishi.mech.waseda.ac.jp with permission)

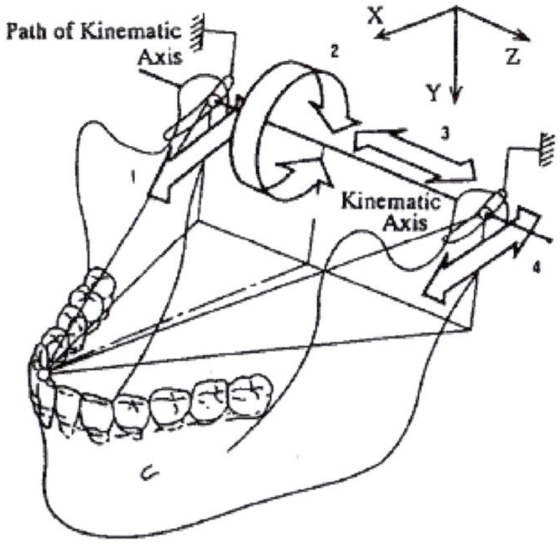

Fig. 1.15 Jaw movements of 3 DOFs (© 2010 TAKANISHI A, Reprinted from www.takanishi.mech.waseda.ac.jp with permission)

A carbon rod passes through the right and left condyles. Forward/backward motion of this rod is constrained on a virtual plane that is inclined 40 degrees from the horizontal plane and is composed of two artificial TMJs. The robot is driven by eleven artificial muscle actuators (AMA); each actuator is made up of a set of a DC motor, an encoder, a wire and a force sensor (Fig. 1.17) [89]. The AMAs were designed to simulate the force of muscular contraction. One end of the tendon is attached on the robotic mandible and the motor pulls the other end. Three 2-DOF bite force sensors were developed and are placed in the right, centre and left part of the tooth arch.

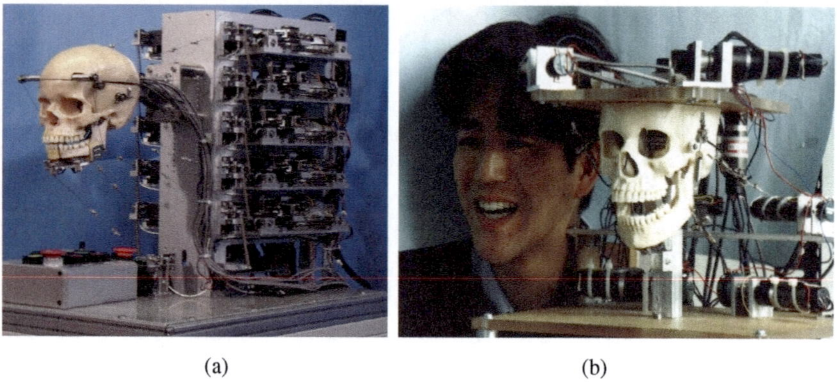

Fig. 1.16 Mastication robot as patient, a) WJ-3 and b) WOJ-1 (© 2010 TAKANISHI A, Reprinted from www.takanishi.mech.waseda.ac.jp with permission)

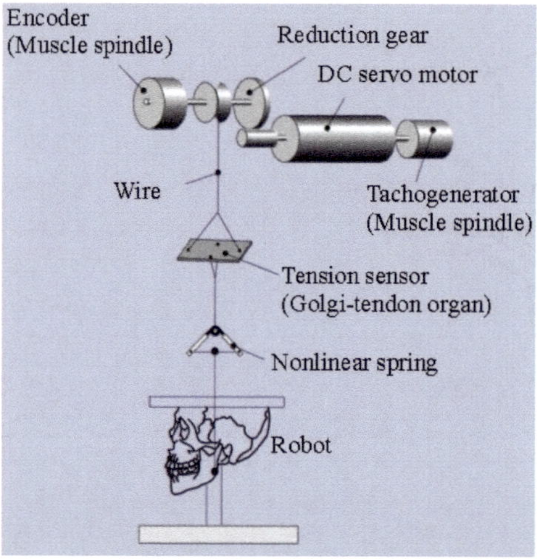

Fig. 1.17 Artificial muscle actuators (© 2010 TAKANISHI A, Reprinted from www.takanishi.mech.waseda.ac.jp with permission)

A control algorithm for adaptive jaw motion using jaw position and biting force was developed, and the experimental trials on chewing ball-shaped cookies were performed [89]. Nonlinearity of the human masticatory muscle was modelled by a non-liner spring mechanism [90] and a nonlinear viscoelastic mechanism [91].

1.5.2 Jaw Simulator

In the 1990s, a JSN series of jaw simulators was developed at Niigata University, Japan. This development was motivated by a desire to reproduce a life-like open-close movement [92-94] and focused on the coordinated activities of several jaw muscles for mastication. The simulator consists of the upper and lower jaws equipped with bilateral occlusal contact and bite-force sensors at the maxillary canine and first molars. The position of the jaw is measured by the rotary encoder and the muscular force is detected by a cable tension sensor via a so-called alpha-gamma linkage.

Each actuator is implemented as a cable-tendon driven by a DC servo motor activated by the EMG data recorded during the open-closing experiment. The TMJ is modeled by rubber bands suspending the condyle housing [92].

The most advanced JSN jaw simulator was the JSN/2A [93]. It involves two DOFs and consists of six muscle actuators; being the masseter, lateral pterygoid, interior pterygoid, digastric, anterior temporalis and post temporalis. To emulate the muscular compliance, an impedance control scheme is employed [94], whereby the occlusal position (measured by the rotary encoder) and the occlusal force (via a bite-force sensor) are fed back and compared with the driving signal. The driving error is then corrected by changing cable tension via the so-called alpha-gamma linkage.

Some experiments were performed on the earlier jaw simulator JSN/1C which has one DOF and three muscular actuators (masseter, lateral pterygoid and diagastric) [92]. From a time series of bite forces at the first molars, it was found that in the early stage of mastication, bilateral occlusal contacts and bite forces are significantly different. These differences are gradually narrowed with the progression of mastication.

1.5.3 Foods Chewing Robot

We began the food chewing robotics research program at Massey University in 2002 with the aim of analytically characterising food texture using a robotic model. The intention was that the robot chewed foods in a human-like fashion – both in terms of mastication movement and chewing force. It was further intended that food particles be collected during the chewing process and food properties evaluated. Much of the research work is the subject of this book and is summarised in the following section.

1.5.3.1 Robotic Model of Linear Actuation

The robot was constructed following a review of the biomechanical findings about jaw structure and muscles of mastication [26, 83, 95]. Each of the major mouth-closing muscles (temporalis, masseter and pterygoid) is represented by a linear actuator. Since the linear actuators are double-acting, the mouth-opening muscles are not required. Muscle origin and insertions are modeled as a spherical joint; the actuators were therefore placed between the mandible and the skull so that each actuator always acted in the direction of the resultant muscular force. This resulted in a robotic platform mechanism, with the mandible being a moving plate, and the skull a ground plate. The dimensions of the robot match the human counterpart. Fig. 1.18 shows the conceptual model, CAD model and static model of the robotic mechanism.

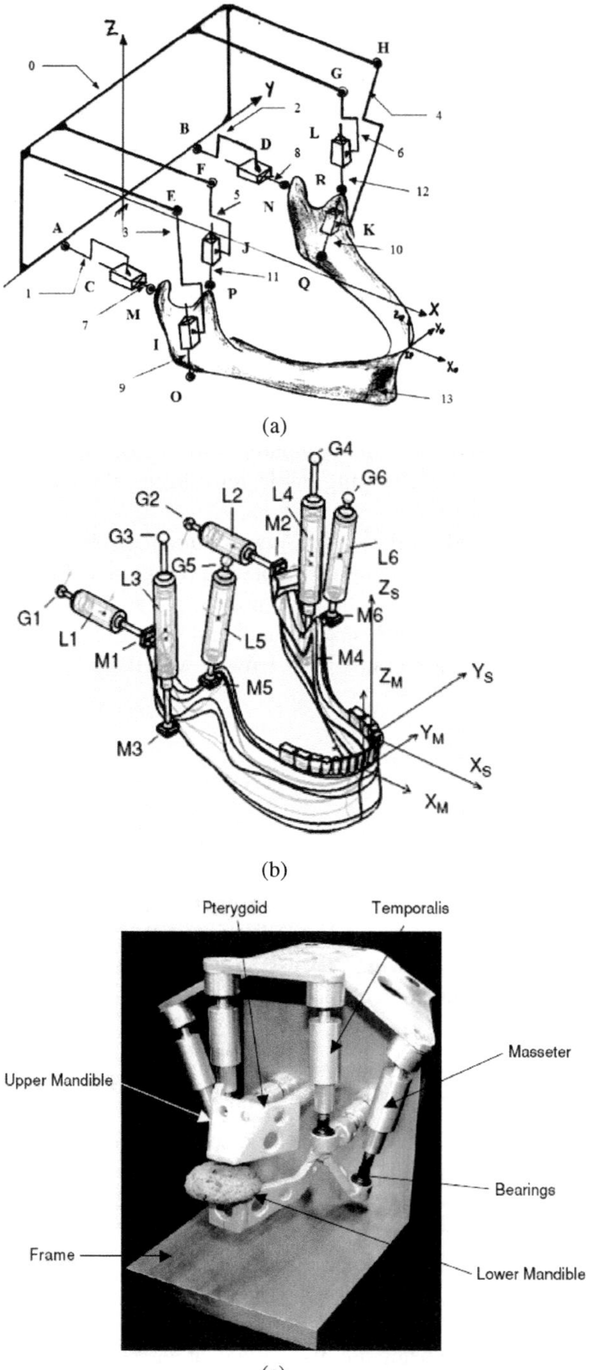

Fig. 1.18 A chewing robot model, (a) conceptual model, (b) CAD model and (c) static model

In the robotic mechanism the points M1 and M2 (Fig. 1.18b) represent right and left condylar points, with each tracing a different trajectory during mandibular movement. This matches both clinical and biomechanical findings which suggest that, during jaw movements, the mandible does not rotate around a fixed condylar axis; rather it rotates around instantaneous axes that continuously change their position in space. The working and balancing condylar points, exhibit different trajectories which themselves vary with the type of food being chewed [31-33].

In the proposed robotic mechanism, there was only one actuator representing one muscle group. To achieve a biologically meaningful range of mandibular movements, the muscle length of the biggest sub-muscle group in the closed-mouth position was used as the actuator length. This gave actuator lengths for masseter, temporalis and lateral pterygoid at 45.6 mm, 52.2 mm and 32.6 mm, respectively [96, 97]. The actuator orientation, which represents the muscle line of action, was obtained using a force-vector calculation in both the frontal and sagittal plane [96, 97]. The insertion coordinates for each actuator were estimated using a scanned mandible.

1.5.3.2 Robotic Model of Crank Actuation

Whereas our efforts were focused on designing and building a robot based on the above model, its realisation was beset by technical challenges. Among them was the fact that the size, motion and power specified for the linear actuator could not be satisfied using existing techniques [97]. Whilst keeping within the specifications for linear actuators derived from the human jaw, a robotic platform of 6RSS parallel mechanism was proposed, as shown in Fig. 1.19. Each RSS linkage is driven by a DC motor and consists of a crank and a coupler. The crank is pivoted to the ground by a revolute joint while the coupler is connected to the crank and the mandible by a ball-socket joint. The coupler size, placement, motion and force requirements were approximated from those specified for the linear actuator in the previous model. The robot was designed as shown in Fig. 1.20a and built in 2005 as photographed in Fig. 1.20b [27].

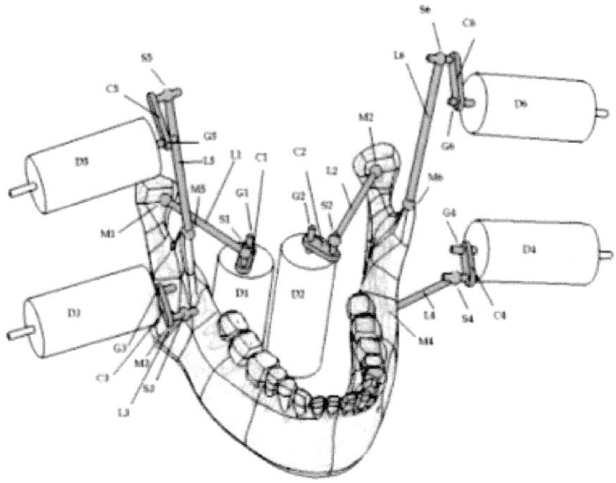

Fig. 1.19 A chewing robot of 6 RSS parallel mechanism.

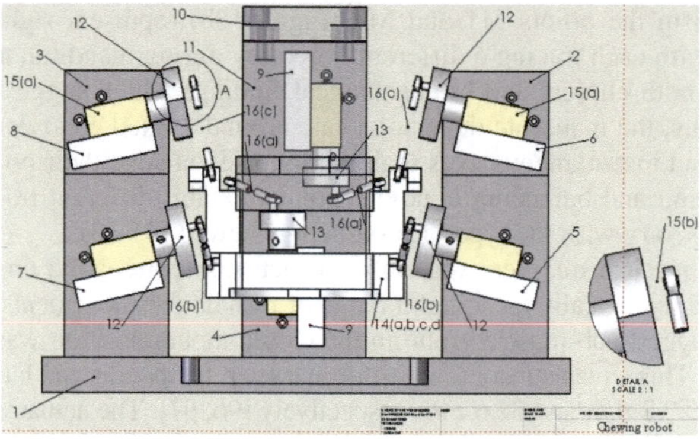

(a)

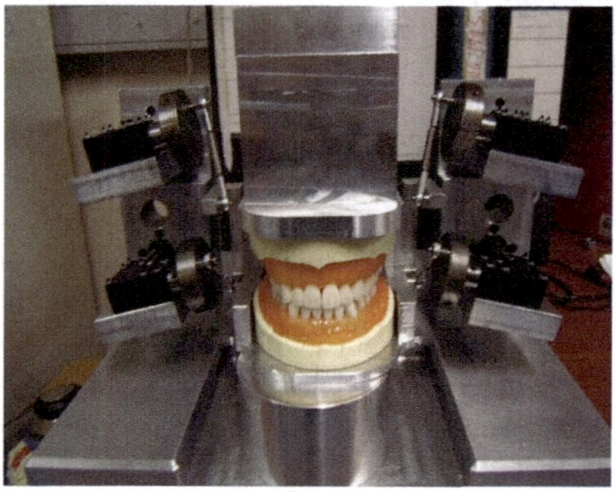

(b)

Fig. 1.20 A chewing robot, (a) the drawing and (b) the photo

The motion control system consists mainly of a six-axis motion control card, DMC-1860 by Galil, which is plugged onto the motherboard of a PC, and two amplifiers, each of which is capable of driving up to four motors using pulse width modulation. The Galil DMC Smart Terminal program provides a means of programming PID control of each axis.

The robot was commanded under PID controllers to reproduce chewing, as shown in Fig. 1.10. A 6 mm thick aluminium plate was inserted between the upper and lower jaws. This experiment was used to test both the ability of the model to follow a trajectory and chewing force limiting ability (i.e. the ability to reproduce those movements used when chewing on a hard food) [97]. To do this, inverse kinematics were performed and the torque was limited to 55Nm at each motor. This torque limit gave rise to a torque of 3.6Nm at the crank, and approximately 250N of force at the link attaching to the jaw. These specifications being close to reported human chewing forces [26].

Fig. 1.21 shows some representative results obtained with actuator 6. A position error can be seen during occlusion as the torque is limited and the jaw locks up and is therefore unable to follow the commanded trajectory. For the rest of chewing cycles, the robot follows the inverse trajectories satisfactorily. The same results hold for all five other actuators.

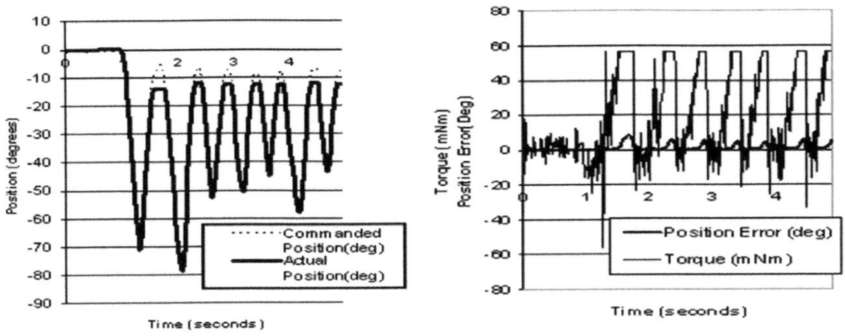

(a) (b)

Fig. 1.21 An actuation example, a) commanded and actual trajectory and b) motor torque and positional error

1.5.4 Other Mastication Robots

A speech robot was developed in 2005, at the University of British Columbia, Canada [28, 29]. Development was aimed at studying the role that jaw movements play in perceiving and understanding face-to-face conversation. This robotic jaw consists of two 3-DOF parallel manipulators for driving the two TMJs, one at each end of the jaw. Unfortunately there has been no literature published to date in relation to this research.

To understand the workings of the human jaw joint, the AJJP (Australian Jaw Joint Project) at the University of Adelaide, Australia built a so-called Mark III chewing machine (http://www.jaw-joint.com/). This is driven by nine pneumatic actuators via stainless steel cables which serve as muscles intended to mimic generic mastication. Fig. 1.22a and b show the jaw construction and pneumatic cable system, respectively. Some modifications were in progress [98], although no other literature out of the AJJP has been found.

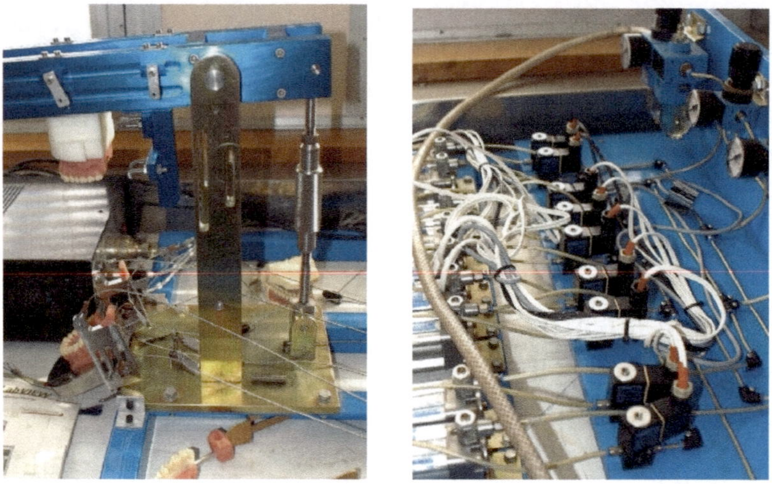

Fig. 1.22 Mark III chewing machine ((© 2005 BOWY C, Burgess D. Reprinted from [98] with permission)

1.6 Computational Intelligence in Mastication

Human mastication is a complex process which is regulated by a brain stem central pattern generator (CPG) [99] and serves to decompose and transport food for further digestion. This process can be studied by way of masticatory physiology [100, 101]. Electromyography (EMG) is used to measure the muscular activity of mastication, rhythmical mandibular movements in relation to opening and closing, salivary secretion, etc. Human mastication is highly specific to both the individual and to the particular food being chewed. The aforementioned masticatory measurements are therefore affected by, firstly factors pertaining to the individual (e.g. dental status, chewing side, age, or gender) secondly, the particular food being chewed (e.g. sample size, hardness or rheological behaviour) [102], and finally, situational factors affecting experimental control (e.g. day-to-day variability, sample supply order, sample control, subject control, specified chewing side, specified chewing cycles, etc) [9, 10, 103].

A robot that chews food depicting human mastication as accurately as possible is obviously preferable. One challenge, among many others in such an effort, relates to how the robot learns variations in food properties and thus adapts its chewing behaviours according to certain performance criteria.

To tackle this challenge, we proposed developing a behaviour-based and on-line learning control methodology for the robot. To this end, two objectives were identified. The first being the synthesis of chewing behaviours, which dealt with the association of jaw movements and chewing behaviours with food characteristics drawn from the available chewing database and/or literature. The second objective was the incorporation into the robotic model of an on-line learning adaptive control of the robot, which dealt with the fusion and memorisation of a set of primitive chewing behaviours and develops on-line learning adaptive control.

Various foods chewing databases were developed by way of experiments and surveys. From there, a set of primitive behaviours was formulated by way of data mining techniques. These rules associated jaw movements (time-functions) with the type of model foods being tested under various conditions, such as an individuals' chewing habits and dentition. The behaviours were formulated so that artificial neural networks could be used to memorise the associations. To command the robot to chew in a normalised way, we would take the aforementioned neural network as a nominal controller. This represented a reflexive action that associated with particular food properties and chewing behaviours. To accommodate unstructured and unpredictable variations in foods being chewed and variations in individual chewing habits/dentitions, we added an on-line learning mechanism to the nominal controller in order to refine jaw movements based on real time sensing of foods being chewed (food tenderness, textures etc). Use of the behavioural control allowed consideration of the fast sensory-reactive nature of human mastication mechanism to be taken into account whilst the use of the neural network was in order to take advantage of incremental/evolutionary learning capability.

Considerable work was done in developing systematic methods for evaluating the capability of a person to effectively chew foods. One such effort was the formal description, in terms of object-oriented method, of the mastication process and contributing attributes [104]. A comprehensive knowledge framework was established to model the mastication process and data mining was used to discover the relationships between masticatory physiology, measurement and various factors affecting mastication.

An object-oriented framework was proposed in order to organise the hierarchies of the mastication knowledge [105]. The framework contains two types of knowledge representation (Fig. 1.23): (1) classes of *objects,* whose slots of attributes characterise the masticatory physiology, measurements, and factors; and (2)

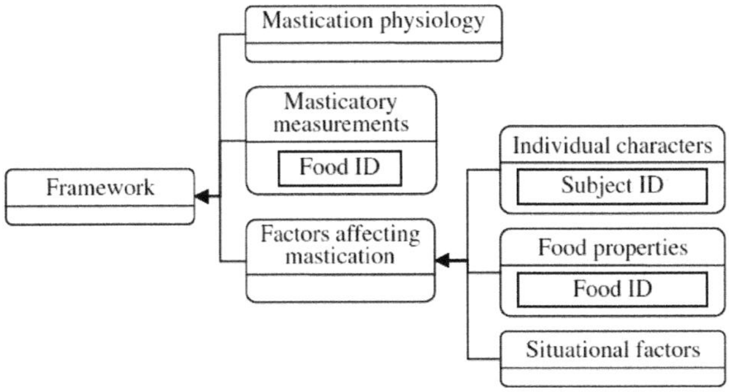

Fig. 1.23 Top class objects for the mastication knowledge framework

rules among the slots of objects to substantiate their relationships. The inheritance links of these objects regulate the measurements and factors, making the foods mastication experiments more controlled and comparable. The object-oriented design, which is the first feature of the framework, is performed using domain knowledge and literature results initially. It is then populated by the knowledge of relationships among the measurements of the mastication and factors affecting the mastication, which is the second feature of the framework. The knowledge is represented by rules about the relationships among the slots (or attributes) of objects. These rules can be discovered from the experimental database following the knowledge discovery in database (KDD) [106].

In [104, 105], a case study was presented to show the procedure of how the hidden knowledge can be discovered, where the database was collected previously while studying the effect of texture on masticatory parameters during the chewing of plastic and elastic model foods [10]. A data mining tool, so-called WEKA [107] was used, and the data mining results were interpreted in relation to food science.

1.7 Discussion

The masticatory system exhibits a distinct suite of kinematics and dynamics. The mandible can be regarded as a rigid body, suspended from the skull via the two TMJs and driven co-ordinately by the six major masticatory muscles by commands from the central nervous system. Kinematically, the mandible is movable in relation to the skull in the three-dimensional space and is constrained by the biological structure of the TMJs. Dynamically, the upper and lower teeth interact with foods by virtue of a highly complex contraction of dynamically varying muscular activities. Consequently, the unique biomechanics of the masticatory system gives rise to a clear roadmap for development of computational and robotic models.

The major challenge lies in dealing with the unusual temporomandibular joint and the masticatory muscles. For the computational models, the modeling of tooth shape would be an additional challenge that has not yet been addressed.

The masticatory process can be measured and used as an objective means to assess firstly, the masticatory system itself, and secondly, its dysfunction, in other words in the diagnosis of pathology of the TMJ. Additionally, it may prove crucial in the development of new clinical modalities in dentistry, as well as in the evaluation of certain food properties.

It is possible to measure the rhythmic jaw movement, chewing force and muscular activities in-vivo on a human subject's masticatory system. Other measurements can also be made on masticated foods after expectoration including properties such as particle size distribution within the food bolus prior to swallowing. Measurements such as these serve as criteria upon which to validate computational masticatory models and/or to assess the performance of masticatory robots.

The computational models of the masticatory system are useful in the testing of various hypotheses for clinical purposes and in order to understand masticatory biomechanics. The models are closely related to masticatory robotics, and may provide significant clues as to their design including for example specifications in relation to the muscular actuation system, jaw movements, biting force and over all in relation to the assembly of the entire robot. Some robots developed may trace their foundation back to the computational models.

Considerable efforts have been made in terms of modelling the muscles of mastication, including origin and insertion coordinates, displacement and force relations plus biological properties. The Hill-type model has been widely adopted in the development of computational models. Due to the efforts made in modelling the masticatory system, a significant body of data in relation to biomechanical parameters has been made available. This, plus measurement of masticatory performance, has made the development of various masticatory robots feasible.

Various types of masticatory robots have been developed since the early 1990s. Both the WY and WJ series have primarily been developed for dental training - the former for assessment of mouth opening and closing in patients and the latter for the assessment of robotic training therapy. Due to the specific objective of the WY robots, their designs did not have to mirror the human counterpart, except in the specific area of the mouth opening. Consequently, few biomechanical findings obtained from these robots have been applicable to human mastication.

WJ robots were partially developed with direct reference to the biomechanics of the human jaw. For example, the mandible designed to have 3-DOF movement, the TMJ was built such that the condyle translates and rotates inside a curved slot and the muscles were replaced by a DC driven motor with a wire-pulling system for compliant contraction. However, there are a few major flaws in these robots. First of all, the human mandible possesses 6 DOFs of motion in a three-dimensional space. Three of these DOFs describe translational displacements and the other three for angular displacements; hence, the 3-DOF WJ robots are unable to simulate complex human jaw movements. Secondly, the human condylar point does not move in a fixed path but instead moves within a narrow-band spatial envelope. This means that the use of a slot containing the TMJ in WJ robots is inappropriate. Thirdly, the muscle actuation system still acts uni-directionally without taking advantage of the double-acting power of an engineering solution; this means that the insertion of the actuation systems between the mandible and the maxilla does not follow its biological human counterpart. Thus, WY and WJ robot series might be good enough for dental training, as evidenced by clinical trials, but may not be appropriate for the analysis of food texture.

The JSN robots have only three or fewer DOFs and apparently have difficulty in reproducing complex human jaw movements. The muscle actuation systems are similar to those used in the WJ robots but their activation using recorded EMG data is significant. The speech robot, if it were converted in order to perform actual chewing of foods, is weak in terms of mechanics as the mandible is driven by two 3-DOF parallel mechanisms at the back of the jaw. This design is like a cantilever with the biting force applied at the rear.

The chewing robot developed at Massey has six DOFs and is able to reproduce any complex human jaw movement, as evidenced by simulations and experimentation. Its design follows biomechanical data such as the placement of the actuators and the ability to achieve "floating" by relaxing the TMJ thereby allowing it to move within an envelope. More importantly, by means of a double-acting, motor-gear system, the robot is in fact more powerful than the human jaw whilst the robotic structure is simplified. The use of a 6RSS parallel mechanism as actuation has made it possible to fit the robotic jaw into a compact space. The initial experiments have shown success in implementation of the jaw trajectory path whilst achieving maximum chewing force (by limiting torque in the motors). This force give-way feature is essential for the robot to chew hard foods adaptively. Although the industrial PID control is sufficient in its current design, the non-linear compliance found in the mastication muscles will need to be incorporated in controls for more realistic food chewing scenarios.

1.8 Summary

The masticatory system, measurements of mastication, together with computational and robotic models of mastication was reviewed. It is hoped that the review benefits the niche field of robotics and generates more interest in the field of mastication from the robotics community. The masticatory system was described in this review in a manner allowing both kinematics and dynamics to be understood as multi-rigid-body systems. Measurements in relation to the mastication process were presented in terms of what will be required to specify and assess both computational models and masticatory robots for various applications. The computational models reviewed were concerned with jaw motion and the forces required for this motion. The models reviewed provide specifications for the design of masticatory robots. The masticatory robots were reviewed in terms of their applications with a focus on their robotic mechanisms and muscular actuation systems. The initial work in computational and artificial intelligence for mastication points out a possible direction along which a chewing robot could be developed which replicates human mastication.

References

1. Bourne, M.C.: Relationship between rheology and food texture. In: Welti-Chanes, J., et al. (eds.) Engineering and food for the 21st century. CRC Press, LLC (2002)
2. Heath, M.R., Prinz, J.F.: Oral processing of foods and the sensory evaluation of texture. In: Rosenthal, A.J. (ed.) Food texture: measurement and perception. Aspen Publishers Inc., Gaithersburg (1999)
3. Ow, P.K.K., Carlsson, G.E., Karlsson, S.: Relationship of masticatory mandibular movements to masticatory performance of dentate adults: a method study. J. Oral Rehab. 25, 821–829 (1998)
4. Nakajima, J., et al.: Masticatory mandibular movements for different food texture related to onomatopoetic words. J. Med. Dent. Sci. 48, 121–129 (2001)
5. Tsuruta, J., et al.: An index for analysing the stability of lateral excursions. J. Oral Rehab. 29, 274–281 (2002)
6. Hayasaki, H., et al.: A calculation method for the range of occluding phase at the lower incisal point during chewing movements using the curved mesh diagram of mandibular excursion (CMDME). J. Oral Rehab. 26, 236–242 (1999)

7. Palla, S., et al.: Jaw tracking and temporomandibular joint animation. In: McNeill, C. (ed.) Science and practice of occlusion. Quintessence Publishing Co. Inc. (1997)
8. Anderson, K., et al.: The effects of bolus hardness on masticatory kinematics. J. Oral Rehab. 29, 689–696 (2002)
9. Peyron, M.A., et al.: Effects of increased hardness on jaw movement and muscle activity during chewing of visco-elastic model foods. Exp. Brain Res. 142, 41–51 (2002)
10. Foster, K., et al.: Effect of texture of plastic and elastic model foods on the parameters of mastication. J. Neurophysiol. 95, 3469–3479 (2006)
11. Tortopidis, D., et al.: The variability of bite force measurement between sessions in different positions within the dental arch. J. Oral Rehab. 25, 681–686 (1998)
12. Koolstra, J.H.: Number crunching with the human masticatory system. J. Dent. Res. 82, 672–767 (2003)
13. Osborn, J.W., Baragar, F.A.: Predicted pattern of human muscle activity during clenching derived from a computer assisted model: symmetric vertical bite force. J. Biomech. 29, 589–595 (1985)
14. Osborn, J.W.: Features of human jaw design which maximize the bite force. J. Biomech. 29, 589–595 (1996)
15. Koolstra, J.H., et al.: A three-dimensional mathematical model of the human masticatory system predicting maximum possible bite force. J. Biomech. 21, 563–567 (1988)
16. Koolstra, J.H., van Eijden, T.M.: Application and validation of a three-dimensional mathematical model of the human masticatory system in vivo. J. Biomech. 25, 175–187 (1992)
17. Koolstra, J.H.: Dynamics of the human masticatory system. Crit. Rev. Oral Biol. Med. 13, 366–376 (2002)
18. Koolstra, J.H., van Eijden, T.M.: Dynamics of the human masticatory muscles during a jaw open-close movement. J. Biomech. 30, 883–889 (1997)
19. Koolstra, J.H., van Eijden, T.M.: A method to predict muscle control in the kinematically and mechanically indeterminate human masticatory system. J. Biomech. 34, 1179–1188 (2001)
20. Hannam, A.C.: Jaw muscle structure and function. In: McNeill, C. (ed.) Science and practice of occlusion. Quintessence Publishing Co, Inc. (1997)
21. Peck, C.C., et al.: Forces resisting jaw displacement in relaxed humans: a predominantly viscous phenomenon. J. Oral Rehab. 29, 151–160 (2002)
22. Röhrle, O., Pullan, A.J.: Three-dimensional finite element modelling of muscle forces during mastication. J. Biomech. 40, 3363–3372 (2007)
23. Takanobu, H., et al.: Integrated dental robot system for mouth opening and closing training. In: Proceedings of the 2002 IEEE International Conference on Robotics & Automation, Washington DC, pp. 1428–1433 (2002)
24. Takanobu, H., Takanishi, A.: Dental robotics and human model. In: Proceedings of the 1st International IEEE EMBS Conference on Neural Engineering, Capri Island, pp. 671–674 (2003)
25. Nakajima, S.I., et al.: Development of 2-D jaw movement simulator (JSN/S1). J. Robotics Mechatronics 10, 499–504 (1998)
26. Xu, W.L., et al.: A robotic model of human masticatory system for reproducing chewing behaviours. IEEE Robotics Automation Mag. 12, 90–98 (2005)
27. Torrance, J., et al.: Motion control of a chewing robot of 6 RSS parallel mechanism. In: Proceedings of International Conference on Autonomous Robotics and Agents, pp. 593–598. Palmerston North, New Zealand (2006)

28. Lin, B.: Chew on this. UBC Reports. Vancouver, BC, Canada, pp. 1–8 (2005)
29. Flores, E., Fels, S.: Design of a 6 DOF anthropomorphic robotic jaw. J. Acoustical Soc. Am. 117, 2547 (2005)
30. Gray, H.: Anatomy of the human body. Lea & Febiger, Philadelphia (1918)
31. Koolstra, J.H.: Dynamics of the human masticatory system. Crit. Rev. Oral Biol. Med. 13, 366–376 (2002)
32. Scapino, K.P.: Morphology and mechanism of the jaw joint. In: McNeil, C. (ed.) Science and practice of occlusion. Quintessence Publishing Co. (1997)
33. Lehman-Grimes, S.K.: A review of temporomandibular disorder and an analysis of mandibular motion. Master of Dental Science Thesis. The University of Tennessee (2005)
34. Koolstra, J.H., van Eijden, T.M.: Three-dimensional dynamical capabilities of the human masticatory muscles. J. Biomech. 32, 145–152 (1999)
35. Lucas, P.W.: Dental functional morphology: how teeth work. Cambridge University Press, United Kingdom (2004)
36. Ashby, M.F., Jones, D.R.H.: Engineering materials, 2nd edn. Butterworth Heinemann, Oxford (1996)
37. Frank, F.C., Lawn, B.R.: On the theory of Hertzian fracture. Proceedings of the Royal Society London Series A, 291–316 (1967)
38. Oralb (2007), `http://www.oralb.com/images/learningcenter/teaching/` (accessed September 3, 2007)
39. Dawson, P.E.: Evaluation diagnosis and treatment of occlusal problems, 2nd edn., Mo. Mosby, St Louis, pp. 85–91 (1989)
40. Dawson, P.E.: Evaluation, diagnosis, and treatment of occlusal problems, 2nd edn., p. 632. Elsevier Health Sciences, Amsterdam (1989)
41. Wynne, W.P.D.: The art of articulation 3(1) (2005)
42. Posselt, U.: Movement areas of the mandible. J. Pros. Dent. 7, 375–385 (1957)
43. Douglass, G.D., DeVreugd, R.T.: The dynamics of occlusal relationships. In: McNeil, C. (ed.) Science and practice of occlusion. Quintessence Publishing Co., Berlin (1997)
44. Ogawa, T., et al.: Different responses of masticatory movements after alternation of occlusal guidance related to individual movement pattern. J. Oral Rehab. 28, 830–841 (2001)
45. Röhrle, O., et al.: From jaw tracking towards dynamic computer models of human mastication. In: IFBME Proceedings of 12th International Conference on Biomedical Engineering, Singapore (2005)
46. Buschang, P.H., et al.: Quantification of human chewing-cycle kinematics. Arch. Oral. Biol. 45, 461–474 (2000)
47. Gibbs, C.H., Lundeen, H.C.: Jaw movements and forces during chewing and swallowing and their clinical significance. Adv. Occlusion., 2–32 (1982)
48. Anderson, D.J.: Measurements of stress in mastication. J. Dent. Res. 41, 175–189 (1956)
49. Gibbs, C.H., et al.: Occlusal forces during chewing – influences of biting strength and food consistency. J. Prosthetic Dent. 46, 561–567 (1981)
50. Wood, G.D., Williams, J.E.: Gnathodynamometer: measuring opening and closing forces. Dent. Update 8, 239–250 (1981)
51. Every, R.F.: Sharpness of teeth in man and other primates. Postilla 143, 1–20 (1978)
52. Helkimo, E., Ingervall, B.: Bite force and functional state of the masticatory system in young men. Swed. Dent. J. 2, 167–175 (1978)

53. Sharkey, P., et al.: Jaw opening forces in human subjects. Br. Dent. J. 156, 89–92 (1984)
54. Stegman, D.F., et al.: Surface EMG models: properties and applications. J. Electromyogr. Kinesiol. 10, 313–326 (2000)
55. Barbenel, J.: The mechanics of the temporomandibular joint–a theoretical and electromyographical study. J. Oral Rehab. 1, 19–27 (1974)
56. Throckmorton, G.S., Throckmorton, L.S.: Quantitative calculations of temporomandibular joint reaction forces-I. the importance of the magnitude of the jaw muscle forces. J. Biomech. 18, 445–452 (1985)
57. May, B., et al.: A three-dimensional mathematical model of temporomandibular joint loading. Clin. Biomech. 16, 489–495 (2001)
58. Gonalez, R., et al.: Review: the use of electromyography on food texture assessment. Food Sci. Tech. Int. 7, 461–471 (2001)
59. Korioth, T., et al.: Three-dimensional finite element stress analysis of the dentate human mandible. Am. J. Phys. Anthropol. 88, 69–96 (1992)
60. Clason, C., et al.: A method for material parameter determination for the human mandible based on simulation and experiment. Comput. Methods Biomech. Biomed. Eng. 7, 265–276 (2004)
61. Ichim, I., et al.: Mandibular stiffness in humans: numerical predictions. J. Biomech. 39, 1903–1913 (2006)
62. van Essen, N., et al.: Anatomically based modeling of the human skull and jaw. Cells Tissues Organs 180, 44–53 (2005)
63. Ichim, I., et al.: Mandibular biomechanics and development of the human chin. J. Dent. Res. 85, 638–642 (2006)
64. Mulder, L., et al.: Biomechanical consequences of developmental changes in trabecular architecture and mineralization of the pig mandibular condyle. J. Biomech. 40, 1575–1582 (2007)
65. O'Connor, C., et al.: Bite force production capability and efficiency in Neanderthals and modern humans. Am. J. Phys. Anthropol. 127, 129–151 (2005)
66. Koolstra, J., van Eijden, T.: The jaw open-close movements predicted by biomechanical modelling. J. Biomech. 30, 943–950 (1997)
67. Gal, G.A., et al.: Wrench axis parameters for representing mandibular muscle forces. J. Dent. Res. 80, 534 (2001)
68. Gal, G.A., et al.: Analysis of human mandibular mechanics based on screw theory and vivo data. J. Biomech. 37, 1405–1412 (2004)
69. Weijs, W., Hillen, B.: Cross-sectional areas and estimated intrinsic strength of the human jaw muscles. Acta Morphol. Neerl. Scand. 23, 267–274 (1985)
70. van Eijden, T.M.G.J., et al.: Architecture of the human jaw-closing and jaw-opening muscles. Anat. Record. 248, 464–474 (1997)
71. Koolstra, J., et al.: Computer-assisted estimation of lines of action of human masticatory muscles reconstructed in vivo by means of magnetic resonance imaging of parallel sections. Arch. Oral Biol. 35, 549–556 (1990)
72. Cattaneo, P., et al.: Using the finite element method to model the biomechanics of the asymmetric mandible before, during and after skeletal correction by distraction osteogenesis. Comput. Methods Biomech. Biomed. Eng. 8, 157–165 (2005)
73. Koolstra, J., van Eijden, T.: Combined finite-element and rigid-body analysis of human jaw joint dynamics. J. Biomech. 38, 2431–2439 (2005)

74. Hirose, M., et al.: Three-dimensional finite-element model of the human temporomandibular joint disc during prolonged clenching. Eur. J. Oral Sci. 114, 441–448 (2006)
75. Tanaka, E., et al.: Dynamic compressive properties of the mandibular condylar cartilage. J. Dent. Res. 85, 571–575 (2006)
76. Koolstra, J., van Eijden, T.: Prediction of volumetric strain in the human temporomandibular joint cartilage during jaw movement. J. Anat. 209, 369–380 (2006)
77. van Loon, J., et al.: Loading of a unilateral temporomandibular joint prosthesis: a three dimensional mathematical study. J. Dent. Res. 77, 1939–1947 (1998)
78. De Zee, M., et al.: Validation of a musculo-skeletal model of the mandible and its application to mandibular distraction osteogenesis. J. Biomech. 40, 1192–1201 (2007)
79. van Eijden, T.M.G.J., et al.: Architecture of the human pterygoid muscles. J. Dent. Res. 74, 1489–1495 (1995)
80. van Eijden, T.M.G.J., et al.: Three-dimensional structure of the human temporalis muscle. Anat. Record 248, 565–572 (1996)
81. Weingartner, T., et al.: Dynamic simulation of the jaw and chewing muscles for maxillofacial surgery. In: Proceedings of IEEE Nonrigid and Articulated Motion Workshop (NRAMW 1997). Puerto Rico, pp. 104–111 (1997)
82. Weingartner, T., et al.: Virtual jaw: a 3D simulator for computer assisted surgery and education. Stud. Health Technol. Inform. 50, 329–335 (1998)
83. Daumas, B., et al.: Jaw Mechanism Modeling and Simulation. Mech. Machine Theor. 40, 821–833 (2005)
84. Takanobu, H., et al.: Mouth opening and closing training with 6-DOF parallel robot. In: Proceedings of the 2000 IEEE International Conference on Robotics & Automation, San Francisco, pp. 1384–1389 (2000)
85. Takanobu, H., et al.: Jaw training robot and its clinical results. In: Proceedings of IEEE/ASME International Conference on Advanced Intelligent Mechatronics (AIM 2003), pp. 932–937 (2003)
86. Okino, A., et al.: A clinical jaw movement training robot for lateral movement training. In: Proceedings of IEEE International Conference on Robotics & Automation, Taipei, Taiwan, pp. 244–249 (2003)
87. Takanobu, H., et al.: Quantification of masticatory efficiency with a mastication robot. In: Proceedings of the IEEE Internal Conference on Robotics & Automation, pp. 1635–1640 (1998)
88. Takanishi, A., et al.: Development of 3 DOF jaw robot WJ-2 as a human's mastication simulator. In: Proceedings of Fifth International Conference on Advanced Robotics, Pisa, pp. 277–282 (1991)
89. Takanobu, H., et al.: Adaptive masticatory jaw motion using jaw position and biting force information. In: Proceedings of IEEE International Conference on Multisensor Fusion and Integration for Intelligent Systems, Las Vegas, pp. 360–365 (1994)
90. Takanobu, H., et al.: Control of rapid closing motion of a robot jaw using nonlinear spring mechanism. In: Proceedings of IEEE/RSJ International Conference on Intelligent Robots and Systems, Pittsburg, pp. 372–377 (1995)
91. Takanobu, H., et al.: Development of a mastication robot using nonlinear viscoelastic mechanism. In: Proceedings of IEEE/RSJ International Conference on Intelligent Robots and Systems, pp. 1527–1532 (1997)

92. Hayashi, T., et al.: Control mechanism of an autonomous jaw-movement simulator JSN/1C during open-close movement. In: Proceedings of 18th Annual International Conference of the IEEE Engineering in Medicine and Biology Society, Amsterdam, pp. 613–614 (1996)
93. Hayashi, T., et al.: Physiological control scheme of jaw simulator JSN/2A for improving reproducibility of open-close movement. In: Proceedings of the First Joint BMES/EMBS Conference, Atlanta, USA., vol. 564 (1999)
94. Hayashi, T., et al.: A physiological control of chewing-like jaw movement for robotized jaw simulator JSN/2A. In: Proceedings of the 22nd Annual EMBS International Conference, Chicago, USA, pp. 730–731 (2000)
95. Pap, J.S., et al.: A robotic human masticatory system – kinematics simulations. Int. J. Intell. Syst. Technol. Appl. 1, 3–17 (2005)
96. Pap, J.S., et al.: Designing a robot based on parallel mechanism to reproduce human chewing behaviour. In: Proceedings of the Joint Conference on Robotics, Munich, Germany (2006)
97. Xu, W.L., et al.: Design of a biologically inspired chewing robot. IEEE Transactions on Industrial Electronics 55, 832–841 (2007)
98. Bowey, C., Burgess, D.: Robotic temporomandibular joint. Final Year Project Report. School of Mechanical Engineering, The University of Adelaide (2005)
99. Dellow, P.G., Lund, J.P.: Evidence for central timing of rhythmical mastication. J. Physiol. 215, 1–13 (1971)
100. Karker, F.R., et al.: Texture of fresh fruit. Hort. Rev. 20, 121–224 (1997)
101. Thexton, A.J.: Mastication and swallowing: a review. Br. Dent. J. 173, 197–206 (1992)
102. Woda, A., et al.: Adaptation of healthy mastication to factors pertaining to the individual or to the food. J. Oral. Rehab. 89, 28–35 (2006)
103. Lassauzay, C., et al.: Variability of the masticatory process during chewing of elastic model foods. Euro. J. Oral Sci. 108, 484–492 (2000)
104. Xu, W.L., et al.: Object-oriented knowledge representation and discovery of human chewing behaviours. Eng. Appl. Artif. Intell. 20, 1000–1012 (2007)
105. Xie, D., et al.: Object-oriented knowledge framework for modeling human mastication of foods. Exp. Sys. Appl. 36, 4810–4821 (2009)
106. Han, J., Kamber, M.: Data mining: concepts and techniques. Morgan Kauffman Publishers, San Francisco (2001)
107. Witten, I.H., Frank, E.: Data mining: practical machine learning tools and techniques, 2nd edn. Morgan Kauffman Publishers, San Francisco (2005)

Chapter 2
Robotic Models of the Masticatory System*

Abstract. In order to quantitatively evaluate the dynamic changes in the texture of foods during chewing in humans, a robotic device is required to enable the reproduction of human chewing behaviour. The first step in designing such a device requires the jaw mechanism to be modeled and analysed through simulations. Following a biological examination of the muscles used in the process of mastication, it was determined that those responsible for chewing movements can be represented by a set of linear actuators. By placing these actuators between the mandible and the skull, according to human biological structure and functionality, a robotic model of parallel mechanism was identified. The physical dimensions and properties of the mechanism were measured from a model replica of the human skull. Simulations for the mandible movements with respect to given muscular actuations, and for the muscular actuations required for actual human chewing patterns, were conducted using the Matlab/SimMechanics toolbox and the SolidWorks/CosmosMotion, respectively.

2.1 A Mechanism Model of the Masticatory System

2.1.1 Jaw Muscles and Movements

More than twenty muscles are involved in the process of human mastication [1, 2]. The temporalis muscle (as shown in Fig. 2.1a) is a large, flat muscle. Its fibres can be divided into two parts: firstly, the anterior fibres that elevate the mandible (lower jaw) and close the mouth and secondly, the posterior fibres that contribute to the complex grinding movement by retraction of the mandible. The pterygoid (as shown in Fig. 2.1b) consists of a family of muscles including the lateral and medial pterygoids. The lateral pterygoids work to protract the mandible and open the mouth, and the medial pterygoids are mostly used to protract the mandible.

---

* Reprinted with modification from the following papers: Daumas B, Xu WL and Bronlund J (2005) Jaw mechanism modelling and simulation. Mech. Machine Theor. 40:821-833, Xu WL, Bronlund J and Kieser J (2005) Choosing new ways to chew, a robotic model of the human masticatory system for reproducing chewing behaviours. IEEE Robotics Automation Mag.12:90-98, and Pap JS, Xu WL and Bronlund J (2005) A robotic human masticatory system: kinematics simulations. Int. J. Intell. Sys. Techno. Appl. 1:3-17, with permission from Elsevier, IEEE and Inderscience.

W. Xu and J.E. Bronlund: Mastication Robots, SCI 290, pp. 35–58.
springerlink.com

The masseter (as shown in Fig. 1c) is a flat, quadrilateral muscle with both deep and superficial component parts. It contributes mostly to the elevation of the mandible (i.e. mouth closing), but also plays a role in protracting the mandible. Underneath the mandible, the hyoid bone supports a muscle set referred to as the suprahyoid muscles (Fig. 2.1d). Among them, the digastric, stylohyoid, mylohyoid, geniohyoid and platysma muscles are involved in mouth opening and subsequently the depression of the mandible.

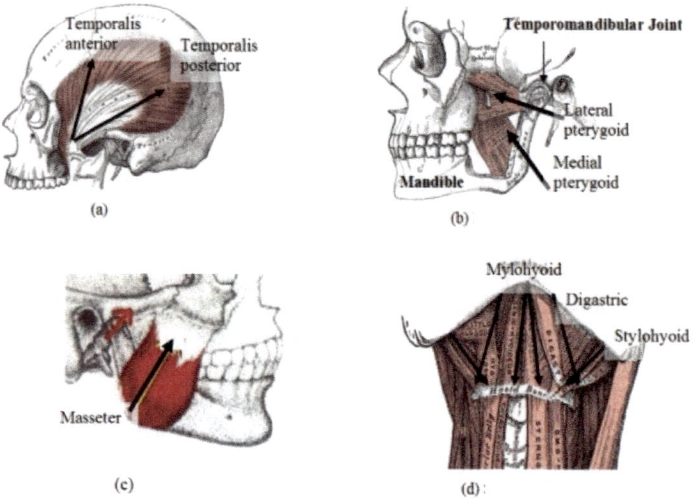

Fig. 2.1 Muscles for mastication (a) left temporalis muscles, (b) left pterygoids muscles, (c) right masseter muscles and (d) suprahyoid muscles. (©1918 Lea & Febiger, Reprinted from [2] with permission).

The mandible, or the lower jaw, is attached to the rest of the skull by muscles by the so-called temporo-mandibular joint or TMJ (as shown in Fig. 2.1b). The mandible therefore, cannot move as a free body in space, as it is biologically constrained by both joints and muscles.

Human chewing behaviour can be described by way of two basic mandibular movements: namely clenching and grinding movements (Fig. 2.2). Clenching consists of the successive elevation and depression of the mandible and uses a variety of muscles, but mostly the masseter and temporalis anterior muscles. Grinding involves almost all of the jaw muscles together with the incisal point (the point between the two lower incisors) with the resulting movement tracing a circle within the horizontal plane. Thus, complex human mastication can be regarded as an aggregate of both clenching and grinding movements.

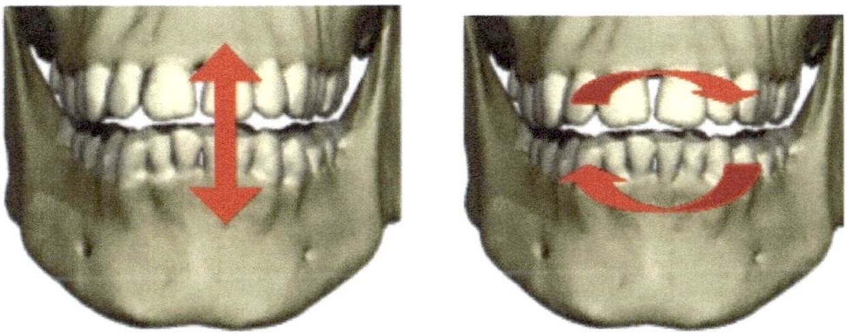

(a) (b)

Fig. 2.2 Two basic chewing movements, (a) basic clenching and (b) basic grinding

2.1.2 Jaw Mechanism Modeling

2.1.2.1 Mandible Reference Points

Reproducing the jaw movements in a 3D space requires an appropriate mechanism. The movability of a mandible can be observed by the trajectories of a small number of reference points. Shown in Fig. 2.3 are the reference points commonly observed: these being the incisal point (IP), kinematic condylar point (i.e., LCP and RCP) and first molar point (i.e., LMP and RMP) [3]. The incisal point is the point between the two lower jaw incisors. The kinematic condylar points are the points at the top of both the left and right processus, these points are simply referred to as the left processus and right processus. These two points are highly constrained in their movement as they must remain positioned within the mandibular fossa. The first molar points are those on the first lower molars. The trajectories of all the five aforementioned reference points should constitute the qualitative specifications in the development of any scientifically suitable human jaw mechanism.

2.1.2.2 Jaw Mechanism Model

A muscle works along a linear axis and actuates only in the contraction phase. In this study linear actuators were used to replace muscles in providing both contraction and depression. As discussed previously, the muscles collectively generate the movement of the mandible which is pivoted at the two TMJs and the mandible movement may be regarded as an aggregate of both clenching and grinding movements. For clenching, the anterior temporalis and masseter muscles are largely responsible for mandible elevation, whereas the suprahyoid muscles together with gravity perform the depression action.

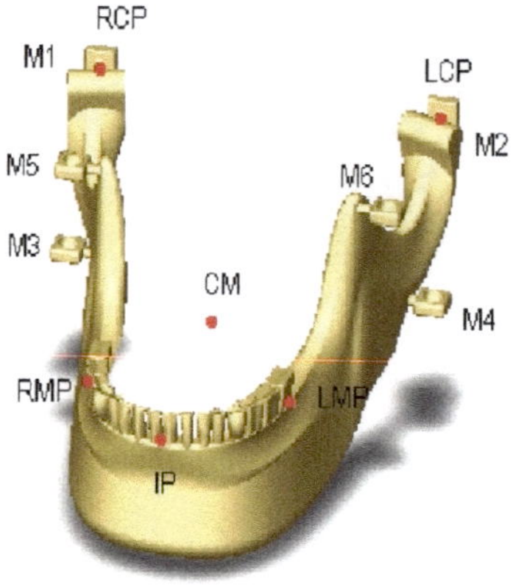

Fig. 2.3 A mandible model and reference points

Four linear actuators were therefore used to reproduce both mandible elevation and depression in place of the anterior temporalis, masseter and suprahyoid muscles. As shown in Fig. 2.4, these actuators are referred to as the Right Masseter, Left Masseter, Right Temporalis and Left Temporalis.

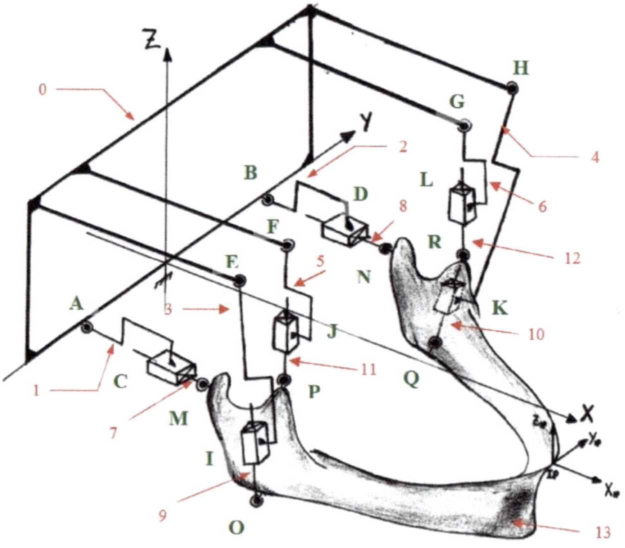

Fig. 2.4 Sketch showing jaw mechanism

In relation to the grinding movement, which is mainly produced by the left and right posterior temporalis muscles, two actuators are placed behind each mandible processus and connect to the TMJs. They are referred to as the Right Grinder and Left Grinder (Fig. 2.4). Each actuator is then attached to the skull and mandible through spherical joints at both ends, resulting in the jaw mechanism model (Fig. 2.4). The spatial model consists of 14 links, 6 prismatic actuators and 12 spherical joints (as itemised in Table 2.1).

Table 2.1 Notations of jaw mechanism constituents

Links	Passive joints	Active joints (actuators)
0: Skull (ground)	A: back right mandibular fossa	C: right grinder
1: Back right grinder	B: back left mandibular fossa	D: left grinder
2: Back left grinder	E: upper right masseter attach	I: right masseter
3: Upper right masseter	F: upper right temporalis attach	J: right temporalis
4: Upper left masseter	G: upper left temporalis attach	K: left temporalis
5: Upper right temporalis	H: upper left masseter attach	L: left masseter
6: Upper left temporalis	M: frontal right mandibular fossa	
7: Frontal right grinder	N: frontal left mandibular fossa	
8: Frontal left grinder	O: lower right masseter attach	
9: Lower right masseter	P: lower right temporalis attach	
10: Lower left masseter	Q: lower left temporalis attach	
11: Lower right temporalis	R: lower left masseter attach	
12: Lower left temporalis		
13: Mandible (end-effector)		

It was essential to verify that the reproduction of human-like chewing movements was achievable in terms of kinematics with this model. Kinematics is concerned with mechanism mobility which can be evaluated using the Kutzbach criterion [4],

$$m = 6(n-1) - 5j_1 - 4j_2 - 3j_3 - 2j_2 - j_5 \qquad (2.1)$$

where m is the number of independent degrees of freedom (DOF) required for the model to be a movable mechanism, n (=14) is the number of links, j_1 (=6), j_2(=0), j_3(=12), j_4(=0) and j_5(=0) refers to numbers 1-, 2-, 3-, 4- and 5-, being each of the 5 DOF joints respectively.

The number of calculated independent DOFs is m=6x13-5x6-3x12=12. Considering there is one local rotational movement associated with each actuator (i.e., between each pair of spherical joints), the final independent DOFs becomes m=12-6=6 - being the number of active joints (actuators). This therefore proves that the reproduction of human chewing movements is possible using the mechanism model as specified.

2.2 Kinematics Simulations in SimMechanics

To analyse the mandible movements given actuation inputs or to find suitable actuation inputs (controls) for a predefined mandible motion, SimMechanics, one of the toolboxes for Matlab Release 13, was used [5]. The toolbox is an object-oriented program able to perform simulations of most motion analysis problems in relation to machines without the need for dynamic equations. The physical assembly of a machine and its physical quantities as well as actuation/control modes must be provided.

2.2.1 Physical Quantities of the Mechanism

To compute a machine model in SimMechanics, all physical attributes must be specified in terms of an inertial matrix expressed at each attributes center of gravity. The mandible, which has a complicated shape, is approximated by a set of five rigidly-welded parallelepipeds. The actuators are represented as two steel cylinders linked by a cylindrical joint.

The approximated jaw mechanism has three body components: the skull (or the ground), the actuators and the mandible (or the end-effector). Each of the three component dimensions and mass properties were either measured, estimated or calculated from a replica human skull. However, the skull (as the ground in the jaw mechanism) does not require a specified mass.

Table 2.2 and Table 2.3 give the coordinates of the points (G_i, i=1,2,…, 6) where the actuators are attached to the ground and the coordinates of the points (M_i, i=1,2,…, 6) where the actuators respectively attach to the mandible.

Each actuator is composed of one hollow and one solid cylinder, the total length of which is the distance between the attachment points M_i and G_i. It is assumed that each cylinder is ¾ of the total length between its attachment points. The hollow cylinder has a 1 cm inner radius and a 2 cm outer radius whilst the solid cylinder is 1 cm in radius. The actuator stroke is either +/- ¼ of its length. Table 2.4 presents the physical properties of the six actuators.

Table 2.2 Ground attaching point coordinates of the mechanism[1]

Description	Origin	X_0 (m)	Y_0 (m)	Z_0 (m)
Attaching the left grinder to the skull	G1	−0.04	0.05	−0.01
Attaching the right grinder to the skull	G2	−0.04	−0.05	−0.01
Attaching the left masseter to the skull	G3	0.015	0.065	−0.015
Attaching the right masseter to the skull	G4	0.015	−0.065	−0.015
Attaching the left temporalis to the skull	G5	0.04	0.05	0.075
Attaching the right temporalis to the skull	G6	0.04	−0.05	0.075

Table 2.3 End-effector attaching point coordinates of the mechanism

Description	Origin	X_0 (m)	Y_0 (m)	Z_0 (m)
Attaching the left grinder to the mandible	M1	0	0.05	−0.01
Attaching the right grinder to the mandible	M2	0	−0.05	−0.01
Attaching the left masseter to the mandible	M3	0.015	0.055	−0.095
Attaching the right masseter to the mandible	M4	0.015	−0.055	−0.095
Attaching the left temporalis to the mandible	M5	0.03	0.05	−0.005
Attaching the right temporalis to the mandible	M6	0.03	−0.05	−0.005

[1] All coordinates are defined in the world coordinates system (WCS), XYZo.

Table 2.4 Actuators properties[2]

Actuator name	Nominal length (m)	Mass of hollow cylinder (kg)	Mass of solid cylinder (kg)	Total mass (kg)	Stroke (m)
Right grinder	0.04	0.219	0.073	0.292	±0.01
Left grinder	0.04	0.219	0.073	0.292	±0.01
Right masseter	0.0806	0.442	0.147	0.589	±0.02
Left masseter	0.0806	0.442	0.147	0.589	±0.02
Right temporalis	0.0806	0.442	0.147	0.589	±0.02
Left temporalis	0.0806	0.442	0.147	0.589	±0.02

Table 2.5 Mandible dimensions and mass properties[3]

Mandible components	GC	GC position in WCS	Length (m)	Width (m)	Height (m)	Mass (kg)
Right processus	A1	[0.015 −0.05 −0.05]	0.03	0.01	0.1	0.027
Right side	A2	[0.07 −0.05 −0.09]	0.07	0.01	0.02	0.0126
Central part	A3	[0.095 0 −0.09]	0.01	0.09	0.02	0.0162
Left side	A4	[0.07 0.05 −0.09]	0.07	0.01	0.02	0.0126
Left processus	A5	[0.015 0.05 −0.05]	0.03	0.01	0.1	0.027

Five parallelepipeds make up the mandible. A parallelepiped is defined by its length (along x-axis), its width (along y-axis) and its height (along z-axis). Table 2.5 gives the physical properties of the mandible.

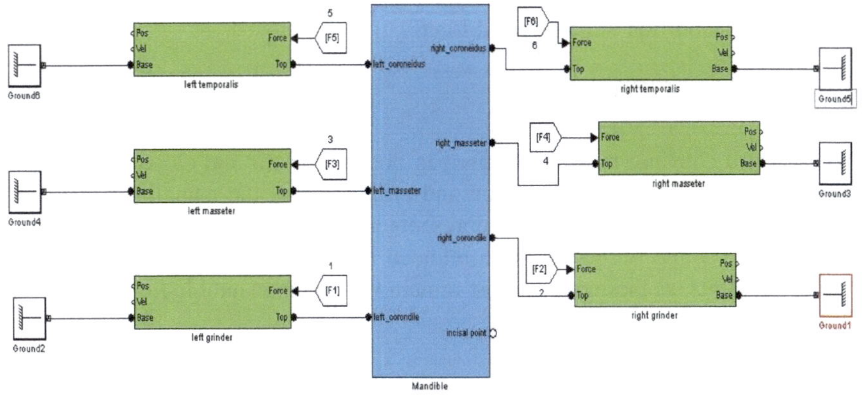

Fig. 2.5 The jaw mechanism in SimMechanics

2.2.2 *Actuation Modes*

The jaw mechanism in SimMechanics is illustrated in Fig. 2.5. The mandible is moved by the six actuators placed between the mandible and the skull (or ground).

[2] Steel density is 7747 kg/m^3.
[3] Human bone density is estimated at 900 kg/m^3.

Two actuating modes are available in SimMechanics, one for specifying pure motion of the actuators and the other for applying a force to the actuators. The former is used to perform an open-loop simulation (mainly in order to determine the trajectories of some of the reference points on the mandible). The latter is used to control the jaw mechanism so as to achieve pre-specified mandibular movements and chewing forces in a closed-loop manner. Both open- and closed-loop simulations of the mechanism were conducted.

For the open-loop simulations, the mandible motion was specified in terms of its position and orientation whilst actuations were calculated via a Simulink generator - this having originally been developed for the Stewart platform mechanism [6]. The inputs of this generator are three time-functions of the mandible position in the x-, y- and z- axis, respectively and three time-functions of the mandible orientation, defined as the angles around the x-, y- and z-axis, respectively. The outputs are the position, velocity and acceleration of all actuators.

The closed-loop simulations require the design of a controller for each of the six actuators. These controllers evaluate the current state of the jaw mechanism and calculate the forces being applied to each actuator.

While the real mastication process involves application of biting forces on foods, the control of pure mandible motion (without taking forces into account) was considered at this stage of the research. In the simulations, each actuator was independently controlled by a PID controller. The desired position was then extracted from the actuation generator and the difference between the desired and measured positions constituted the error to be corrected. The error from proportional and integral actions and the velocity measurements were used to generate the derivative portion of the applied force.

2.2.3 Results and Analysis

The desired chewing movements used in both the open- and closed-loop simulations were specified by the position and orientation of the mandible (as given in Table 2.6). The ideal movement represents a banana-shaped trajectory of the incisal point [7] and occurs within a reduced range of movement at the kinematic condylar points, as these points must remain within the mandible fossa.

Table 2.6 Desired chewing movement

Generator input	Symbol	Signal type	Time-function
Roll angle of the mandible	Xang	Constant	0
Pitch angle of the mandible	Yang	Sinusoid	$15 + 15\sin(2\pi t)$
Yaw angle of the mandible	Zang	Sinusoid	$5\sin(2\pi t + \pi/3)$
X position of the mandible	Xpos	Sinusoid	$0.005\sin(2\pi t + \pi/3)$
Y position of the mandible	Ypos	Constant	0
Z position of the manidble	Zpos	Constant	0

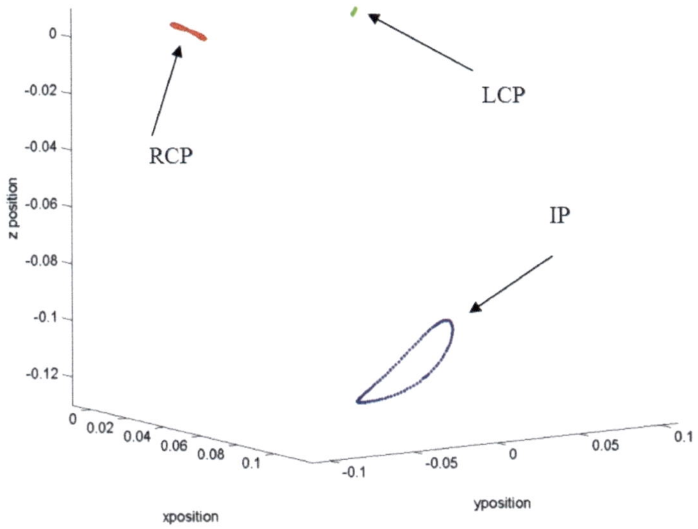

Fig. 2.6 Trajectories of the jaw reference points IP, LCP and RCP under open-loop control

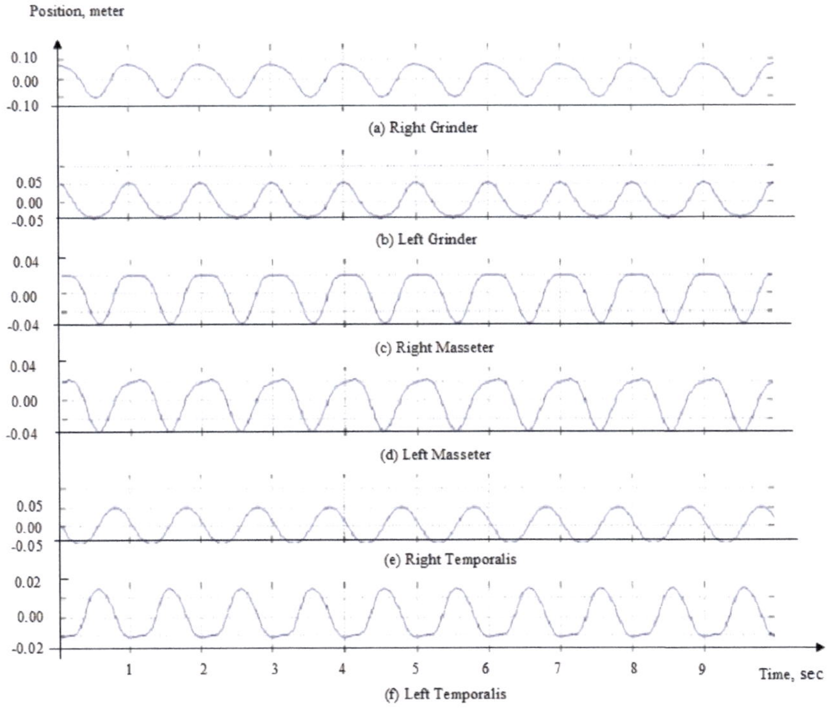

Fig. 2.7 Time variations of six muscular actuations

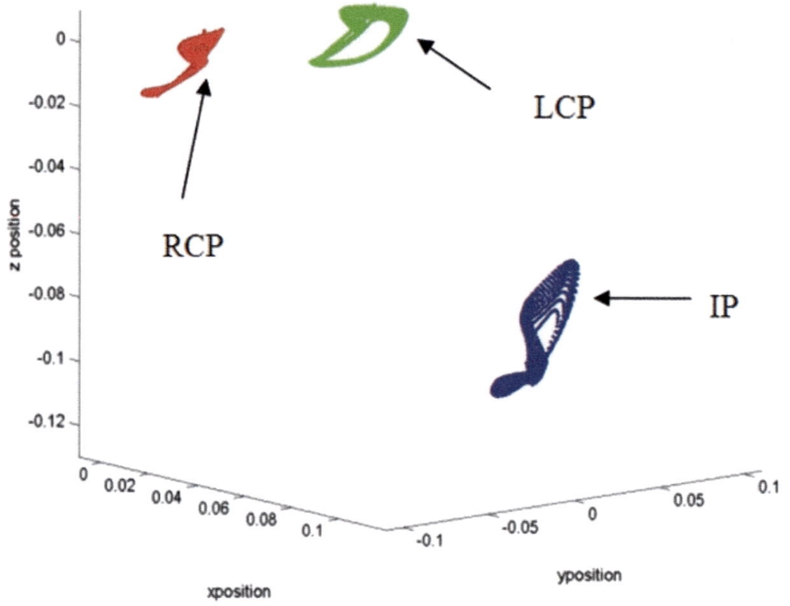

Fig. 2.8 Trajectories of the reference points IP, LCP and RCP under closed-loop control

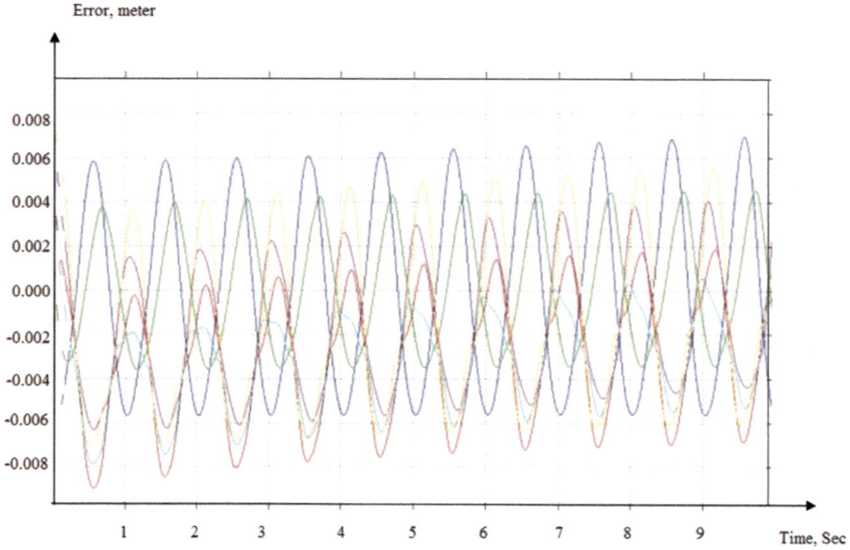

Fig. 2.9 Position errors in the six actuations

Fig. 2.6 shows the trajectories of the mandible reference points under open-loop control. The results revealed that the jaw mechanism reproduced a promising mandible motion in terms of the shape of the trajectories. No further quantitative analysis was possible as data on human chewing of various food samples was not available at this stage of the project research.

Fig. 2.7 presents the time variations of the six actuations which were determined using inverse kinematics from the Stewart platform mechanism [6].

For the closed-loop simulations, various sets of the PID controller gains were trialled. There was no significant difference in the trajectory tracking performances of the jaw mechanism under control. Fig. 2.8 shows the trajectories of the three reference points with the gains K_p=2000, K_i=200 and K_d=200 for the six independent PID controllers. It was found that the reproduced trajectories did not mirror those that were desired and the jaw mechanism was in fact unstable. This finding was further confirmed by the error that occurred in actuating positions (Fig. 2.9). These should ideally have diminished over time.

Although the trajectory tracking performance might be improved through carefully tuning the three PID gains, the six independent PID controllers would never be capable of correcting this control problem. This is due to fact that the jaw mechanism is a highly coupled, nonlinear, multi-input and multi-output system. Further work must therefore be conducted in relation to modeling, stability analysis, and in the design of a multivariable controller. This would allow the development of a controller capable of that tracking a specified trajectory.

2.3 A Robotic Model of the Masticatory System

2.3.1 The Robotic Model

According to the aforementioned biomechanics information, the mandible can be regarded as a rigid body, suspended from the skull through both TMJs and driven coordinately by three muscles controlled by the central nervous system. Functionally, the mandible is movable in relation to the skull in a 3D space. Movement is constrained by the biological structure of the TMJs and the jaw muscles together with the interaction between the individual's dentition and foodstuffs.

It seems technically challenging to mechanically replicate each of these distributed muscles specified in the robotic model. One approach to solving this problem is the use of a linear cylindrical actuator in place of a group of muscles. An actuator can act bi-directionally with both ends attached to the skull and the mandible via spherical joints. This enables the actuating force to reliably be exerted in the same direction as the resultant muscle forces. This results in a robotic model (as shown in Fig. 2.10). L_1 and L_2 refer to the actuators for the right and left pterygoid externus, respectively; L_3 and L_4 refer to the right and left temporalis, respectively;

L_5 and L_6 refer to the right and left masseter, respectively; finally G_i and M_i (i=1,2, …, 6) denote the muscles' origin and insertion locations on the skull and the mandible, respectively.

The pterygoid externus actuators (L_1 and L_2) are placed posterior to the TMJs (at M_1 and M_2) for ease of placement. This is made possible due to the actuators capability to act in a bi-directional manner. The pterygoid internus are not present in the model because they only play a minor role in the closing and opening of the mouth. Their absence helps avoid a redundant affect upon the actuations (due to their use in place of muscles) and consequently eliminates any over-constraints that may prevent the mandible from being movable.

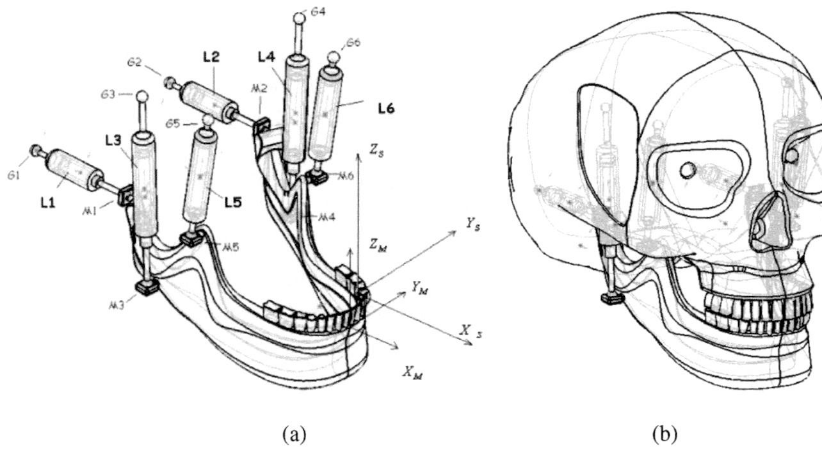

Fig. 2.10 A robotic model in SolidWorks, (a) nomenclature and coordinate systems and (b) the robot covered by the skull

In the model, points M_1 and M_2 represent the right and left condylar points, respectively. Each of these points is capable of tracing a different trajectory when the mandible moves. This replicates both clinical and biomechanical findings that, during jaw movement, the mandible does not rotate around a fixed condylar axis, but rather around an instantaneous axis that continuously changes its position in space [8, 9, 10]. The working and balancing condylar points exhibit different trajectories that vary in relation to the type of food being chewed [11]. The forces within the moving spherical joints, M_1 and M_2, are always the normal to the condylar paths, if friction is not taken into consideration.

2.3.2 Physical Quantities of the Model

The robotic model proposed above can be viewed as a variant version of the Stewart platform robot due to the presence of both a moving and a fixed plate. The proposed model is a 6-DOF robot and consists mainly of the skull (or the ground), the six cylindrical actuators and the mandible (or the end-effector). The mandible has an irregular shape and is approximated using a replica human skull (as shown in Fig. 2.11). The skull, as the immovable component of the model, does not require physical parameter specification. The locations of both actuators' insertions on the mandible and positions on the skull are approximated from the same replica skull (as shown in Fig. 2.10 and Fig. 2.11). The coordinates of these actuators are given in Table 2.7 and Table 2.8. The actuators represent a cylindrical joint that allows both translation along, and rotation about the axis M_iG_i (i=1, 2,…, 6).

The reference points that can be used for the description of human chewing behaviour, are incisal point (IP), right molar point (RMP), left molar point (LMP), right condylar point (RCP), and left condylar point (LCP). They are shown in Fig. 2.11 and their coordinates are given in Table 2.9. The centre of mass (CM) point is determined in accordance with a bone density of 900 kg/m3 having been assigned to the mandible. The two main coordinate systems used in the model are: the skull system ($oxyz_s$), which is fixed on the skull and to which the mandible movements refer, and the mandible system ($oxyz_m$), which is fixed on the mandible and whose initial location (when the mouth is closed) differs from $oxyz_s$ by only a translation of -9.1 mm along the z-axis.

Table 2.7 Coordinates of the locations where the actuators are attached on the skull in $oxyz_s$ (unit: mm)

Attaching points	*Initials*	X_S	Y_S	Z_S
Joint between skull and L1	G1	–158.5	–41.91	25.4
Joint between skull and L2	G2	–158.5	41.91	25.4
Joint between skull and L3	G3	–85	–48.75	85.5
Joint between skull and L4	G4	–85	48.75	85.5
Joint between skull and L5	G5	–53.8	–38.75	82.5
Joint between skull and L6	G6	–53.8	38.75	82.5

Table 2.8 Coordinates of the locations where the actuators are attached on the mandible in $oxyz_m$ (unit: mm)

Attaching points	*Initials*	X_M	Y_M	Z_M
Joint between jaw and L1	M1	–101.344	–42	34.5
Joint between jaw and L2	M2	–101.344	42	34.5
Joint between jaw and L3	M3	–80.844	–44.75	–1
Joint between jaw and L4	M4	–80.844	44.75	–1
Joint between jaw and L5	M5	–62.044	–38.75	31.5
Joint between jaw and L6	M6	–62.044	38.75	31.5

Table 2.9 Reference points' coordinates in the mandible system (unit: mm)

Reference points	*Initials*	X_M	Y_M	Z_M
Incisor point	IP	0	0	0
Right molar point	RMP	–25.734	–24.324	–0.766
Left molar point	LMP	–25.734	24.324	–0.766
Right condylar point	RCP	–101.344	–42	34.5
Left condular point	LCP	–101.344	42	34.5
Centre of mass	CM	–41.272	–0.003	4.181

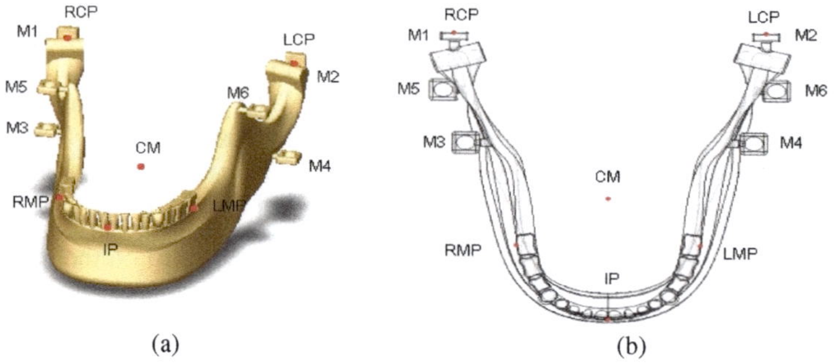

Fig. 2.11 The mandible, the actuators attaching points and reference points (a) 3D model and (b) the top view wire frame model

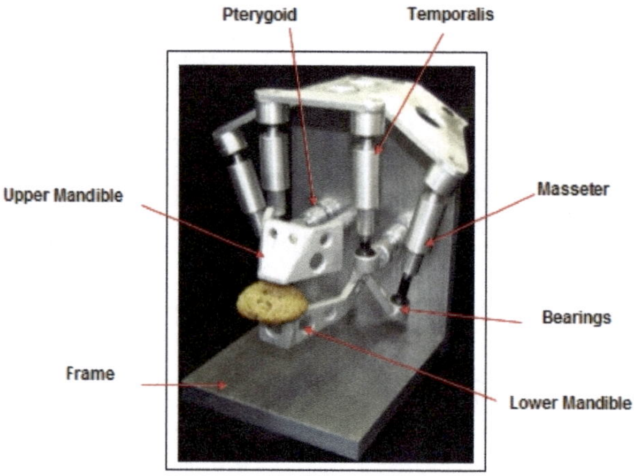

Fig. 2.12 A kinematical model of the robot

Fig. 2.12 is a photo of the final physical model based on the aforementioned design. The gap between the two mandibles allows for the absence of the upper and lower sets of teeth.

The model is made entirely of aluminium and the spring-loaded cylindrical joints ensure the model maintains its equilibrium. Fig. 2.13 illustrates the robotic model performing clenching and grinding movements. It should be noted that some of the actuators' attachments were shifted slightly from their corresponding biological locations in order that the physical interference between the actuators could be eliminated. It should be further noted that the proposed model is not a typical platform robot. This is due to the fact that the two plates, one for the moving lower jaw and the other for the non-moving frame involving the upper jaw, are connected in such a way that the attachment points on either plate do not lie in a single plane. Consequently, a theoretical framework accounting for dynamics requires development.

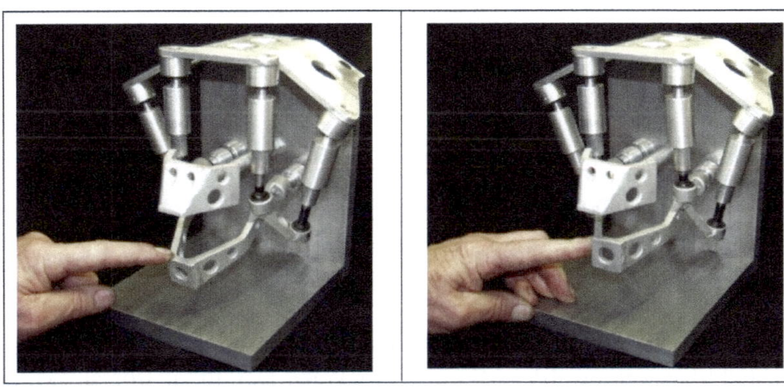

Fig. 2.13 The model performing clenching and grinding movements

2.4 Kinematics Simulation in SolidWorks/CosmosMotion

The robotic model developed in this research allows for two types of simulations to be performed. Firstly, forward kinematics, for generating the masticatory patterns by specified actuations of the actuators and secondly, inverse kinematics for determining the muscular actuations required for a desired masticatory pattern.

CosmosMotion, embedded within SolidWorks, was the tool used to simulate and animate robotic kinematics. Extensive simulations have been conducted with respect to the specified actuations of various continuous time functions, and the specified masticatory movements of different chewing behaviours. Two of these simulation scenarios are described below.

2.4.1 Forward Kinematics Simulation

This simulation was used to produce a chewing pattern in the 3-D space and to determine the ranges of the reference points on the mandible. The initial specifications of this robot were as follows:

The incisor point was placed at $[x \quad y \quad z]^T = [-12 \quad 0 \quad -30]^T$ within the skull system $oxyz_s$, the pterygoid actuators L1 and L2 were displaced by 4.45 mm from the their lowest end points and the coincident x-z plane of $oxyz_s$ and $oxyz_m$.

Being sophisticated in terms of accommodating multiple time-functions for chewing, the muscular actuations were able to be approximated by a series of harmonic functions. This study employed the following harmonic time function in order to drive the six actuators and demonstrated the capability of the model to simulate masticatory patterns. It should be noted however, that this particular set of functions is not associated with any realistic chewing behaviours.

$z = 4\sin(0.5t + \pi/2)$, for pterygoid externus actuators L_1 and L_2
$z = \sin(0.5t)$, for temporalis actuators L_3 and L_4
$z = 4\sin(0.5t - 0.083\pi)$, for right masseter actuator L_5
$z = 4.175\sin(0.5 - 0.1\pi)$, for left masseter actuator L_6

where z is the actuator's displacement around its initial position.

The spatial trajectories of IP, RMP, LMP, RCP and LCP in the frontal, sagittal and horizontal planes are shown in Fig. 2.14 and 2.15. It was found that the lateral, inferior-superior and anterior-posterior ranges of the incisors are 8, 25 and 20 mm, respectively, and the lateral, inferior-superior and anterior-posterior ranges of the RCP are 8, 2 and 8 mm, respectively.

The condylar movements were consistent with clinical and biomechanical findings [13, 14] and the incisal point moved within the limits of the Posselt envelope [12]. The shape of the chewing path appears to be one from a subject chewing on the left-side of the jaw. However, it should be noted that the periodic actuations were arbitrarily chosen for simulation purposes only and do not reflect actual human chewing processes. This was particulary true in terms of the temporal information in relation to chewing.

Fig. 2.16 and Fig. 2.17 show the side and frontal views of the jaw with the reference point trajectories superimposed.

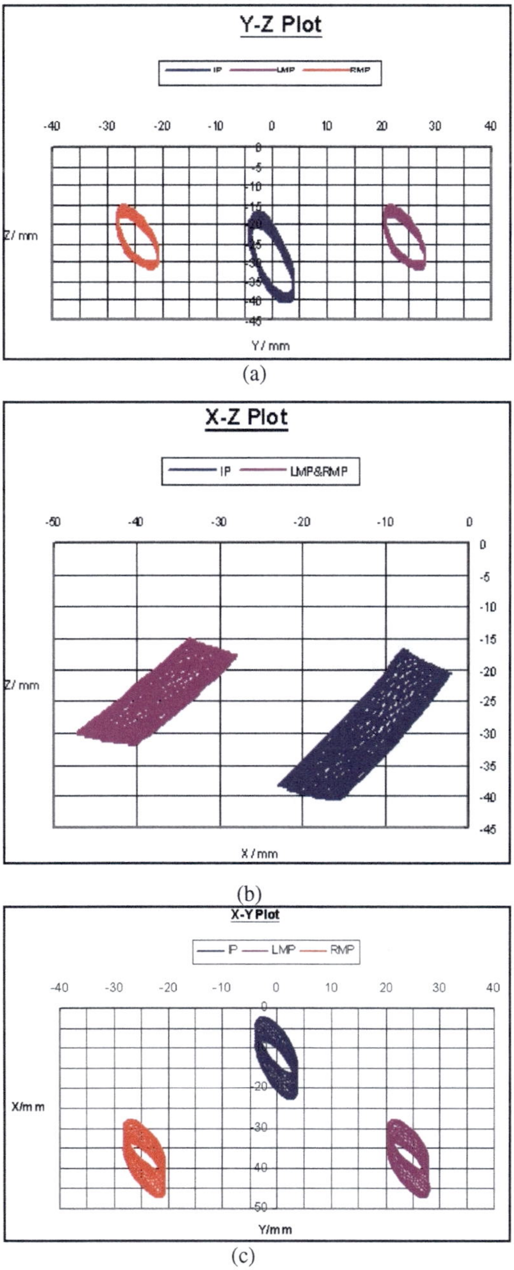

Fig. 2.14 IP, RMP and LMP trajectories, (a) the frontal plane, (b) the sagittal plane and (c) the horizontal plane

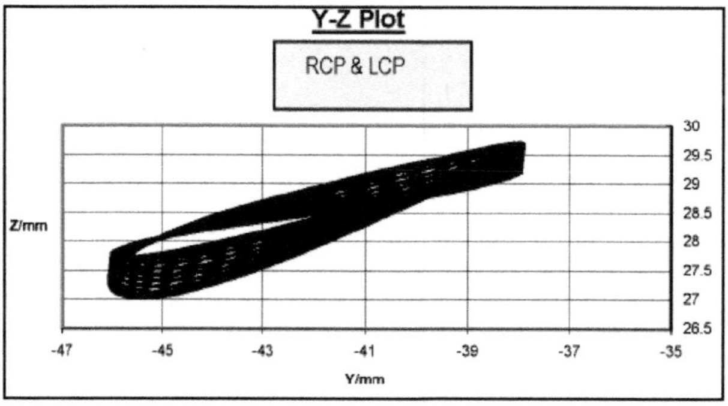

(a)

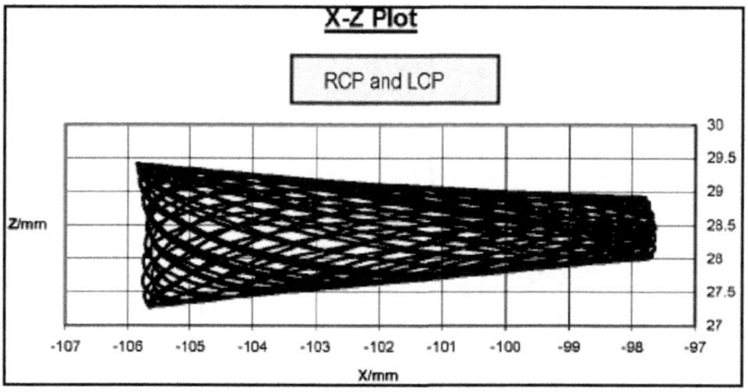

(b)

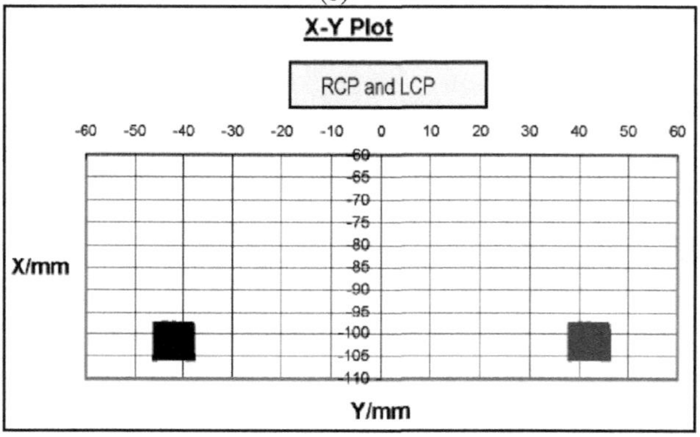

(c)

Fig. 2.15 RCP and LCP trajectories, (a) the frontal plane, (b) the sagittal plane, and (c) the horizontal plane

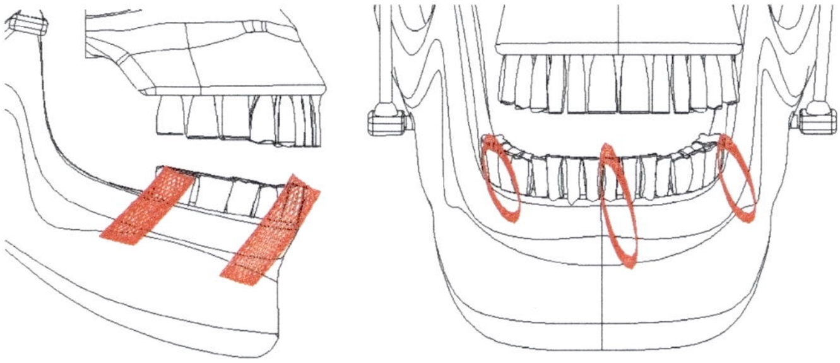

(a) (b)

Fig. 2.16 Jaw with IP, RMP and LMP trajectories, (a) side view and (b) frontal view

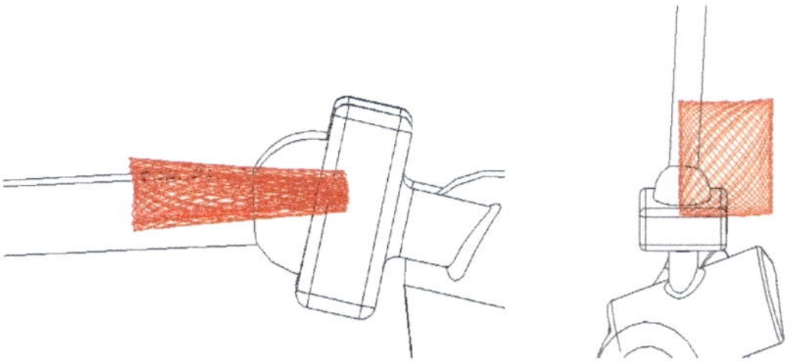

(a) (b)

Fig. 2.17 TMJ with RCP and LCP trajectories, (a) side view and (b) frontal view

2.4.2 Inverse Kinematics Simulation

Data for this simulation was gathered from the recorded masticatory sequence of a subject (provided by INRA, France). The subject chewed on the right-side of the jaw and the test food exhibited hard elastic properties [15]. The only historical data available from the simulation was that of the y and z coordinates of IP in $oxyz_s$ (as plotted in Fig. 2.18 and Fig. 2.19). This data alone, however, is insufficient to enable the mandible movements to be fully specified. In order to do so, it must be taken into account that the mandible is further constrained as follows: the middle point of the two condyles M_1 and M_2 moves along a straight line in the *x-z* plane and 25.4 mm distant from the origin of $oxyz_s$. Fig. 2.20 and Fig. 2.21 show the trajectories of IP, RMP and LMP of the mandible, and RCP and LCP of the TMJ with respect to a specified motion.

It can be seen from the lateral movements (Fig. 2.18 and Fig. 2.19) that the subject chewed on the right-side of the jaw and performed three peaks of the lateral excursion movements. These peaks can be explained by the fact that the tongue works to collect or re-locate the food bolus within the mouth to an accurate position. It can also be seen from the superior-inferior movements that for first couple of chewing cycles the mouth openings reached a maximum of 30 mm. During the next few chewing cycles the maximum openings reached approximately 25 mm and for the remainder of the chewing process the maximum opening reached was 20 mm.

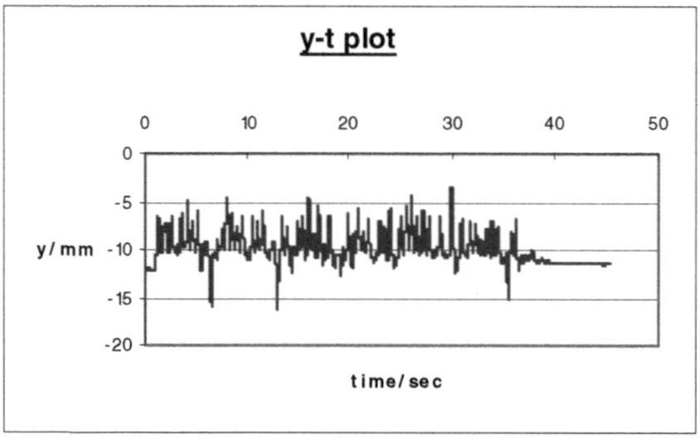

(a)

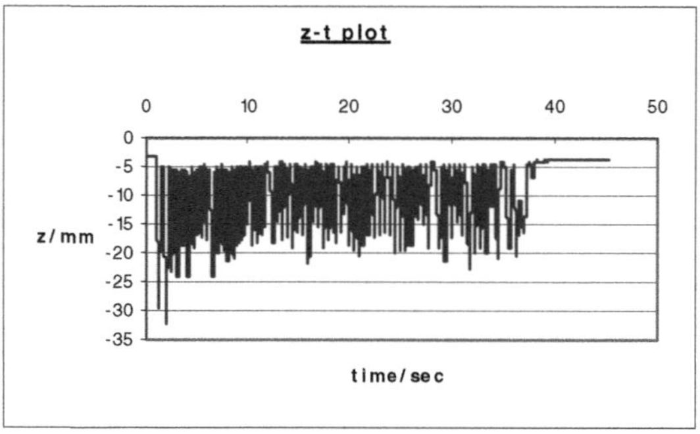

(b)

Fig. 2.18 Temporal chewing trajectories of IP from experiments, (a) lateral movement and (b) superior-interior movement

In a simulation such as this, the six actuators must be actuated to follow the time functions given in Fig. 2.22. This is required in order to reproduce an actual real human chewing pattern. It can be seen that the maximum strokes of the six actuators are approximately 18.7, 15.4, 2.4, 4.5, 9.4 and 12.7 mm, respectively.

The two pterygoid actuators must be more displaced than the two temporalis actuators and even more displaced than the two masseter actuators. The maximum pterygoid actuations occur in the first few chewing cycles, and for most of the chewing process their actuations were less than 9.3 and 7.7 mm, respectively. The two masseter actuations were smaller relatively and more uniform over the entire chewing period. Like the pterygoid actuators, the temporalis actuators experience maximum displacements in the first few chewing cycles but exhibit fairly small actuations of less than approximately 4.7 and 6.3 mm, respectively. The results of these actuations revealed an average chewing cycle of 1.9 seconds. Other time related chewing parameters could also be determined if required.

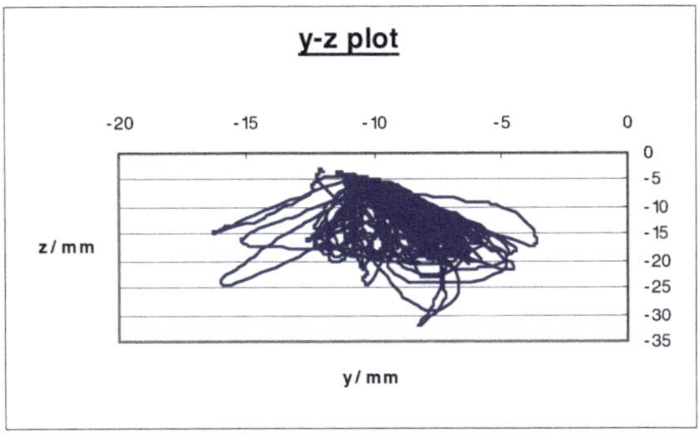

Fig. 2.19 Spatial chewing trajectories of IP in the frontal plane

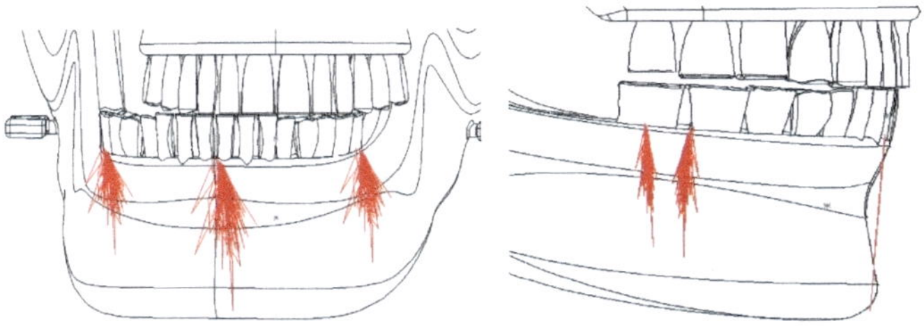

(a) (b)

Fig. 2.20 IP, RMP and LMP trajectories for a person, (a) frontal view and (b) side view

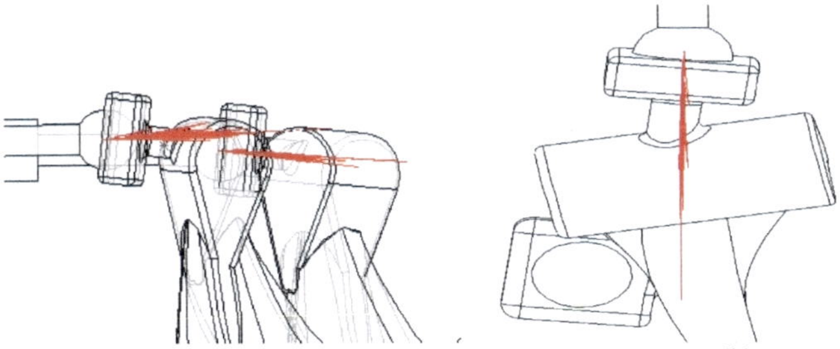

(a) (b)

Fig. 2.21 RCP and LCP trajectories, (a) right view and (b) top view

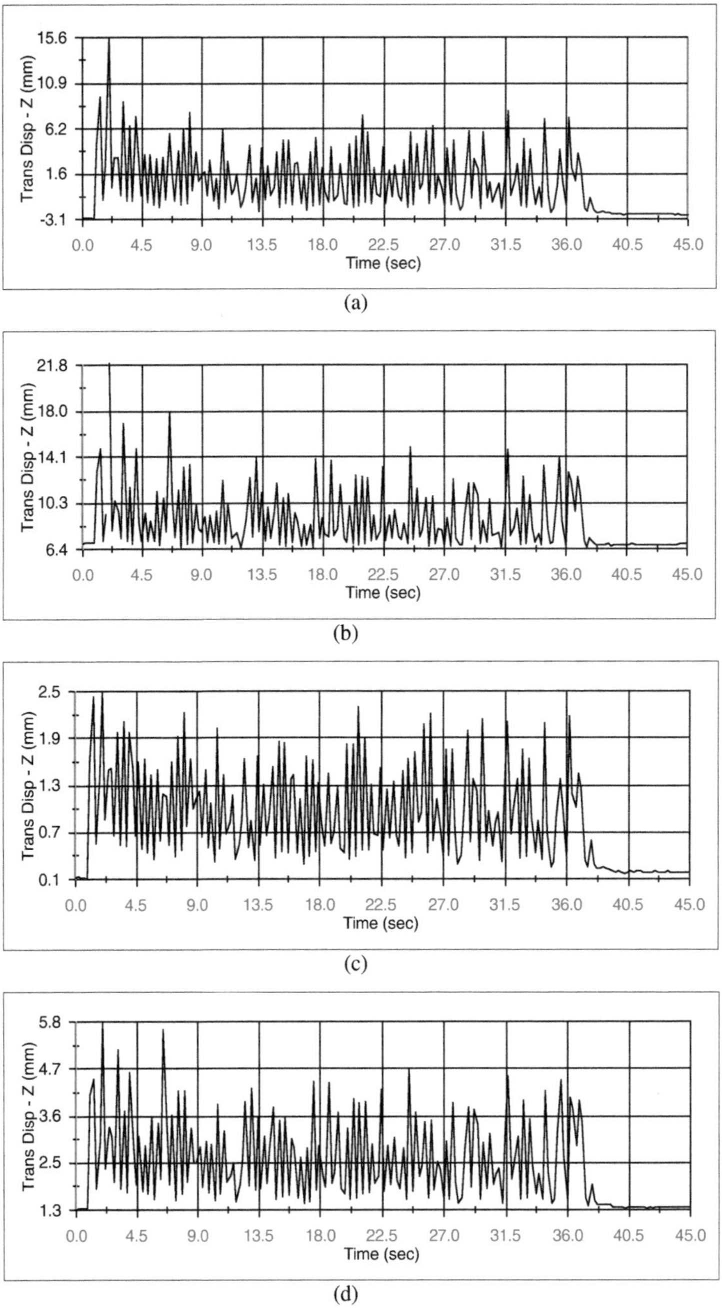

Fig. 2.22 Actuations required for specified chewing patterns, (a) right pterygoid actuator, (b) left pterygoid actuator, (c) right masseter actuator, (d) left masseter actuator, (e) right temporalis actuator, and (f) left tempolaris actuator

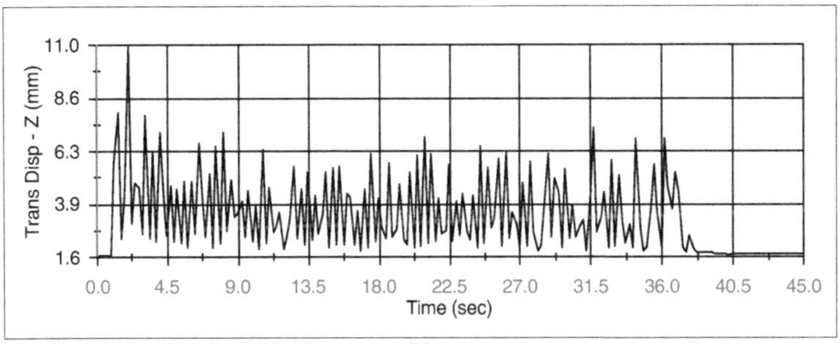

(e)

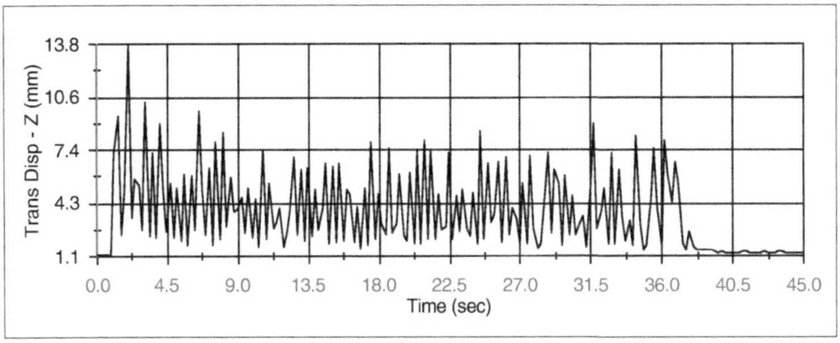

(f)

Fig. 2.22 (*continued*)

2.5 Summary

In order to reproduce human chewing movements a robotic model was developed based upon previous biomechanical findings of the human mastication system. The model allowed the performance of two different kinds of simulations. Firstly, it allowed the specification of actuator displacement in order to animate mandible movements. Secondly, it enabled the use of measurements from actual chewing processes as input data in order to generate the actuations required to accurately reproduce human chewing movements. The robotic model was validated by extensive simulations in Matlab/SimMechanics and SolidWorks/CosmosMotion.

Whilst the work presented in the chapter is preliminary, it does provide a viable robotic model upon which further research and development work can be conducted. As the robotic model is not a typical parallel mechanism, the issues that require further research include those in the areas of kinematics, dynamics, robotic force-motion control, and mechatronics design.

References

1. Palastanga, N.: Anatomy and human movement: structure and function. Butterworth-Heinemann, Oxford (1998)
2. Gray, H.: Anatomy of mastication. Lea & Febiger, Philadelphia (1918)
3. Nishigaw, K., et al.: Current and future technologies in jaw movement analysis. Lecture Slides. School of Dentistry, University of Tokushima, Japan (2003), `http://www.dent.tokushima-u.ac.jp/hotetu2/WCB/WCB01.html` (accessed May 19, 2003)
4. Shigley, E., Uicker, J.J.: Theory of machines and mechanisms. McGraw-Hill, New York (1980)
5. The MathWorks, SimMechanics user's guide. Version 1.1. Matlab Release 13 (2002)
6. Smith, N., Wendlandt, J.: Creating a Steward platform model using SimMechanics, The Matheworks - MATLAB Digest, 10 (2002)
7. Takanishi, A., et al.: Development of 3 DOF jaw robot WJ-2 as a human's mastication simulator. In: Fifth International Conference on Advanced Robotics, vol. 1, pp. 277–282 (1991)
8. Hannam, A.C.: Jaw muscle structure and function. In: McNeill, C. (ed.) Science and practice of occlusion. Quintessence Publishing Co., Inc. (1997)
9. Palla, S., et al.: Jaw tracking and temporomandibular joint animation. In: McNeill, C. (ed.) Science and practice of occlusion. Quintessence Publishing Co. Inc. (1997)
10. Hayasaki, H., et al.: A calculation method for the range of occluding phase at the lower incisal point during chewing movements using the curved mesh diagram of mandibular excursion (CMDME). J. Oral Rehab. 26, 236–242 (1999)
11. Anderson, K., et al.: The effect of bolus hardness on masticatory kinematics. J. Oral Rehab. 29, 689–696 (2002)
12. Tsuruta, J., et al.: An index for analyzing the stability of lateral excursions. J. Prosthodontic Dentistry. 7, 375–385 (2002)
13. Nakajima, J., et al.: Masticatory mandibular movements for different food texture related to onomatopoetic words. J. Med. Dent. Sci. 48, 121–129 (2001)
14. Scapino, R.P.: Morphology and mechanism of the jaw joint. In: McNeill, C. (ed.) Science and practice of occlusion, Quintessence Publishing Co. (1997)
15. Peyron, M.A., et al.: Effects of increased hardness on jaw movement and muscle activity during chewing of visco-elastic model foods. Exp. Brain Res. 142, 41–51 (2002)

Chapter 3
Mastication Robot of Linear Actuation*

Abstract. The aforementioned robotic model of the masticatory system was refined in this section of the research, utilising substantially more data. This chapter presents both the design of, and data collected from such a masticatory robot of linear actuation. As a linear actuator was used to represent the muscle groups required for human mastication, the spatial placement between the mandible (moving plate) and the maxilla (ground plate) allowed both the line of action and attachment sites of each muscle to be followed. The design requirements for each actuation system were specified in terms of actuation range, velocity and acceleration, plus actuation force. These specifications were determined via firstly, inverse kinematics analysis, by application of simulation software, and secondly via jaw force analysis, by the application of Pythagorean Theorem. The end design of the physical linear actuation is made up of a rotary motor, a gear reduction train and a lead-screw. The challenges in relation to building an entire mastication robot are also discussed.

3.1 A Refined Robotic Model

Several research teams have focused on the architecture of masticatory muscle groups in order to provide a set of data that can be used in the biomechanical modelling of the masticatory system [1, 2]. For example, using eight research cadavers, van Eijden *et al.* [3] plotted the 3D coordinates of the muscle attachment sites in order to determine the spatial positioning of the line of action of various masticatory muscle groups in the closed-mouth position. They differentiated *inter alia* the superficial and deep masseter, the anterior and posterior temporalis, and the inferior and superior lateral pterygoid. Their spatial orientations are shown in Fig. 3.1 where the action lines are projected onto the frontal and sagittal planes. Notation is represented as follows: 1 represents the superficial masseter; 2 the deep masseter, 3 the anterior temporalis, 4 the posterior temporalis, 5 the inferior lateral pterygoid and 6 the superior lateral pterygoid. Table 3.1 gives the right side attachment coordinates, the total muscle lengths and physiological cross-sectional areas of the main masticatory muscle groups.

* Reprinted with modification from Xu WL, Pap JS and Bronlund J (2008) Design of a biologically inspired parallel robot for foods chewing. IEEE Trans. Industrial Electron. 55:832-841, with permission from IEEE.

W. Xu and J.E. Bronlund: Mastication Robots, SCI 290, pp. 59–89.
springerlink.com

In the refined robotic mechanism, there is only one actuator representing one muscle group. To achieve the movement range of the mandible the muscle length of the biggest sub-muscle group in the closed-mouth position is taken as the actuator length. As a result, the actuator lengths for masseter, temporalis and lateral pterygoid were 45.6, 52.2 and 32.6 mm, respectively.

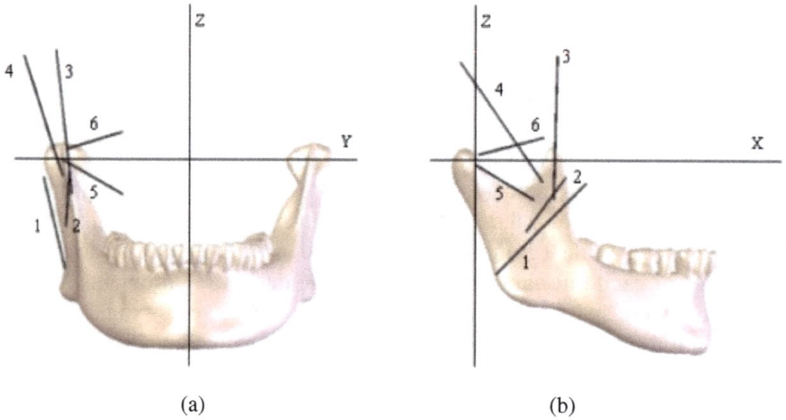

Fig. 3.1 Spatial orientations of the sub-muscle action lines in (a) frontal and (b) sagittal planes

Table 3.1 Coordinates of the muscle attachment points (mm), total muscle lengths (mm) and physiological cross-sectional areas (cm^2) in the closed-mouth position

	Origin			*Insertion*[1]			*Total muscle length (mm)*	*Physiological cross-sectional area (cm^2)*
	x	*y*	*z*	*x*	*y*	*z*		
Masseter								
Superficial part	37.9	-52.3	-9.1	6.2	-44.9	-40.7	45.6	6.82
Deep part	30.7	-43.4	-7	16.7	-44.6	-26	25.3	3.49
Temporalis								
Anterior part	27.7	-48	34.4	26.4	-42.6	-14.8	49.7	7.7
Posterior part	-6.2	-59.3	32.3	22.4	-45.8	-8.8	52.2	5.55
Lateral pterygoid								
Inferior part	19.4	-24.3	-14.9	-0.9	-46.1	-3.1	32.6	2.82
Superior part	23	-24.2	6.4	0.2	-44.1	0.5	31.3	0.95

[1] Insertion: mandibular attachment.

The actuator orientation (which represents the muscle line of action) can be obtained using a force-vector calculation in the frontal and sagittal planes as follows. The magnitude and the direction of the sub-muscle forces define the resultant muscle orientation, as shown in Fig. 3.2. The resultant maximum muscle force is calculated using a unit of 30 N/cm^2 of PCS [4]. For example, a muscle of 10 cm^2 PCS can produce a force of 10 x 30 N = 300 N. Therefore, using the figures in Table 3.1, the actuator orientations in the frontal and sagittal planes are calculated as given in Table 3.2.

The insertion coordinates for each actuator were estimated using a scanned human study model mandible and the attachment locations of each muscle group are given in Table 3.1. The origin coordinates were then determined using the calculated resultant muscle line of action and the insertion coordinates. Table 3.3 lists the attachment coordinates and the total length of the three actuators.

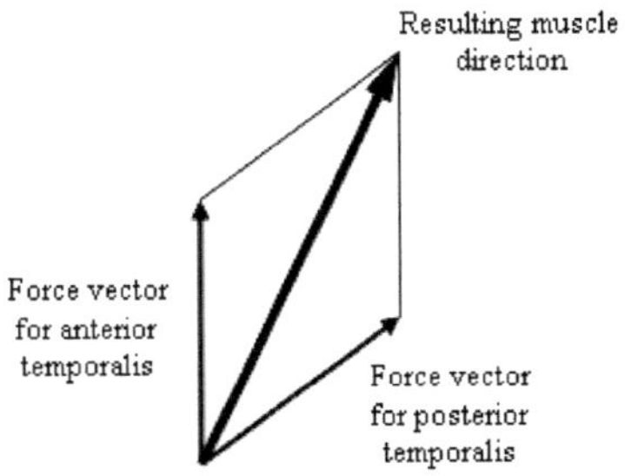

Fig. 3.2 Diagram of force vector analysis using the temporalis muscle group as an example

Table 3.2 Determined frontal and sagittal angle of the masseter, temporalis and lateral pterygoid actuators

	Frontal angle (°)	*Sagittal angle (°)*
Masseter	9.95	42.02
Temporalis	11.28	13.61
Lateral pterygoid	72.47	70.84

The use of the above biomechanical analysis, results in the robotic mechanism as shown in Fig. 3.3. Where S_i and M_i (i=1, 2, .., 6) denote the muscles' origins and insertion locations on the skull and the mandible, respectively. In order to determine the mechanism motion, two coordinate systems were established. Firstly, the skull system $oxyz_S$, fixed onto the skull and the mandible system $oxyz_M$, fixed onto the mandible. The mandible system differed from the skull system by $[x\ y\ z]^T = [78.85\ 0\ -41.87]^T$.

Table 3.3 Determined attachment coordinates (mm) and total actuator length (mm)

	Origin			*Insertion*			*Total actuator length (mm)*
	x	*y*	*z*	*X*	*y*	*z*	
Masseter							
right side	48.2	-52.5	-14.7	18	-44.7	-47.7	45.6
left side	48.2	52.5	-14.7	18	44.7	-47.7	45.6
Temporalis							
right side	13.3	-53.2	51.8	25.6	-43	2	52.2
left side	13.3	53.2	51.8	25.6	43	2	52.2
Lateral Pterygoid							
right side	21.5	-17.5	-12.4	-0.9	-41.1	-3.1	32.6
left side	21.5	17.5	-12.4	-0.9	41.1	-3.1	32.6

In order to accurately depict human mandibular movement, at least three reference points on the mandible are required. The following points are typically chosen: the incisor point (IP), the left molar point (LMP), the right molar point (RMP), the left condylar point (LCP), and finally the right condylar point (RCP) (see Fig. 2.11). The coordinates of these points within the mandibular coordinate system $OXYZ_M$, are given in Table 3.4.

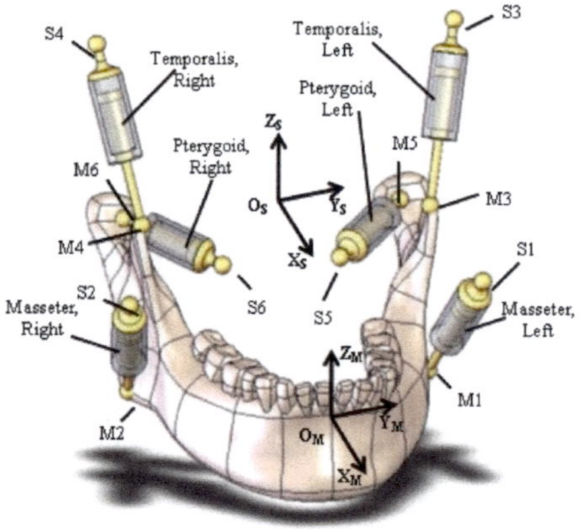

Fig. 3.3 A refined robotic mechanism in SolidWorks

Table 3.4 Coordinates of reference points in the mandible system (unit: mm)

Reference points	*Initials*	x_M	y_M	z_M
Incisor point	IP	0	0	0
Left molar point	LMP	-35.2	26.37	12.26
Right molar point	RMP	-35.2	-26.37	12.26
Left condylar point	LCP	-79.75	41.1	38.77
Right condylar point	RCP	-79.75	-41.1	38.77

3.2 Motion Design of the Actuation

The trajectories of the mandible reference points in real-life human chewing can be recorded during the chewing of a sample food. These can be reproduced by the masticatory robot through programming the six actuations. Thus, typical real-life chewing trajectories may be used in order to determine the actuations required in the robotic model. This was achieved via the use of inverse kinematics. the principles of which were extended so as to specify the actuators' motion properties in terms of other variables such as, displacement, velocity, and acceleration.

Inverse kinematics was simulated using CosmosMotion embedded into SolidWorks. In the previous chapter Fig. 2.18 and Fig 2.19 present the time history of the lateral and superoinferior movements of the incisor point during the chewing of a sample food. The test food displayed hard elastic properties. It can be seen that the subject chewed on the left side except for three peaks of the lateral excursion movement, which were performed to the right side. This movement can be explained due to the work of the tongue, which influences the chewing cycle by collecting and/or relocating the food bolus to an accurate position within the mouth. Furthermore, it can be seen in the superoinferior movement that chewing amplitudes vary between 30 mm and 20 mm for the beginning and the end of the chewing cycle, respectively.

To perform the inverse kinematics, a full specification of the masticatory sequence in 3D space is required. Therefore, in addition to the 2-D information that was measured while a subject chewed a sample food, specifications as to the anteroposterior movement of the IP and the angular movement of the mandible, about all three axes, was also required. It was found that the sagittal movement while chewing, occurred within an angle ranging between 0° to 30° [5].

In the simulation for the design of the actuation motion, a sagittal angle of 15° was chosen in order to represent the anteroposterior movement. Each data point involving the antero-posterior movement was then determined. As the mandibular angular movement about the x- axis is usually very small, it vanished in the simulation. To specify the mandibular angular movement about the y- and z- axis, the middle point of the segment connecting LCP and RCP (at $[x\ y\ z]^T$=[-79.75 0 38.77]^T in $oxyz_M$) was guided allowing movement in the sagittal plane through the origin of the skull system. The traced paths of the reference points in the horizontal, frontal and saggittal planes are shown in Fig. 3.4. The incisor trajectory was measured initially and moved within the Posselt envelope [6]. The condylar movements matched previous clinical and biomechanical findings [7]. The jaw movements are similar to those described in [3].

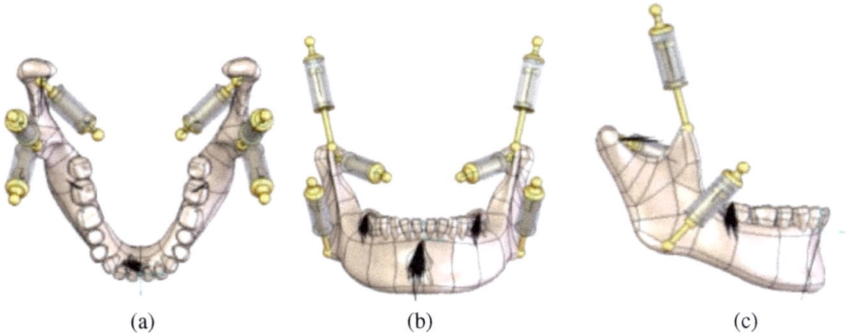

Fig. 3.4 SolidWorks model with IP, LMP, RMP, LCP and RCP trajectories in (a) horizontal, (b) frontal, and (c) sagittal views

Table 3.5 Determined actuator properties and their maximum angular movement in the frontal and sagittal planes

	Actuator properties			*Mandibular angular movement*	
	Displacement (mm)	*Velocity (mm/s)*	*Acceleration (mm/s²)*	*Frontal angle (°)*	*Sagittal angle (°)*
Masseter					
Left	7	40	500	2	4
Right	7	40	600	5	6
Temporalis					
Left	17	75	1000	4	12
Right	14	65	1000	4	8
Lt. Pterygoid					
Left	8	40	400	14	17
Right	8	40	450	10	15

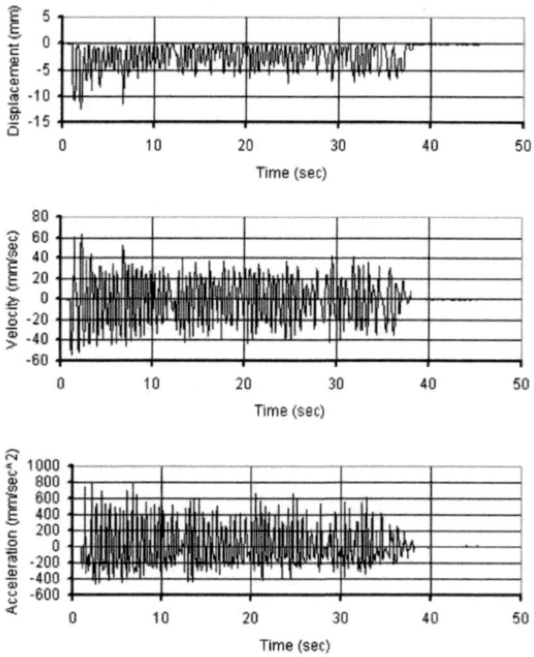

Fig. 3.5 Right temporalis actuation required for the specified chewing pattern: displacement, velocity and acceleration

The time functions of displacement, velocity, and acceleration for each of the six actuators (which must be followed to reproduce the given chewing pattern) were obtained. Table 3.5 gives the maximum motion properties of the actuators and their maximum angular movement, in both the frontal and sagittal planes.

The time functions generated during the simulation for the right temporalis actuator, for example, are depicted in Fig. 3.5. It was found that the temporalis actuator exhibited the largest displacement, velocity and acceleration in the first few chewing cycles. Higher velocity and acceleration occur whilst the mouth is open rather than closed. The maximum angular movements of the mandible in the sagittal and horizontal planes were 24° and 6°, respectively.

3.3 Force Design of the Actuation

Besides masticatory movement, a further important factor to be taken into consideration when designing a masticatory robot is the force applied whilst chewing. This force occurs between the teeth of the mandible and the maxilla, and is applied in horizontal, vertical and antero-posterior directions. The biting and chewing forces of a representative group of people were measured in [8]. The findings were as follows: the average bite force at the incisor was 40% of the force at the molars; the chewing force at the incisors was approximately 47% of that at the molars, and the average chewing force was 52% of the average bite force at the molars and 60% at the incisors.

These results have been confirmed by further research. By way of example, Wood and Williams [9] and Anderson [10] found that the maximum biting forces between the upper and lower molars were usually in the range 500–700 N, whereas the maximum mouth opening forces generally did not exceed 150 N. Furthermore results of this research found that the forces applied to a single tooth during chewing of foods such as biscuits, carrots, and cooked meat were in the range 70-150 N [11]; and the total biting forces across all contacting teeth for dentate individuals were in between 190-260 N [12].

Consequently, it can be concluded that the biting/chewing forces increase as the chewed food becomes hard and that chewing forces are always less than the maximum biting force. This indicates that a chewing force of 260 N across all contacting teeth should be considered in the design specifications of a masticatory robot.

The calculation of the required actuator forces is complicated due to both the number of muscle groups, and the fact that these forces occur within a 3D space. As demonstrated in Fig. 3.6, the 3D jaw model can in fact be simplified and represented by a 2D model, when actuation forces are balanced by occlusal forces [1]. In the saggital plane, the force equilibrium is achieved by the torque about the hinge axis at point A, i.e.

$$\left(-F_{Temp,z}\cdot a\right)+\left(-F_{Mass,z}\cdot b\right)+\left(F_{Occ}\cdot c\right)=0 \tag{3.1}$$

where $F_{Temp,z}$ is the force of the temporalis muscle group actuator in z- direction, $F_{Mass,z}$ is the force of the masseter muscle group actuator in z- direction, F_{Occ} the occlusal force, and a, b and c are the normal distances to the hinge axis. By Pythagorean Theorem, $F_{Temp,z}$ and $F_{Mass,z}$ can be calculated by:

$$F_{Temp,z} = F_{Temp}\cdot\cos\beta\cdot\cos\delta \tag{3.2}$$

and

$$F_{Mass,z} = F_{Mass} \cdot \cos\alpha \cdot \cos\gamma \tag{3.3}$$

respectively, where α and β are the angles in the sagittal plane, and δ and γ the angles in the frontal plane.

According to [13], the force ratio of the pterygoid and masseter muscle groups is 0.81 as follows:

$$f_r = \frac{F_{Temp}}{F_{Mass}} = 0.81 \tag{3.4}$$

Substituting Eq. (3.2) to (3.4) into Eq. (3.1) yields:

$$\left(-\left(\frac{0.81 \cdot \cos\beta \cdot \cos\delta}{\cos\alpha \cdot \cos\gamma}\right) \cdot F_{Mass,z} \cdot a\right) + \left(-F_{Mass,z} \cdot b\right) + \left(F_{Occ} \cdot c\right) = 0 \tag{3.5}$$

and $F_{Mass,z}$ is found as follows:

$$F_{Mass,z} = \frac{\left(F_{Occ} \cdot c\right)}{\left(\left(\frac{0.81 \cdot \cos\beta \cdot \cos\delta}{\cos\alpha \cdot \cos\gamma} \cdot a\right) + b\right)} \tag{3.6}$$

Given a = 0.0265 m, b = 0.0189 m, c = 0.047 m, α = 42.02 °, β = 13.61 °, δ = 11.28 °, γ = 9.95 °, and F_{Occ} = 260 N, $F_{Mass,z}$ = 260.78 N by Eq.(6), and F_{Mass} = 356.38 N by Eq. (3.3). Applying Eq. (3.2) and (3.4) yields F_{Temp} = 288.67 N and $F_{temp,z}$ = 275.15 N, respectively.

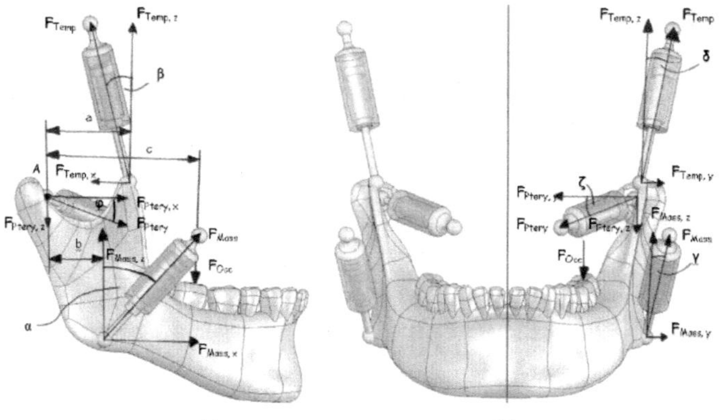

(a) (b)

Fig. 3.6 Two-dimensional model of the jaw with applied muscle and occlusal forces, (a) sagittal view and (b) frontal view

To obtain the pterygoid muscle actuation force, the force equilibriums in *x*- and *y*- directions can be determined, as follows:

$$F_{Ptery,x} + F_{Mass,x} - F_{Temp,x} = 0 \tag{3.7}$$

and

$$-F_{Ptery,y} + F_{Mass,x} + F_{Temp,x} = 0 \tag{3.8}$$

in which:

$$F_{Mass,x} = F_{Mass,z} \cdot \sin\alpha \tag{3.9}$$

$$F_{Temp,x} = F_{Temp,z} \cdot \sin\beta \tag{3.10}$$

$$F_{Mass,y} = F_{Mass,z} \cdot \sin\gamma \tag{3.11}$$

$$F_{Temp,y} = F_{Temp,z} \cdot \sin\delta \tag{3.12}$$

Use of Eqs. (3.7) to (3.12) yields $F_{Ptery,x}$ = 109.81 N, and $F_{Ptery,y}$ = 98.88 N. By using the Pythagorean Theorem, the actuator force F_{Ptery} for the lateral pterygoid muscle group can be calculated by:

$$F_{Ptery} = \sqrt{\left(F_{Ptery,y} \cdot \cos\varsigma\right)^2 + \left(F_{Ptery,x}\right)^2} \tag{3.13}$$

With ζ = 17.53 °, F_{Ptery} = 144.74 N.

Since all the forces are calculated for the actuators on both the left and right sides of the mandible combined, F_{Mass}, F_{Temp} and F_{Ptery} need to be halved. Consequently, the required actuator forces are F_{Mass} = 178.19 N, F_{Tema} = 144.34 N and F_{Ptery} = 72.37 N. As the above calculation is an estimation, the final requirement of the actuation forces is set at F = 200 N for all six actuators, this takes into account a certain safety factor allowance.

3.4 Design of the Actuation System

3.4.1 Basic Design Requirements

A uniform actuation system, in terms of both its transmission and energy system, was required in order to ensure that the design, control and maintenance of the masticatory robot were simple. To this end, the maximum values of all actuation systems in relation to linear velocity, linear acceleration, angular movement, and linear force were chosen. Only the muscle lengths in the closed-mouth position and the maximum displacement changes were individually considered. Table 3.6 summarises the required actuator properties. It is noted that the compliance and damping characteristics of the muscles have not been specified. This will be done during design of the robot's compliant motion control, in conjunction with specifying chewing force.

Table 3.6 Required actuator properties.

Length (*mm*)	
Masseter	45.6
Lateral Pterygoid	32.6
Temporalis	52.2
Displacement (*mm*)	
Masseter	10
Lateral Pterygoid	10
Temporalis	20
Linear velocity (*mm/s*)	75
Linear acceleration (mm/s^2)	1000
Linear force (*N*)	200
Angular movement (°)	20

3.4.2 Types of Actuation

Extensive research was conducted in order to find an electromechanical 'off-the-shelf' actuator. Actuators possessing small dimensions, a fast speed and strong torque, were difficult to obtain on the market. The specifications of various 'off-the-shelf' products were compared with the required actuator properties listed in Table 3.6. By way of an example, comparison was made between required actuator properties and the 21000 Series Size 8 linear actuators produced by Haydon Switch & Instrument, Inc., Waterbury, USA. This actuator is small in size, but cannot produce adequate torque and speed. The same issue was found with the Series 7220 produced by ACD & D Limited. Other actuators, such as the M-05 series from Nook Industries Inc., Cleveland, USA, or the S12 series from Power Drives Inc., Pittsburgh, USA, can produce enough torque, but are too slow and their dimensions too large.

No actuator found met the set requirements adequately. In response, various design concepts for a linear actuator were investigated. The following two actuator design concepts seemed realisable. First, it seemed possible to design an actuator, where the gearbox and the motor were mounted sideways to the ball screw unit. This concept is widely used for linear actuators and, therefore, became the first design. Secondly, the concept that an actuator based on a slider-crank mechanism was considered. This mechanism however, does not enable movement similar to that of the muscle line of action, but has the advantage that it is a simple design. The following gives an overview of the main ideas considered.

Double ball screw system. This concept involves a double ball screw system with two bi-directional threads (Fig. 3.7). If the spindle is turning in one direction, then both ball nuts will move either, away from, or towards the gear. This design is associated with various problems. The gearbox and the motor have to be fixed to the spindle so that the actuator is able to orientate in the desired position. To fix the actuator to the mandible and skull, two spherical joints can be mounted on both ball nuts. However, the outer diameter of the spherical joint could prove problematic due to spatial requirements.

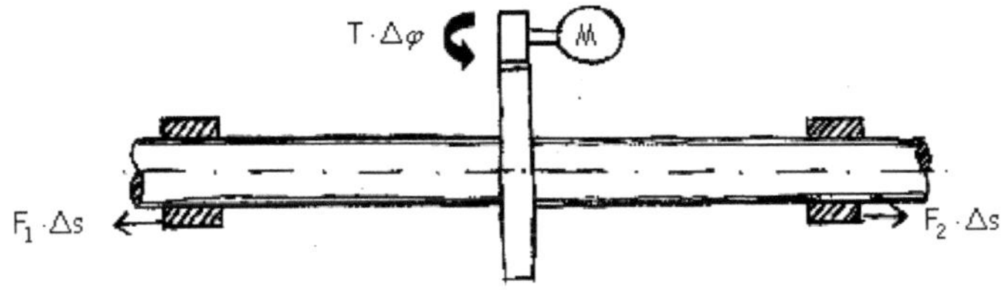

Fig. 3.7 Double ball screw system

Reverse linear motion by using two racks and one pinion. This solution involves two racks and a pinion (Fig. 3.8).

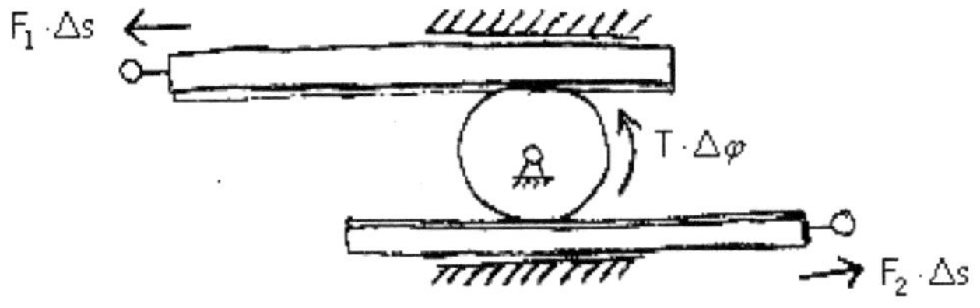

Fig. 3.8 Reverse linear motion using two racks and one pinion

While the pinion is turning in one direction, both racks are either moving away from one or towards one another. One problem lies in ensuring that both racks remain parallel to each other. Another issue concerns the angle between the line of action and the racks. This must be considered when developing control algorithms of the actuator. The required torque can be achieved using a gearbox.

Ball screw, gearbox and motor in line. This concept represents the most commonly used actuator. The motor, gearbox and ball screw systems are all in line

(Fig. 3.9). However, due to this arrangement, the overall length of the actuator will be larger than required. A spherical joint, mounted on the ball nut, and a universal joint, mounted to the motor end, could be used for mounting the actuator to the mandible and skull. The gearbox and motor could be positioned parallel to the spindle shaft and would reduce the size of the actuator.

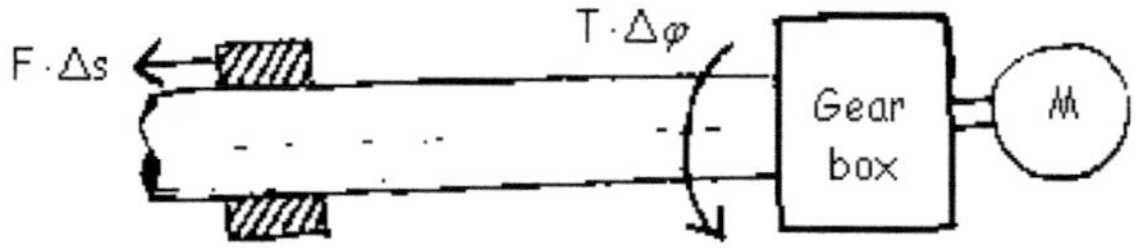

Fig. 3.9 Design concept with ball screw, gearbox and motor in line

2-bar linkage mechanism. This concept involves a 2-bar linkage and a two pinion transmission system (Fig. 3.10). Both bars are connected to each other via a hinge joint. The motor and the smaller pinion are fixed at the middle of one bar, whereas the bigger pinion is connected to the second bar at the hinge. The big gear and the second bar move together. One possible drawback occurs due to vibrations caused by the weight of the motor.

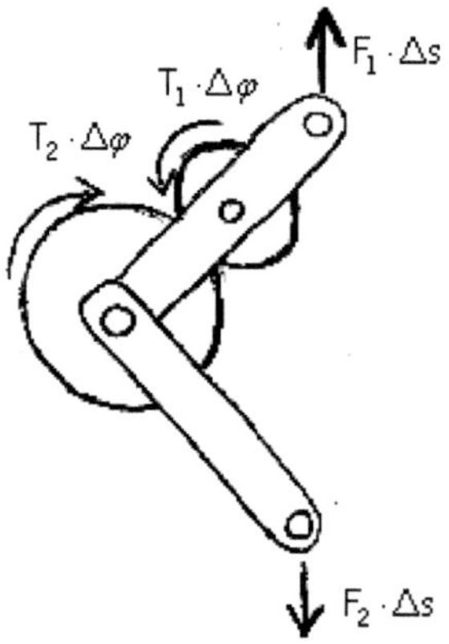

Fig. 3.10 2-bar linkage mechanism driven by two pinions

Slider-crank mechanism. This design concept works like the piston in a car engine (Fig. 3.11). The slider is connected via a coupler to a crank. By turning the crank, the slider moves up and down along the guide-way. For the masticatory robot, the slider would be fixed on the mandible. This is a simple solution, but does not work in the muscle line of action.

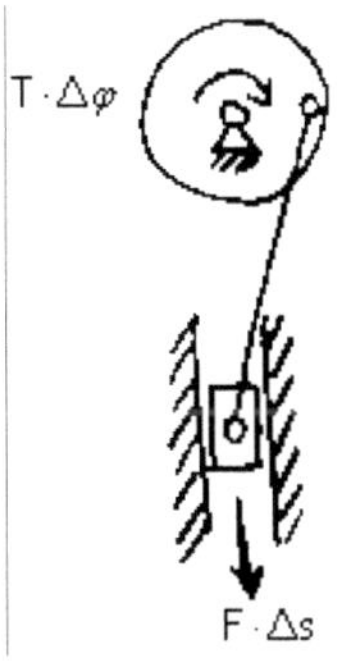

Fig. 3.11 Slider-crank mechanism

5-bar linkage mechanism. This solution is elaborate. The 5-bar linkage mechanism was originally developed to allow for the vertical dunking of a blender (Fig. 3.12). Given a certain rotation of the crank, the blender is able to make a vertical movement. However, this concept is hard to realise given the requirement for three-dimensional movements.

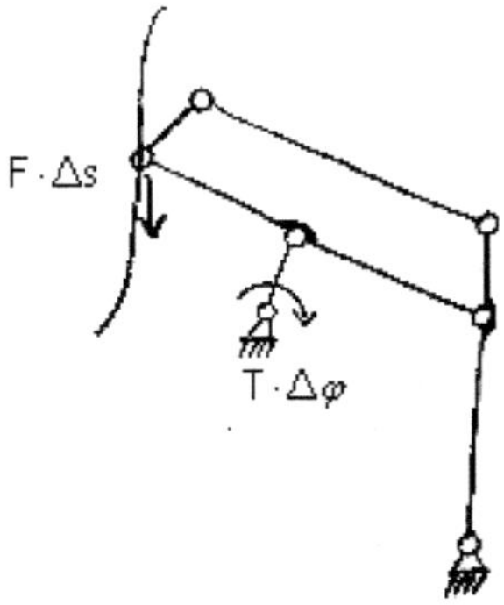

Fig. 3.12 5-bar linkage mechanism

3.4.3 Actuator Control

The purpose of developing a mastication robot is to reproduce human chewing behaviour. This can be achieved by implementing actuation patterns by following the time-function of displacement. The basic structure for actuator control can be open-loop or closed-loop. With an open-loop-controlled actuator, position and velocity change are possible. However, this kind of actuator provides no feedback allowing the control of mechanical output. Disturbances or properties such as introduced forces, friction, backlash, or electromagnetic hysteresis, can all influence mechanical output. Due to a high degree of precision being required, a feedback-controlled actuator is necessary. This feedback can be achieved by using a sensor in the form of an encoder attached to the rear motor shaft. Fig. 3.13 shows the basic system of a feedback-controlled electromechanical actuator, which is referred to as the servo-system.[14]

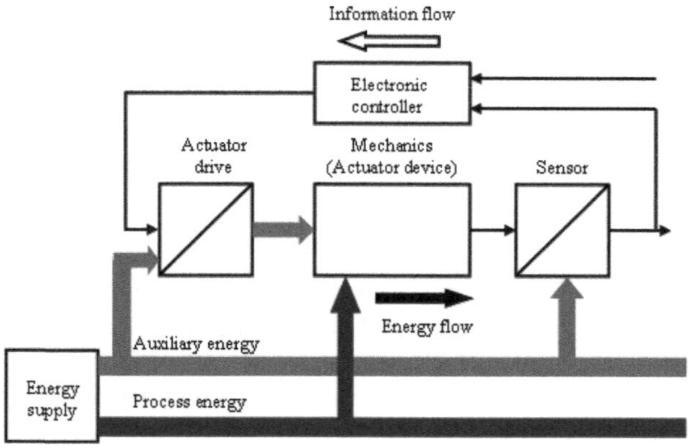

Fig. 3.13 Electromechanical actuator as feedback controlled mechatronic system

Fig. 3.14 DMC-1860 series motion control card and amplifier AMP-19540

The Galil motion control system allows the independent control of all six actuators by using one DMC-1860 series motion control card and two amplifiers AMP-19540 (Fig. 3.14). The control card provides motor control for standard servo motors, brushless servo motors and stepper motors. Therefore, one of these actuation systems had to be used. The amplifier can drive four servo motors, each up to 500 Watts.

3.4.4 *Physical Design*

This section discusses the development of the physical design of a ball screw unit linear actuator. The design was performed in SolidWorks.

3.4.4.1 Ball Screw System

An electrical motor generates a rotary motion. This motion does not satisfy the requirements of an actuator device either in terms of demands with regards to linear motion, rotational velocity and torque. These demands can be satisfied by using a ball screw system and a gear reduction, which enable the actuator to perform according to the desired motion. In order to choose motor and gear ratios according to actuator performance specifications, the transmission system had to be determined first.

Rotary movement can be transformed into linear movement by using a ball screw system. It should be noted that the lead specification of the ball screw influences the required torque and motor speed. Various parameters, such as the permissible rotational speed and axial load, together with a static safety factor, needed to be verified in order to ensure that the ball screw system was able to handle performance requirements. To satisfy the specific requirements (as shown in Table 3.6), the ball screw system needed to be as short as possible, precise, and have a stroke of at least 20 mm. The smallest ball screw (ball spindle) system sourced was the BNK 0401-3 system, THK Co., Ltd., Japan (Fig. 3.15).

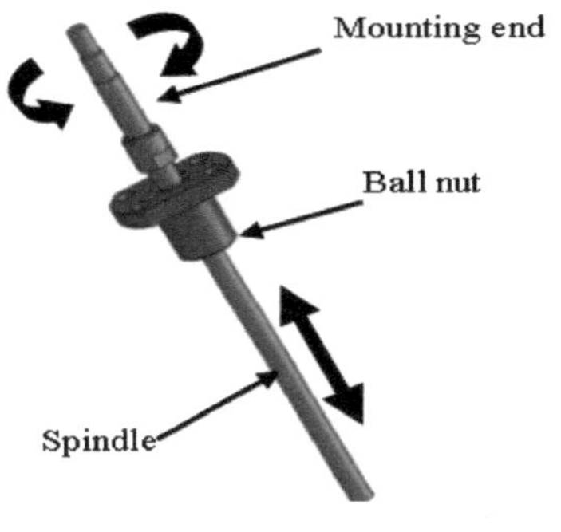

Fig. 3.15 Ball screw system BNK

The ball screw system consists of a spindle and a ball nut. The spindle has a 4 mm thread, a 1 mm lead, a 20 mm stroke, and a total length of 77 mm. One end of the spindle was fully supported with a mounting unit, whilst the other end was free.

To verify that the correct ball screw system had been selected, the maximum rotational speed and the maximum axial load on the spindle shaft required calculation. The following calculations were carried out according to the standard ball screw system catalogue from THK Co., Ltd. [15].

Maximum axial load

The maximum rotational speed $n_{Spindle}$ [min^{-1}] of the spindle can be determined:

$$n_{Spindle} = \frac{v_{Spindle} \cdot 60}{R} \tag{3.14}$$

where $v_{Spindle}$ is the required linear velocity [m/s] and R the lead of the spindle [m] with, $v_{Spindle}$ = 0.075 m/s (Table 3.6) and R = 0.001 m, $n_{Spindle}$ = 4500 min^{-1} is given. The maximum axial force F_{axial} [N] can be described as the sum of the maximum constant force F [N] and the force due to acceleration F_{acc} [N]:

$$F_{axial} = F + F_{acc} \tag{3.15}$$

According to Newton's second law, F_{acc} is:

$$F_{acc} = m_F \cdot a \tag{3.16}$$

where m_F represents the mass induced due to force [kg] and a represents the linear acceleration [m/s^2]. m_F can be described as:

$$m_F = \frac{F}{g} \tag{3.17}$$

where g is the acceleration due to gravity [m/s^2]. With, F = 200 N (Table 3.6) and g = 9.81 m/s^2, gives m_F = 20.39 kg. With a = 1 m/s^2 (Table 3.6), Eq. (3.16) yields F_{acc} =20.39 N. According to Eq. (3.15), F_{axial} = 220.39 N.

Permissible rotational speed

The permissible rotational speed of the spindle is determined based upon the critical speed together with a so-called DN value. At high speeds, the spindle produces resonance, due to the frequency characteristics of the spindle shaft, which may make operation impossible. The shaft speed should therefore be set at a level below the resonant point (critical speed). The permissible rotational speed n_{cs} based upon critical speed [min^{-1}] can be calculated by:

$$n_{cs} = \frac{60 \cdot \lambda_1{}^2}{2 \cdot \pi \cdot l_u^2} \cdot \sqrt{\frac{E \cdot 10^3 \cdot I}{\gamma \cdot A}} \cdot f_{s1} \tag{3.18}$$

where λ_1 is the mounting coefficient [-], l_u the unsupported spindle length [mm], E the Young's modulus [N/mm^2], I the geometrical moment of inertia [mm^4], γ the specific density [kg/mm^2], A the screw-shaft cross sectional area [mm^2] and f_{s1} the safety factor. The moment of inertia I can be calculated:

$$I = \frac{\pi}{64} \cdot d_k^4 \tag{3.18}$$

where d_k is the screw shaft thread diameter [mm]. With, $\lambda_1 = 1.875$, $l_u = 54$ mm, $E= 2.06 \cdot 10^5$ N/mm^2, $d_k = 3.4$ mm, $\gamma = 7.85 \cdot 10^{-6}$ kg/mm^3, $f_{s1} = 0.8$ [15], the critical speed is found, $n_{cs} = 40104.60$ min^{-1}.

The permissible rotational speed n_{DN} based on the DN value [min^{-1}] an can be calculated by:

$$n_{DN} = \frac{x_{DN}}{d_{BC} \cdot 60} \tag{3.19}$$

where x_{DN} is the DN value [mm/s] and d_{BC} is the ball centre diameter [mm]. With, x_{DN}= 70,000 mm/s [14], $d_{BC} = 4.15$ mm, it gives $n_{DN} = 16867.47$ min^{-1}. Since, n_{cs} and n_{DN} are larger than the required value of $n_{Spindle}$, the chosen ball screw system met requirements.

Permissible axial force

Where an axial force is exerted upon the ballscrew, the screw shaft should be determined with consideration given to the buckling force and the permissible tensile-compressive force, which may exert yielding stress on the shaft. The ballscrew should not buckle under the maximum compressive force applied in its axial direction. The buckling force F_b [N] can be calculated by:

$$F_b = \frac{\lambda_2 \cdot \pi^2 \cdot E \cdot I}{(l_u)^2} \cdot f_{s2} \tag{3.20}$$

where λ_2 is a mounting coefficient, and f_{s2} the safety factor. When $\lambda_2 = 0.25$, and f_s=0.5 [15] the resultant buckling force is $F_b = 571.71$ N. In this case, the maximum force F_{axial} is below the buckling force F_b.

The permissible tensile-compressive force F_{tc} [N] of the screw shaft can be calculated by:

$$F_{tc} = \delta \cdot \frac{\pi \cdot (d_k)^2}{4} \tag{3.21}$$

where δ is the permissible tensile-compressive stress [N/mm^2]. With, $\delta = 147$ N/mm^2, the formula above yields $F_{tc} = 1334.64$ N. It was found that the maximum force F_{axial} is below the permissible tensile-compressive force of F_{tc}. Hence, the ball screw met the actuator force requirements.

Static safety factor

If the ball screw receives an excessive load or a significant impact (e.g. crushing of food), localised and permanent deformation develops between the raceway and balls. This deformation can hinder the movement of the ball screw. Therefore, the basic static load rating C_{0a} [N] was taken as the permissible axial load. C_{0a} is normally equal to the permissible static axial load of the ball screw. According to the operation conditions of the actuator, a static safety factor f_{s3} was considered. Therefore, the permissible static axial load $F_{a,max}$ can be described as:

$$F_{a,.\max} = \frac{C_{0a}}{f_{s3}} \tag{3.22}$$

With, C_{0a} = 420 N and f_{s3} = 2 [15], the formula above yields $F_{a,max}$=210 N. The permissible static axial load $F_{a,max}$ is bigger than the required static axial load F = 200 N.

According to the aforementioned calculations, it was confirmed that the ball screw system chosen was appropriate for the specified requirements.

3.4.4.2 Motor, Gear Reduction, and Encoder Selections

Motor type selection

To find a suitable motor-gear reduction combination, the maximum required torque and speed of the motor needed to be determined for various gear reductions, to make comparisons possible. The equations used to make these calculations are presented in the following sections. A number of factors had to be considered before making the final choice of a brushless DC-Servomotor produced by Minimotor SA, Switzerland.

Maximum motor torque

In order that the actuator would operate under different loadings induced while chewing, the calculations for the motor torque were divided in torque under constant load T_c [Nm], torque under acceleration T_{acc} [Nm] and torque under deceleration T_{dec} [Nm]. All three determined values had to be smaller than the maximum possible torque of the chosen motor.

The torque T_c under constant load [Nm] can be calculated by:

$$T_c = \frac{1}{i} \cdot T_{spindle} \tag{3.23}$$

where $T_{Spindle}$ is the required torque on the spindle shaft before gear reduction [Nm] and i represents the reduction ratio between $n_{driving}$, the rotation on the motor shaft, and n_{driven}, the rotation on the spindle shaft, where i is determined as follows:

$$i = \frac{n_{driving}}{n_{driven}} \tag{3.24}$$

$T_{Spindle}$ can be described as:

$$T_{Spindle} = \frac{F \cdot R}{2 \cdot \pi \cdot \mu} \cdot f_{s1} \tag{3.25}$$

where μ is the spindle efficiency and f_{s1} is the safety factor. With, F = 200 N, μ =0.95 (95%), f_{s1} = 1.5, the formula above yields $T_{Spindle}$ = 0.05 Nm. The safety factor f_{s1} was included in order to ensure that the motor could handle unexpected torques and to take into consider any unknown efficiency factor in relation to the gear drive.

The torque T_α throughout acceleration [Nm] can be described with the rotational form of Newton's second law:

$$T_\alpha = I_e \cdot \alpha \tag{3.26}$$

where I_e is the entire mass moment of inertia [kgm^2] about the longitudinal axis through the mass centre, and α is the angular acceleration [s^{-2}]. I_e can be calculated by:

$$I = I_F \cdot \left(\frac{1}{i}\right)^2 \cdot 10^{-6} + I_{Spindle} \cdot \left(\frac{1}{i}\right)^2 \cdot 10^{-6} + I_{driven} \cdot \left(\frac{1}{i}\right)^2 + I_{driving} \tag{3.27}$$

where I_F is the mass moment of inertia [kgmm^2] due to the axial force F_{axial}, $I_{Spindle}$ the mass moment of inertia of the spindle [kgmm^2], I_{driven} the entire mass moment of inertia [kgmm^2] of the driven parts and $I_{driving}$ the entire mass moment of inertia [kgmm^2] of the driving parts. The mass moments of inertia I_{driven} and $I_{driving}$, such as those of the gears, did not significantly impact upon the calculations and were, therefore, not included. Using Eq. (3.17), the mass moment of inertia due to F_{axial} can be calculated as:

$$I_F = m_F \cdot \left(\frac{R}{2 \cdot \pi}\right)^2 = 0.52\, kgmm^2 \tag{3.28}$$

To calculate the mass moment of inertia of the spindle $I_{Spindle}$, the spindle shape was approximated with a cylinder shape. Therefore, $I_{Spindle}$ can be described by:

$$I_{Spindle} = \frac{m_c \cdot d_c^2}{8} \tag{3.29}$$

where m_c is the cylinder mass [kg] and d_c the cylinder diameter [mm]. The cylinder mass m_c can be calculated:

$$m_c = \frac{\pi \cdot d_c^2 \cdot l_c \cdot \gamma}{4} \tag{3.30}$$

where l_c is the spindle length [mm]. With, d_c = 5 mm (approximated thickness), l_c=77 mm (according to the spindle datasheet), the formula above yields $I_{Spindle}$ = 0.04 kgmm^2.

The torque T_{acc} under acceleration is the sum of torque T_c under constant speed [Nm] and torque T_α through acceleration [Nm] as follows:

$$T_{acc} = T_c + T_\alpha \tag{3.31}$$

and the torque under deceleration T_{dec} is the difference between each as follows:

$$T_{dec} = T_c - T_\alpha \tag{3.32}$$

Maximum speed

The maximum speed n_{max} of the motor [min^{-1}] is equal to the product of the gear reduction i and the maximum speed $n_{Spindle}$ of the spindle [min^{-1}]:

$$n_{\max} = i \cdot n_{Spindle} \tag{3.33}$$

Due to the fact that T_c, T_{acc}, T_{dec}, and n_{max} are all dependent upon the gear reduction i, Table 3.7 gives the required torque values for various gear reductions. The comparison between the required properties in Table 3.7 and various motor datasheets of brushed and brushless DC-motors showed that the brushless Minimotor DC-servomotor 2036 024 B with a torque of 4.9 mNm best suited the requirements of the linear actuator.

Table 3.7 Torque vs. gear ratio for the linear actuator

			Solution									Chosen
	Symbol	Unit	1	2	3	4	5	6	7	8	9	specification
Gear Reduction	i	-	100	75	50	25	20	15	10	5	1	**10.2**
Mass moment of inertia	J	kgm^2	6.E-11	1.E-10	2.E-10	9.E-10	1.E-09	2.E-09	6.E-09	2.E-08	6.E-07	5.E-09
Torque due to acceleration	T_α	Nm	6.E-08	1.E-07	2.E-07	9.E-07	1.E-06	2.E-06	6.E-06	2.E-05	6.E-04	5.E-06
Torque (constant speed)	T_c	mNm	0.5	0.7	1.0	2.0	2.5	3.3	5.0	10.0	50.0	4.9
Torque (acceleration)	T_{acc}	mNm	0.5	0.7	1.0	2.0	2.5	3.3	5.0	10.0	50.6	4.9
Torque (deceleration)	T_{dec}	mNm	0.5	0.7	1.0	2.0	2.5	3.3	5.0	10.0	49.4	4.9
Speed	n_{max}	min^{-1}	420000	315000	210000	105000	84000	63000	42000	33600	25200	16800

Gear reduction selection

The next step involved selection of the right transmission system with the gear reduction i = 10. The motor could have been mounted in line, parallel or inclined to the ball screw system. The in line mounting would have increased the actuator length and was, therefore, not chosen. For parallel and incline mounting, the most commonly used principles are those of gear or belt drives.

Gear drives can transmit rotary motion with spur gears (parallel shafts) and bevel gears (inclined shafts). The larger gear is often called the spur gear and the smaller gear the pinion. Gears can be distinguished between spur gears, gears with axial teeth, and helical gears (i.e. gears with helical teeth). To produce a large gear ratio, several stages of gear pairs are required.

Other possibilities include planetary gear drives and harmonic drives. These drives could be sourced with the required gear ratio however, the motor input shaft and the output shaft of the planetary gear drive are inclined. To achieve parallel mounting, an additional transmission system would be necessary. Harmonic drives are widely used for high-precision motions with a gear ratio between 50 and 320. Furthermore, these drives require an additional transmission system to shift the motor and ball screw system from being in line. The additional transmission system would require a 1:1 gear ratio. However, to ensure that the motor shaft and the spindle shaft are far enough apart, two gears with a large diameter would be needed. This would increase the overall dimensions of the actuator and could potentially interfere with other parts of the actuator.

Belt drives consist of a pair of pulleys and a belt for the transmission of motion and are usually used to accommodate large distances. However, the gear ratio is limited to about three in order to maintain an adequate arc of contact [14]. Therefore, at least three stages are necessary. Another disadvantage is the low nominal input speed of less than 6000 RPM.

After considering all these facts, a custom made gearbox with spur gears was considered the most suitable solution. To achieve a reduction of $i = 10$, three gear stages are recommended in order to ensure that the overall dimensions of the gear system was not too large. The entire gear reduction i can be described as the product of the three stage reductions i_x ($x = 1, 2, 3$):

$$i = i_1 \cdot i_2 \cdot i_3 \tag{3.34}$$

and each stage reduction can be calculated by:

$$i_x = \frac{n_1}{n_2} = \frac{t_2}{t_1} \tag{3.35}$$

where n_1 and t_1 are the speed and the number of teeth of the pinion gear, and n_2 and t_2 of the spur gear, respectively.

US Company, SDP/SI offered a good range of gears. When choosing the pinion and spur gear, it had to be ensured that the same module m_x (x = 1, 2 , ..., 6) was present. To fulfil the required gear reduction, the following gears were chosen:

A1B1MY04010 ($m_1 = 0.4$, $t_1 = 10$), A1B1MY04024 ($m_2 = 0.4$, $t_2 = 24$), 2x S10T05M020S0303 ($m_{3,5} = 0.5$, $t_{3,5} = 20$), S10T05M034S0303 ($m_4 = 0.5$, $t_4 = 34$), and S10T05M050S0303 ($m_6 = 0.5$, $t_6 = 50$).

Fig. 3.16 shows the gear train where gear 1 is mounted on the motor shaft and gear 6 on the spindle shaft. Combining Eq. (3.34) and Eq. (3.35) gives the total gear ratio of 10.2, i.e.

$$i = \frac{t_2}{t_1} \cdot \frac{t_4}{t_3} \cdot \frac{t_6}{t_5} = 10.2 \tag{3.36}$$

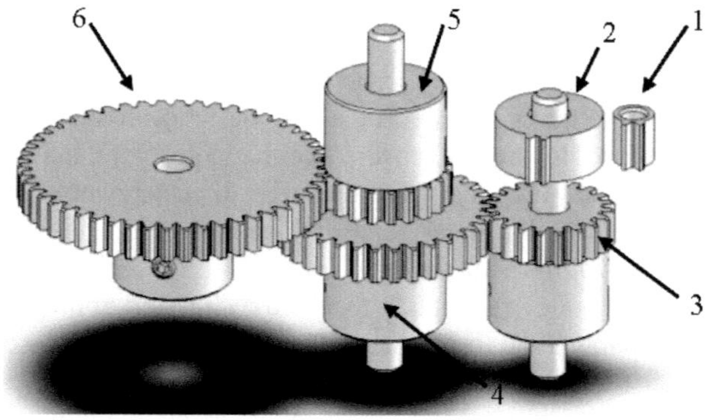

Fig. 3.16 Gear reduction system for the linear actuator in SolidWorks

Due to the fact that the new gear ratio was slightly different to the previously chosen one, the motor torques and speed had to be recalculated in order to verify that the motor could achieve the requirements given the new gear ratio. Using Eq. (3.24) to Eq. (3.34), T_c, T_{acc}, and T_{dec} were all found to be approximately 4.9 mNm and n_{max} 42840 min^{-1}. The comparison of these values with Fig. 3.17, revealed that the data points (n_{max}, T) were lying inside the recommended area required for continuous operation. Thus, the motor chosen met the actuator design requirements.

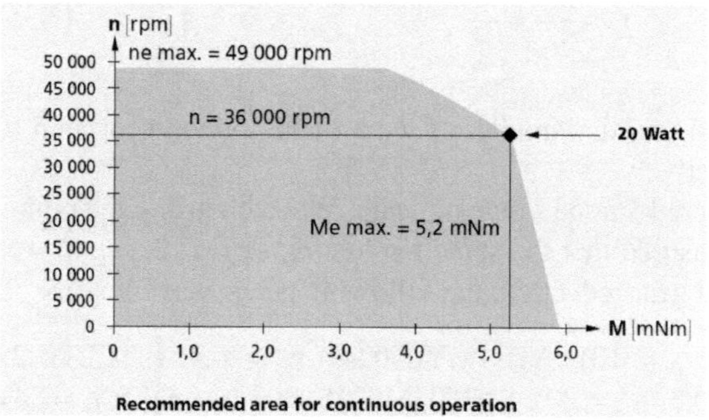

Fig. 3.17 Motor performance diagram for brushless DC-Servomotor 2036 024 B from Minimotor SA, Switzerland

Encoder selection

The required encoder resolution B [-] (counts per turn) can be described by:

$$B = \frac{R \cdot i}{f} \tag{3.37}$$

where f is the minimum required feed accuracy [mm] of the entire system. With, $f = 0.1\ mm$, the above formula yields a required encoder resolution $B = 102$. The incremental shaft encoder HEDS 5540 A, produced by Minimotor SA, Switzerland, has a resolution of $B = 500$. This shaft encoder, in combination with the brushless DC-Servomotor 2036 024 B, enabled specification and control of both, shaft velocity and direction of rotation, together with positioning.

3.4.4.3 Actuator Assembly

The actuator was designed according to the aforementioned calculations together with appropriate part selection. The steps involved in the design process were as follows:

- Choosing a spindle support unit,
- Designing a gearbox,
- Testing the gearbox for deformation, *and*
- Designing the joint systems.

Fig. 3.18 shows the actuator assembly as a SolidWorks model and the manufactured actuator.

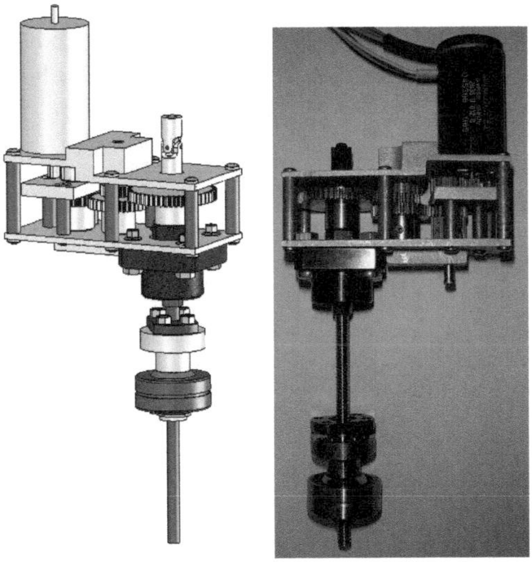

(a) (b)

Fig. 3.18 Actuator assembly: (a) model in SolidWorks, and (b) constructed actuator

Spindle support unit

THK Co. offered a large range of support units for the chosen ball screw system BNK0401-3. The support unit FK 4 was chosen because of its size. The unit used two single row angular contact bearings, which were mounted face to face, so that the load lines converged towards the bearing axis [16]. This bearing orientation allowed the accommodation of the axial loads to act in both directions.

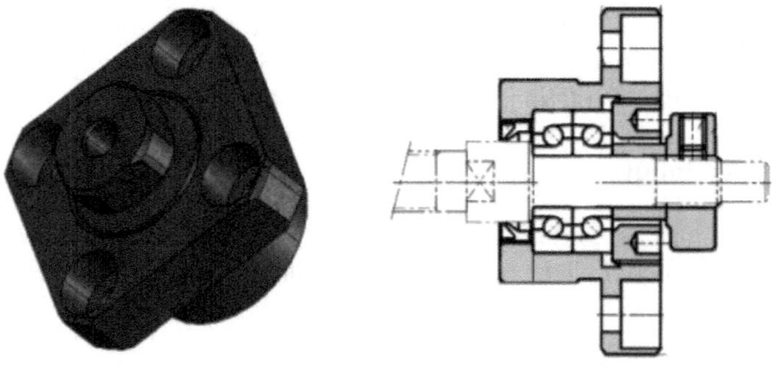

(a) (b)

Fig. 3.19 Support unit FK4: (a) isometric perspective, and (b) cut in the sagittal plane

Gearbox design

The gearbox comprised an upper, middle and lower plate, all manufactured out of stainless steel (Fig. 3.20). The lower plate was joined by 6 pins to the upper plate, whilst the lower plate was joined to the middle plate by three pins. Both ends of these pins were drilled and reamed to a smaller diameter than that of the body diameter.

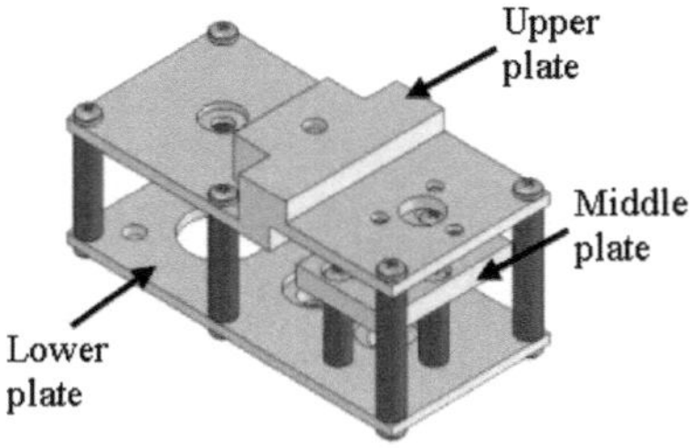

Fig. 3.20 Gearbox assembly in SolidWorks

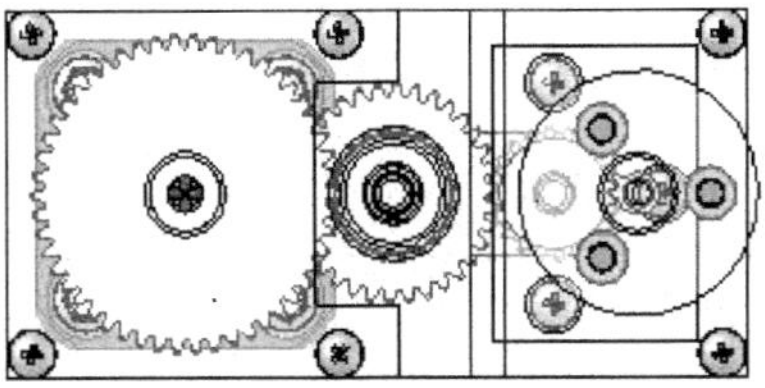

(a)

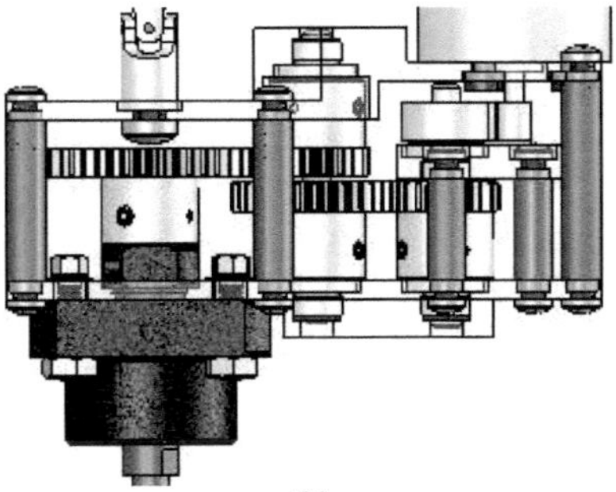

(b)

Fig. 3.21 Designed gearbox in SolidWorks in (a) horizontal and (b) frontal view

The pin ends were placed into reamed holes in the plates so as to operate, not just as parts to join the plates, but also as dowels to ensure that everything was aligned correctly. Inside these pins were drilled and taped holes. These allowed the pins to be screwed onto the plates, as seen in Fig. 3.21.

To ensure that the gear shafts were parallel and could turn with a minimum of friction, two sintered bearings were used for both of the middle shafts. To fix these bearings, mounting holes were included in the plates. These bearings were mounted slightly higher than the plates, and therefore allowed for the smooth running of the gears when in a vertical position. The gears would otherwise have scratched on the plate surface, causing high friction. To ensure that the motor was mounted in the correct position, a reamed hole was inserted into the upper plate in order to locate the motor and its flange. For further support, the motor was fixed with three screws. The ball screw support unit was mounted with four screws.

Gearbox testing

To ensure that the designed gear box was strong enough to resist deformations due to applied forces during chewing, a FEM analysis was performed. The gearbox, as seen in Fig. 3.20, was solid meshed with 18,917 elements of size 2.2 mm. The model was restrained on the mounting position by the ball screw support unit on the lower plate (which had four screw holes). A static force of 300 N, located on the universal joint mounting hole in the upper plate, and was applied normal to it.

An actuator applies dynamic forces (over various frequencies) to the gearbox during the chewing process. A static force 1.5 times higher than the theoretical maximum applied force was required in order to consider the deformation due to dynamic influences. The displacement plot, including the maximum displacement, is shown in Fig. 3.22.

It can be seen that the maximum resultant displacement was $d = 0.069\ mm$ and was therefore smaller than the required displacement accuracy of 0.1 mm, demonstrating that the gearbox was designed appropriately.

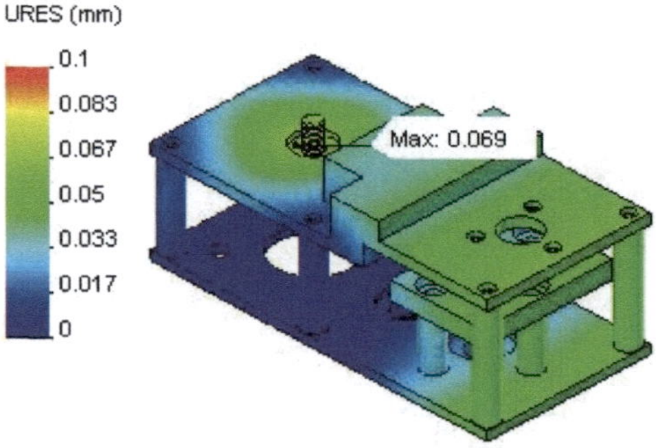

Fig. 3.22 Material displacement of the gearbox due to 300 N static force

Joint systems

Two joints were required to fix and guide the actuator. The first joint was a standard steel universal joint from RS New Zealand (designated item number 689-215) [16], see Fig. 3.18. This joint allowed two DOF and was fixed with an M3x6 screw. The second joint was spherical with three DOF. The bearing was fixed onto a mounting unit, which was in turn fixed onto the ball nut. The mounting unit could either be positioned before the ball nut by using spherical bearing PB12, or after the ball nut by using spherical bearing PB8 (Fig. 3.23).

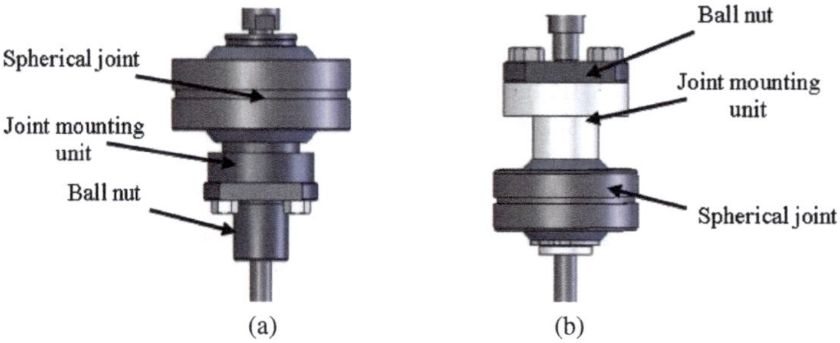

Fig. 3.23 Different mountings for the spherical joints: Mounting (a) before the ball nut with bearing PB12, and (b) after the ball nut with bearing PB8

Both bearings were purchased from THK Co., Japan. In both positions, an external retaining ring fixed the spherical bearing onto the mounting unit. The placement of the mounting unit before the ball nut, reduced the distance between the two joints, but expanded the outer diameter of the spherical joint to the size of the spindle flange. Placement after the ball nut produced the opposite effect. The first mounting position enabled an actuator length of 53 mm, the specified length for the temporalis muscle, whereas the second mounting position produced 73 mm.

3.4.4.4 Design Related Problems

The design of the linear actuator had the following drawbacks:

Backlash between gears. Backlash causes inaccuracy and should be considered in the control of the actuator. A turning test revealed that the inaccuracy was about 2°. Therefore, a second encoder, mounted on the spindle shaft, could improve the control.

Open spindle end. Due to an error, the ball nut could be wound past the spindle end and resulted in the destruction of the ball screw system by releasing the little balls inside. To prevent this, a metal ring with rubber should be mounted by a screw on the spindle end of the actuator. The rubber would absorb the reaction force caused by contact between the ball nut and the metal ring and would also ensure that the balls did not jam.

Mounting problems due to the outer diameter of the spherical bearing. The orientation of the actuator made mounting of the spherical joint (with its large outer diameter) difficult. The actuator orientation should be re-evaluated in order to solve this problem.

Actuator size. It was possible to achieve the specified actuator length for the temporalis muscle, but the length specifications for the masseter and lateral pterygoid could not be achieved.

3.5 Design of Mastication Robot

The sections above give the physical design of a linear actuator. This actuator was designed in order to replicate the temporalis muscle group. This design was successful in relation to; the defined actuator length in the closed mouth position, stroke, kinematic properties (velocity and acceleration), and actuator force. However, there were several problems related to building a masticatory robot. These problems included; Firstly, the actuator length for the masseter and lateral pterygoid muscle group was unachievable, secondly, difficulties were encountered in attachment of the actuator to the mandible, and thirdly, the actuator spindle and mandible collided.

This section introduces a model with longer actuators for the masseter and lateral pterygoid muscle groups. An alternative approach, which may provide a solution to the problem in relation to actuator attachment, is also outlined.

3.5.1 Actuator Assembly

Due to the actuators that were physically feasible being longer than initially specified, a simplified assembly model with extended actuator lengths was built. To make comparisons between the original and the amended model possible, the actuator lengths for the two muscle groups were changed to the length of the temporalis muscle group. Due to the spatial positioning of the two actuators in relation to the mandible, together with the actuator dimensions, it was necessary to turn them about 180° to the muscle line of action in the closed mouth position. Fig. 3.24 shows the robot assembly with the altered orientation plus the lengths of the masseter and temporalis actuators.

Table 3.8 lists the new origin coordinates and actuator lengths. The robotic model, as shown in Fig. 3.3, was updated with the new actuator properties and re-simulated according to Section 3.2.

The position and orientation of the actuators for the temporalis muscle group were not altered and therefore design values were unchanged. Table 3.9 lists the new maximum actuator properties and angular movement for the masseter and lateral pterygoid muscle groups.

It was found that the design values for the masseter muscle group actuators have changed only slightly, which was probably due to the fact that the masseter muscle group length remained almost the same. The only significant change was lower accelerations in comparison to those given in Table 3.5 The new masseter origins did not cause many differences to the previous model. For the pterygoid muscle group actuators, all properties determined were larger. The actuator positions behind the mandible caused a larger stroke. However, the resultant stroke, angular movement and acceleration were all still compatible with the actuator design as given in Section 3.4.

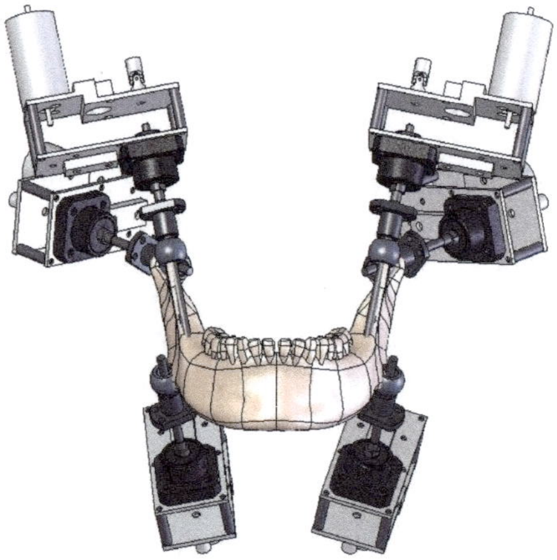

Fig. 3.24 Robot assembly with altered masseter and lateral pterygoid actuators

Table 3.8 New origin coordinates and actuator lengths in the closed mouth position for the masseter and lateral pterygoid muscle groups

	Origin			Total actuator
	x	y	z	length (*mm*)
Masseter				
right side	-16.94	-35.68	-85.90	52.2
left side	-16.94	35.68	-85.90	52.2
Lateral Pterygoid				
right side	-36.83	-78.96	11.75	52.2
left side	-36.83	78.96	11.75	52.2

Table 3.9 New determined actuator properties for masseter and lateral pterygoid muscle groups (maximum values)

	Actuator properties			Angular movement	
	Displacement (*mm*)	Velocity (*mm/s*)	Acceleration (*mm/s*^{2})	Frontal angle (°)	Sagittal angle (°)
Masseter					
Left	4	30	1000	5	2
Right	5	30	1000	7	2
Lt. Pterygoid					
Left	16	130	3000	6	9
Right	19	130	3000	11	9

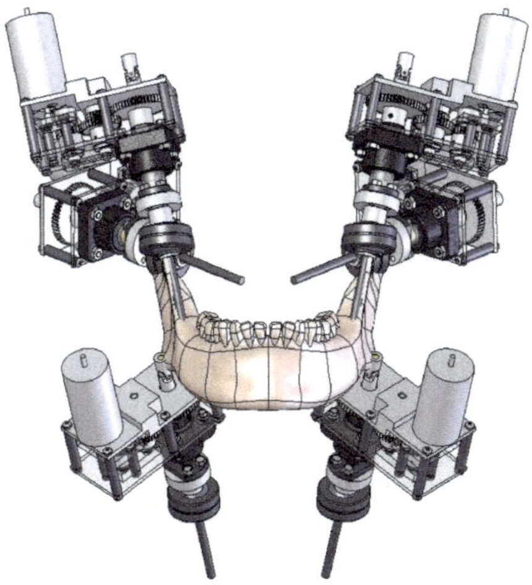

Fig. 3.25 Robot assembly with turned masseter muscle group actuators

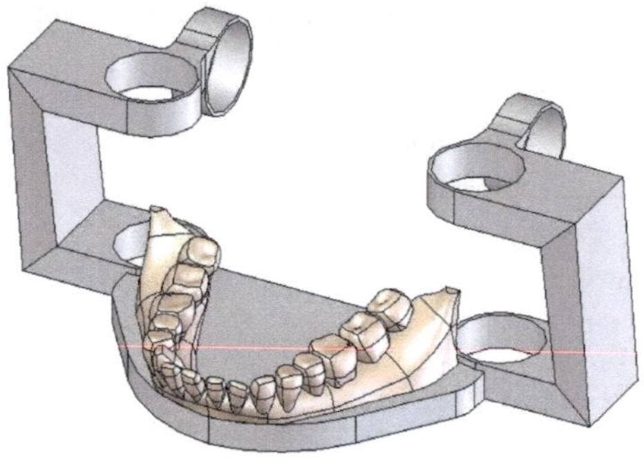

Fig. 3.26 A design concept for the mechanical jaw under consideration

3.5.2 Actuator Mounting

Significant problems were encountered during the assembly of the masticatory robot in relation to spindle-mandible collisions. These resulted in restriction of mandibular movement. Due to the actuators being positioned close to each other and to the long spindle ends, the spindle ends of each actuator collided with the mandible. Turning the actuators around thereby fixing the actuator to the universal joint at the mandible and the spherical joint to the skull, solved the collision problem theoretically.

A turning of either the lateral pterygoid or temporalis muscle group actuators was not possible, due to their close proximity to each other (Fig.3.25). Another problem encountered related to the actuator-mandible mountings. Due to the diameter of the spherical joints, fixing the actuator to the mandible was difficult to achieve. Fig. 3.26 shows the conceptual design of a mechanical jaw taking into account the original actuator orientation. It can be seen that due to the diameters of the spherical joints and the positions of each being so close, the manufacture of a mechanical jaw would be difficult to achieve.

At this stage of the research, it was hard to imagine that the problems relating to spatial requirements could be solved using the current actuator design.

3.6 Summary

This chapter introduced an actuator design for use in the construction of a masticatory robot. This was required due to the fact that no off-the-shelf product could be sourced that met the outlined requirements. One physical actuator model for the temporalis muscle group was built successfully with the specified actuator properties having been taken into consideration. However, actuator lengths for the masseter and lateral pterygoid muscle groups were not achievable. Furthermore, the masseter and lateral pterygoid actuator origins had to be altered as a result of the actuators' dimensions and their close proximity to the mandible. To investigate the resultant influences of these alterations, a new simulation with the updated actuator origins and lengths was carried out. The results revealed that the lateral pterygoid actuator properties necessitated a new design. Furthermore, design related obstacles, such as mounting problems and mandible-spindle collision, revealed that a different approach in relation to actuator design was essential.

References

1. Osborn, J.W., Baragar, F.A.: Predicted pattern of human muscle activity during clenching derived from a computer assisted model: symmetric vertical bite force. J. Biomech. 29, 589–595 (1985)
2. Koolstra, J.H., van Eijden, T.M.: Dynamics of the human masticatory muscles during a jaw open-close movement. J. Biomech. 30, 883–889 (1997)
3. van Eijden, T.M.G.J., et al.: Architecture of the human jaw-closing and jaw-opening muscles. Anat. Record 248, 464–474 (1997)
4. van Eijden, T.M.G.J.: Personal communication. Department of Functional Anatomy, Academic Centre for Dentistry, Amsterdam, The Netherlands (2005)
5. Buschang, P.H., et al.: Quantification of human chewing-cycle kinematics. Arch. Oral Biology 45, 461–474 (2000)
6. Posselt, U.: Movement areas of the mandible. J. Pros. Dent. 7, 375–385 (1957)
7. Helkimo, E., Ingervall, B.: Bite force and functional state of the masticatory system in young men. Swed. Dent. J. 2, 167–175 (1978)
8. Scapino, R.P.: Morphology and mechanism of the jaw joint. In: McNeill, C. (ed.) Science and practice of occlusion, Quintessence Publishing Co. (1997)
9. Sharkey, P., et al.: Jaw opening forces in human subjects. Br. Dent. J. 156, 89–92 (1984)
10. Wood, G.D., Williams, J.E.: Gnathodynamometers: measuring opening and closing forces. Dent. Update 8, 239–250 (1981)
11. Anderson, D.J.: Measurements of stress in mastication. J. Dent. Res. 1. 41, 175–189 (1956)
12. Gibbs, C.H., et al.: Occlusal forces during chewing - influences of biting strength and food consistency. J. Prosthet. Dent. 46, 561–567 (1981)
13. Gray, H.: Anatomy of the human body. Lea & Febiger, Philadelphia (1918)
14. Isermann, R.: Mechatronic systems: Fundamentals. Springer-Verlag London Ltd., London (2003)
15. THK, Linear motion (2005), `guide.www.thk.com`
16. SKF, SKF - General catalogue (2003)
17. RS, The catalogue. RS New Zealand (2005)

Chapter 4
Mastication Robot of Crank Actuation*

Abstract. A life-sized masticatory robot with a 6RSS parallel mechanism is discussed within this chapter. The robot is intended to simulate human chewing in order to evaluate food properties during the chewing process. A robotic mechanism is proposed, the kinematic parameters of which are defined according to biomechanical research findings together with actual measurements of the human masticatory system. In order that a given mandibular trajectory was able to be tracked, the closed-form solution to the inverse kinematics of the robot is found for joint actuations whereas differential kinematics is derived in Jacobian matrices. Major specifications of the robot are presented which include details of the motion control system employed. Experimental results for free chewing, soft-food chewing, and hard-food chewing are reported. Each of these food groups were simulated using foam and hard objects. The crank actuations and driving torques required for the chewing of various foods are compared and finally, a simulation model in SimMechanics is presented.

4.1 The 6RSS Parallel Mechanism

Chapter 3 presented a mechanism for the development of a life-sized mastication robot model in accordance with biomechanical research findings in relation to the physiological structure of the human jaw and muscles of mastication (Fig. 3.3). The mechanism was driven solely by replicating the mouth-closing muscles (temporalis, masseter and pterygoid muscle groups) via double-acting linear actuators placed between the mandible and maxilla via ball-socket joints. The dimensions and specifications for the robot were defined according to human life-size equivalents and the masticatory patterns to be reproduced.

Because existing technologies were challenged in production of such a small robot, a new mechanism, the 6RSS parallel mechanism (Fig. 4.1), attracted our attention. The mechanism and the mandible are actuated by six RSS linkages, with the couplers being used in place of the linear actuators as shown in the model in Fig.3.3. Each RSS linkage was designed so that the coupler aligns itself with the

---

* Sections 4.1~4.4 and 4.8 are reprinted from Xu WL, Torrance J, Chen BQ, Potgieter J, Bronlund J and Pap JS (2008) Kinematics and experiments of a life-sized masticatory robot for characterizing food texture. IEEE Trans. Ind. Electron. 55:2121-2132, with permission from IEEE.

W. Xu and J.E. Bronlund: Mastication Robots, SCI 290, pp. 91–128.
springerlink.com

line of action of the human muscles. Special attention was given to the linkage transmission angles in order to produce efficient delivery of the force exerted by the couplers. Utilising computer tomography, the specifications of the mandible were replicated using measurements obtained from the jaw of a human cadaver supplied by the School of Dentistry at the University of Otago, New Zealand. The 3-D coordinates of the attachment/insertion positions of the major masticatory muscles were obtained from a sample of eight cadavers [1]. It should be noted however, that the new model is indicative only and rigorous proof of the equivalence of the two models needs further study.

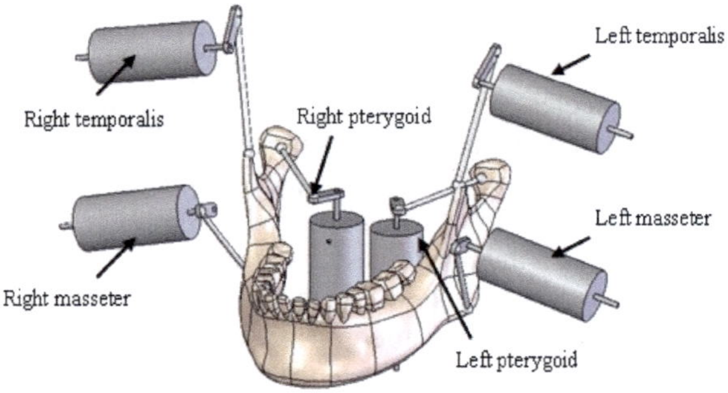

Fig. 4.1 The 6RSS parallel mechanism

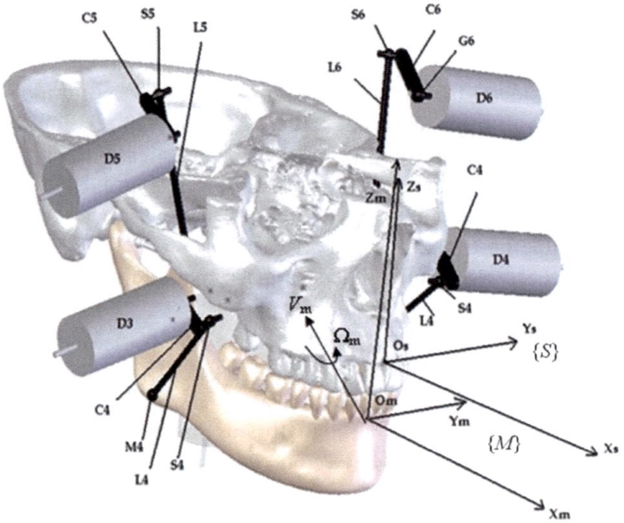

Fig. 4.2 Coordinate systems of the robot

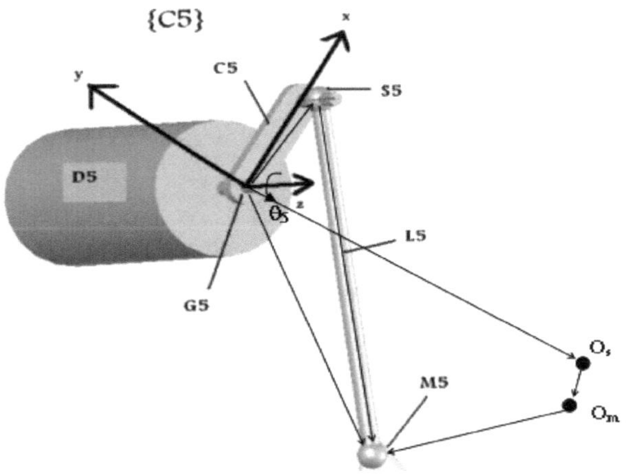

Fig. 4.3 Coordinate system of an RSS linkage

As illustrated in Fig. 4.2 and Fig. 4.3, each linkage consists of a driver unit (D_i, i = 1,2,…6), a crank (C_i, i = 1,2,…6), and a coupler link (L_i, i = 1,2,…6). The three joints are a revolute; at the actuator ground point where the driving shaft connects to the crank (G_i, i = 1,2,…6), a spherical joint where the crank joins to the coupler (S_i, i = 1,2,…6), and a second spherical joint where the coupler joins to the mandible (M_i, i = 1,2,…6).

This new robotic model differs from the Waseda Jaw (WJ) robots (as discussed in Chapter 1) in both its mechanism and biological foundation. Specifically, it has six DOFs which are sufficient to enable all possible masticatory trajectories in a 3-D space. Furthermore, this model employs a double-acting actuator for each group of muscles and attaches the actuators onto the mandible and skull via an RSS linkage specified in accordance with the muscle positions recorded from the human cadaver sample. Lastly the model relaxes the TMJ allowing for free movement within its envelope.

4.2 Coordinate Systems and Kinematical Parameters

In order to develop the robotic mechanism, a frame $oxyz_S$ (or skull frame) and a frame $oxyz_M$ (or mandible frame) must be established based upon the human skull and mandible, respectively (Fig. 4.2). The mandible frame is located at the bottom of the mandibular incisor and has a sagittal plane (x-z plane), a frontal plane (y-z plane) and a horizontal plane (x-y plane). The position of the mandible relative to the skull can be determined by way of homogenous transform ${}^{S}_{M}\mathrm{T}$ [2] as follows:

$$ {}^{S}_{M}T = \begin{bmatrix} c\alpha c\beta & c\alpha c\beta s\gamma - s\alpha c\gamma & c\alpha s\beta c\gamma + s\alpha s\gamma & {}^{S}O_{mx} \\ s\alpha c\beta & s\alpha s\beta s\gamma + c\alpha c\gamma & s\alpha s\beta c\gamma - c\alpha s\gamma & {}^{S}O_{my} \\ -s\beta & c\beta s\gamma & c\beta c\gamma & {}^{S}O_{my} \\ 0 & 0 & 0 & 1 \end{bmatrix} \tag{4.1} $$

in which ${}^{s}o_m = ({}^{s}o_{mx}\ {}^{s}o_{my}\ {}^{s}o_{mz})^T$ is the origin of $\{M\}$ with respect to $\{S\}$ and (γ, β, α) are roll-pitch-yaw rotational angles about the skull's *x*-, *y*- and *z*- axes, respectively. These represent the 3-D motion of the mandible depicted in $\{S\}$. The completed robot (as shown in Fig. 4.10) displays the following configuration for the closed mouth position: $({}^{s}o_{mx}\ {}^{s}o_{my}\ {}^{s}o_{mz})^T = (-6.09\ -0.07\ -20.17)^T$ and $(\gamma, \beta, \alpha) = (0^o, 3^o, 1.5^o)$.

Each RSS linkage is described in its own frame, or crank frame. Fig. 4.3 illustrates crank frame 5 or $\{C5\}$ for linkage 5, with the origin being at the pivotal point G_5, the *x*-axis being along $G_5\ S_5$, the *z*-axis running through the rotational axis of the crank and the *y*-axis completing the system. Crank frame $\{Ci\}$ is expressed in relation to the skull frame $\{S\}$:

$$ {}^{S}_{Ci}\boldsymbol{T} = \begin{bmatrix} c\alpha_i c\beta_i c\theta_i - s\alpha_i s\theta_i & -c\alpha_i c\beta_i s\theta_i - s\alpha_i c\theta_i & c\alpha_i s\beta_i & {}^{S}G_{xi} \\ s\alpha_i c\beta_i c\theta_i + c\alpha_i s\theta_i & -s\alpha_i c\beta_i s\theta_i + c\alpha_i c\theta_i & s\alpha_i s\beta_i & {}^{S}G_{yi} \\ -s\beta_i c\theta_i & s\beta_i s\theta_i & c\beta_i & {}^{S}G_{zi} \\ 0 & 0 & 0 & 1 \end{bmatrix} \tag{4.2} $$

where the upper left *3x3* matrix is a *z-y-z* Euler rotational matrix with angles in order α_i, β_i, and θ_i measured in frame $\{Ci\}$ and ${}^{S}\boldsymbol{G}_i = ({}^{S}G_{xi}\ {}^{S}G_{yi}\ {}^{S}G_{zi})^T$ is the origin of frame $\{Ci\}$ in the frame $\{S\}$.

Table 4.1 Parameters for six linkage frames

	Frame {C1}	Frame {C2}	Frame {C3}	Frame {C4}	Frame {C5}	Frame {C6}
Gx	-70.28	-70.92	-38.50	-41.63	-53.36	-56.36
Gy	-14.04	10.42	-60.91	58.89	-58.69	55.74
Gz	6.72	6.75	16.31	16.47	66.31	66.46
α	0.00	1.97	90.18	-87.99	89.14	-87.86
β	2.99	2.50	80.09	80.21	75.03	77.81
θ	θ_1	θ_2	θ_3	θ_4	θ_5	θ_6

Parameters α_i, β_i and ${}^{S}\mathrm{G}_i$ are deterministic by the robot and angle θ_i is the only variable for actuation. The values of all the deterministic parameters for the robot are given in Table 4.1. The initial crank angles (with zero values as the reference) are defined at the mouth closed configuration. Given in Table 4.2 are: the coordinates of coupler point M_i (i=1,2,…, 6) in the mandible frame {M}, the coordinates of the other couple point S_i (i=1,2,…, 6) in its linkage frame {C_i}, and the length of each coupler S_iM_i (i=1,2,…, 6).

Table 4.2 Other kinematic parameters for the built robot

	M1	M2	M3	M4	M5	M6	S1	S2	S3	S4	S5	S6
x	-79.75	-79.75	-60.85	-60.85	-53.25	-53.25	10.00	10.00	10.00	10.00	15.00	15.00
y	-41.10	41.10	-44.70	44.70	-43.00	43.00	0.00	0.00	0.00	0.00	0.00	0.00
z	38.77	38.77	-5.83	-5.83	43.87	43.87	0.00	0.00	0.00	0.00	0.00	0.00
Coupler	l_1=33.90		l_2=33.90		l_3=49.90		l_4=49.90		l_5=52.26		l_6=52.26	
Crank	c_1=10		c_2=10		c_3=10		c_4=10		c_5=15		c_6=15	

4.3 Inverse Kinematics

The inverse solution is used to find a set of actuation angles, θ_i (i=1,2, .., 6), given the movement of the mandible in relation to the skull, i.e., ${}^S_M\mathrm{T}$ in Eq. (4.1). The inverse solution for the robot can be derived according to the inverse kinematics approach [3].

On one hand, vector $\boldsymbol{G_iM_i}$ (as shown in Fig. 4.3) can be found explicitly as follows:

$$\boldsymbol{G_iM_i} = \boldsymbol{G_iO_s} + \boldsymbol{O_sO_m} + \boldsymbol{O_mM_i} \tag{4.3}$$

where the first term is determined by the location coordinates of G_i on the skull (Table 4.1); the second term is given by the prescribed mandible movement with respect to the skull, Eq.(4.1); and the third term is determined by the location of M_i on the mandible (Table 4.2).

On the other hand, vector $\boldsymbol{G_iM_i}$ can also be expressed by:

$$\boldsymbol{G_iM_i} = \boldsymbol{G_iS_i} + \boldsymbol{S_iM_i} \tag{4.4}$$

in which $\boldsymbol{G_iS_i}$ is the crank vector and $\boldsymbol{S_iM_i}$ is the couple vector.
Rearranging Eq. (4.4) gives rise to:

$$\boldsymbol{S_iM_i} = \boldsymbol{G_iM_i} - \boldsymbol{G_iS_i} \tag{4.5}$$

and further squaring the norm of the above vector equation yields:

$$\|\boldsymbol{S_iM_i}\|^2 = \|\boldsymbol{G_iM_i}\|^2 - 2\boldsymbol{G_iM_i} \bullet \boldsymbol{G_iS_i} + \|\boldsymbol{G_iS_i}\|^2 \tag{4.6}$$

in which $\|\boldsymbol{S_iM_i}\|=l_i$ is the coupler length (as given in Table 4.2), $\|\boldsymbol{G_iS_i}\|=c_i$ is the crank length (also as given in Table 4.2), and $\|\boldsymbol{G_iM_i}\|$ is calculated by Eq. (4.3).

Letting $r_i = -(l_i^2 - c_i^2 - \|\boldsymbol{G}_i\boldsymbol{M}_i\|^2)/2$, which is known up to this point, Eq. (4.6) can be re-arranged as:

$$\boldsymbol{G}_i\boldsymbol{M}_i \bullet \boldsymbol{G}_i\boldsymbol{S}_i = r_i \tag{4.7}$$

According to Eq. (4.3), vector $\boldsymbol{G}_i\boldsymbol{M}_i$ is calculated in {S}:

$$\boldsymbol{G}_i\boldsymbol{M}_i = \begin{bmatrix} GM_{xi} \\ GM_{yi} \\ GM_{zi} \end{bmatrix} = -{}^{S}\begin{bmatrix} G_{xi} \\ G_{yi} \\ G_{zi} \end{bmatrix} + {}^{S}\begin{bmatrix} O_{mx} \\ O_{my} \\ O_{mz} \end{bmatrix} + {}_{M}^{S}\boldsymbol{R}\, {}^{M}\begin{bmatrix} M_{xi} \\ M_{yi} \\ M_{zi} \end{bmatrix} \tag{4.8}$$

in which all the terms are known as per Table 4.1, 4.2 and Eq. (4.3).

Vector $\boldsymbol{G}_i\boldsymbol{S}_i$ is expressed in {S}:

$$\boldsymbol{G}_i\boldsymbol{S}_i = \begin{bmatrix} GS_{xi} \\ GS_{yi} \\ GS_{zi} \end{bmatrix} = {}_{C_i}^{S}\boldsymbol{R}\, {}^{C_i}\begin{bmatrix} S_{xi} \\ 0 \\ 0 \end{bmatrix} = S_{xi} \begin{bmatrix} c\alpha_i c\beta_i c\theta_i - s\alpha_i s\theta_i \\ s\alpha_i c\beta_i c\theta_i + c\alpha_i s\theta_i \\ -s\beta_i c\theta_i \end{bmatrix} \tag{4.9}$$

in which S_{xi}, α and β are the known robotic parameters provided in Table 4.1 and 4.2.

Putting Eq. (4.8) and (4.9) into Eq. (4.7) results in:

$$a_i \cos\theta_i + b_i \sin\theta_i = c_i \tag{4.10}$$

in which:

$$a_i = GM_{xi} c\alpha_i c\beta_i + GM_{yi} s\alpha_i c\beta_i - GM_{zi} s\beta_i$$

$$b_i = -GM_{xi} s\alpha_i + GM_{yi} c\alpha_i$$

$$c_i = r_i / S_{xi}$$

$$i = 1,2,\ldots,6.$$

Eventually, the two inverse solutions for each crank angle are found explicitly [2]:

$$\theta_i = \tan^{-1}(b_i / a_i) \pm \tan^{-1}(\sqrt{a_i^2 + b_i^2 - c_i^2} / c_i) \tag{4.11}$$

In order to control the robot and thereby enable a mandible trajectory to be followed, the above inverse solutions were calculated. These were firstly performed offline. Each crank angle was then sampled and fed into the motion control unit, and the crank angle was then servo-controlled in real time.

4.4 Singularity Analysis

From an RSS linkage (refer to Fig. 4.3), the velocity of point S_i on the coupler in frame $\{Ci\}$ can be found by:

$$^{Ci}\boldsymbol{V}_{S_i} = -(^{Ci}\boldsymbol{GS}_i \times {}^{Ci}z_i)\dot{\theta}_i \tag{4.12}$$

where $^{Ci}\boldsymbol{V}_{S_i}$ is the velocity of S_i in $\{Ci\}$; $^{Ci}\boldsymbol{GS}_i$ is the vector in $\{Ci\}$; $^{Ci}\boldsymbol{GS}_i\times$ is a *3x3* skew-symmetrical matrix of the respective vector; $^{Ci}z_i$ is the axis of crank i in $\{Ci\}$; and $\dot{\theta}_i$ is a scalar quantity of crank angular velocity.

Projecting this velocity along the coupler yields a scalar V_{ms_i}:

$$V_{ms_i} = {}^{Ci}\boldsymbol{MS}_i^T \bullet {}^{Ci}\boldsymbol{V}_{S_i} = -{}^{Ci}\boldsymbol{MS}_i^T \bullet (^{Ci}\boldsymbol{GS}_i \times {}^{Ci}z_i)\dot{\theta}_i \tag{4.13}$$

in which $^{Ci}\boldsymbol{MS}_i$ is a vector from point M_i to S_i in $\{Ci\}$.

Given linear velocity $^{M}\boldsymbol{V}_m$ and angular velocity $^{M}\boldsymbol{\Omega}_m$ of the mandible, expressed in $\{M\}$, the velocity of point M_i on the coupler can be found (Fig. 4.2):

$$^{M}\boldsymbol{V}_{M_i} = {}^{M}\boldsymbol{V}_m - {}^{M}\boldsymbol{O}_m\boldsymbol{M}_i \times {}^{M}\boldsymbol{\Omega}_m \tag{4.14}$$

where $^{M}\boldsymbol{V}_{M_i}$ is the velocity of M_i in $\{M\}$; $^{M}\boldsymbol{O}_m\boldsymbol{M}_i$ is the vector from O_m to M_i; and $^{M}\boldsymbol{O}_m\boldsymbol{M}_i\times$ is a *3x3* skew-symmetrical matrix of the respective vector.

Projecting this velocity along the coupler yields a scalar V_{ms_i}:

$$\begin{aligned} V_{ms_i} &= {}^{M}\boldsymbol{MS}_i^T \bullet {}^{M}\boldsymbol{V}_{M_i} = {}^{M}\boldsymbol{MS}_i^T \bullet (^{M}\boldsymbol{V}_m - {}^{M}\boldsymbol{O}_m\boldsymbol{M}_i \times {}^{M}\boldsymbol{\Omega}_m) \\ &= {}^{M}\boldsymbol{MS}_i^T \bullet {}^{M}\boldsymbol{V}_m - {}^{M}\boldsymbol{MS}_i^T \bullet {}^{M}\boldsymbol{O}_m\boldsymbol{M}_i \times {}^{M}\boldsymbol{\Omega}_m \end{aligned} \tag{4.15}$$

Because both points S_i and M_i are on coupler *i* and the coupler is a rigid body, the two scalars in Eq. (4.13) and (4.15) are identical, i.e.:

$$-{}^{Ci}\boldsymbol{MS}_i^T \bullet (^{Ci}\boldsymbol{GS}_i \times {}^{Ci}z_i)\dot{\theta}_i = {}^{M}\boldsymbol{MS}_i^T \bullet {}^{M}\boldsymbol{V}_m - {}^{M}\boldsymbol{MS}_i^T \bullet {}^{M}\boldsymbol{O}_m\boldsymbol{M}_i \times {}^{M}\boldsymbol{\Omega}_m \tag{4.16}$$

Because the robot involves six RSS linkages, repeating Eq. (4.16) for i=1,2,…, 6, and assembling them yields:

$$\begin{bmatrix} -{}^{C_1}\boldsymbol{MS}_1^{\,T}\bullet({}^{C_1}\boldsymbol{GS}_1\times{}^{C_1}z_1) & 0 & \cdots & 0 \\ 0 & -{}^{C_2}\boldsymbol{MS}_2^{\,T}\bullet({}^{C_2}\boldsymbol{GS}_2\times{}^{C_2}z_2) & \cdots & 0 \\ \vdots & \vdots & \cdots & \vdots \\ 0 & 0 & \cdots & -{}^{C_6}\boldsymbol{MS}_6^{\,T}\bullet({}^{C6}\boldsymbol{GS}_6\times{}^{C_6}z_6) \end{bmatrix}\bullet\begin{bmatrix}\dot{\theta}_1\\ \dot{\theta}_2\\ \vdots\\ \dot{\theta}_6\end{bmatrix}$$
$$=\begin{bmatrix} {}^{M}\boldsymbol{MS}_1^{\,T} & -{}^{M}\boldsymbol{MS}_1^{\,T}\bullet{}^{M}\boldsymbol{O}_m\boldsymbol{M}_1\times \\ {}^{M}\boldsymbol{MS}_2^{\,T} & -{}^{M}\boldsymbol{MS}_2^{\,T}\bullet{}^{M}\boldsymbol{O}_m\boldsymbol{M}_2\times \\ \vdots & \vdots \\ {}^{M}\boldsymbol{MS}_6^{\,T} & -{}^{M}\boldsymbol{MS}_6^{\,T}\bullet{}^{M}\boldsymbol{O}_m\boldsymbol{M}_6\times \end{bmatrix}\begin{bmatrix}{}^{M}\boldsymbol{V}_m\\ {}^{M}\boldsymbol{\Omega}_m\end{bmatrix} \tag{4.17}$$

Denoting $\dot{\boldsymbol{\Theta}}=(\dot{\theta}_1\,\dot{\theta}_2\ldots\dot{\theta}_6)$ and $\dot{\boldsymbol{X}}=({}^{M}\boldsymbol{V}_m\quad{}^{M}\boldsymbol{\Omega}_m)^T$ the above equation becomes:

$$\boldsymbol{J}_\theta\dot{\boldsymbol{\Theta}}=\boldsymbol{J}_x\dot{\boldsymbol{X}} \tag{4.18}$$

in which $\boldsymbol{J}_\theta$ and $\boldsymbol{J}_x$ are called inverse and forward kinematic Jacobian matrices, respectively [4]. They can be calculated easily once the inverse solution has been made available.

Eq. (4.18) shows the relationship between the crank angular velocity ($\dot{\boldsymbol{\Theta}}$) and the mandibular velocity ($\dot{\boldsymbol{X}}$). When either of the two Jacobian matrices is singular, the robot is in a singular configuration where one or more degrees of freedom are uncontrollable instantaneously.

Letting $\|\boldsymbol{J}_\theta\|=0$ easily yields the singular configurations at which a coupler $\boldsymbol{MS}_\mathrm{i}$ is normal to the y-axis of frame {Ci}, i.e. coupler $\boldsymbol{MS}_\mathrm{i}$ and crank $\boldsymbol{GS}_\mathrm{i}$ are folded in a line (Figure 4.3). In motion planning, the robot should be commanded to avoid this singular configuration. To avoid this issue, each crank angle in the designed robot is restricted within a small range, and consequently, this singularity is not present.

Each row of $\boldsymbol{J}_x$ is made up of a vector of coupler $\boldsymbol{MS}_\mathrm{i}$ and a vector that is normal to both $\boldsymbol{O}_m\boldsymbol{M}_i$ and $\boldsymbol{MS}_\mathrm{i}$. Hence, as long as the couplers are not parallel to each other, each row of vectors will be independent and consequently the forward Jacobian $\boldsymbol{J}_x$ will never become singular. The robot (see Fig. 4.10) was designed in order that this singularity was absent.

4.5 Design of the Robot

4.5.1 Framework Design

The entire framework of the robot consisted of four parts: bottom plate, left, middle and right mounting units, as shown in Fig. 4.4. The overall dimensions of the bottom plate were chosen in such a way that it was large enough to accommodate all parts of the entire chewing robot mechanism. Due to its overall dimensions, aluminium was chosen as the material as it gave the whole system an acceptable level of stability and minimised weight.

To make assembly of the mechanism easier, a cutaway was placed underneath the mandible. The three mounting units were each aligned with two dowels and each was fixed to the bottom plate with four cap screws. This prevented the three units from moving. The left and right mounting units supported the left and the right masseter and temporalis actuators, respectively.

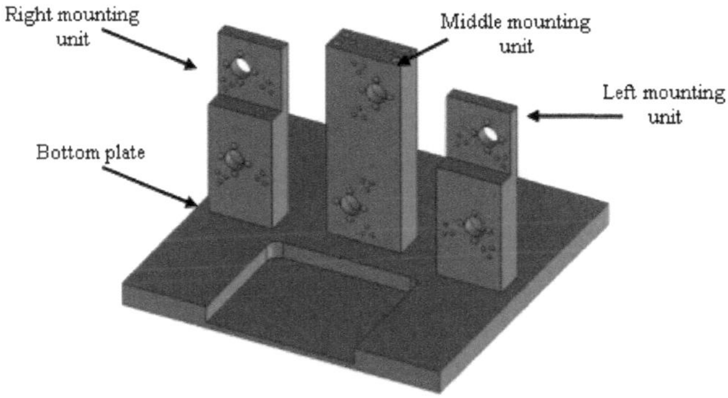

Fig. 4.4 Framework of the chewing robot with bottom plate and three mounting units

The middle mounting unit supported the left and right lateral pterygoid actuators together with the upper jaw. As a result, the thickness of this unit was increased, (relative to the other two units) in order to provide adequate strength. The motor units (gearbox, motor, and encoder) were mounted to the three mounting units with four M3 x 6 cap screws. These screws were embedded in the mounting units to prevent them from buckling and to prevent any movement of the motor unit. By using bevel gearboxes with a gear ratio of 1:1 in front of each motor unit, it was possible to orientate all six motor units in the same direction. This provided a more compact looking model (Fig. 4.5).

The bevel gearboxes had steel gears and shafts and ran in ball and sintered bearings. The housing mouldings were 30% glass filled nylon and provided a greater degree of strength in comparison to normal nylon housings. The motor gearbox (with a 66:1 gear ratio) had a maximum permissible axial load on the output shaft of 140 N. This was below the force required during the chewing process. Therefore, the bevel gearbox in front of the motor unit also had a supporting function and allowed for the absorption of all axial loads which might occur. The output shafts of the motor units and the input shafts of the bevel gearboxes were connected to each other with shaft couplers. Gearboxes and shaft couplers were purchased from MFA/ Comodrills, United Kingdom. Each bevel gearbox was fixed with nine M2 screws to the bevel gearbox mounting unit.

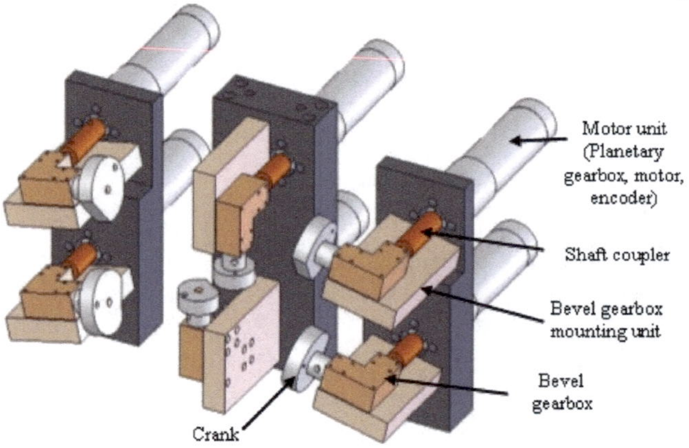

Fig. 4.5 Left, right and middle mounting units with assembled motor units and bevel gearboxes

4.5.2 *Lower and Upper Jaw*

Accurate denture morphology was required in order to enable comparisons to be made with the human chewing system. Therefore, important points on the model jaw (i.e., reference points, attachment points and denture morphology) needed to accurately represent an average human jaw. In order to achieve this, a basic study model from Trimut Corporation, Kyoto, Japan, was used to provide realistic morphology. This model included the important 2:1 tooth relationship between mandible and maxilla, and provided valid contact points (Fig. 4.6).

The teeth were constructed of melamine (a type of plastic) and were considered strong enough for the first chewing tests. For more advanced testing, two layered composite teeth with natural hardness were also available from the same company.

The structure of the lower mechanical jaw consisted of three aluminium parts: base, left and right ramus, as shown in Fig. 4.7. Where ML, MR, TL, TR, PL and PR represent the left masseter, right masseter, left temporalis, right temporalis, left lateral pterygoid and right lateral pterygoid, respectively. Each ramus was fixed to the base with four M4 cap screws and two 4 mm dowels. The dowels ensured a very good lining between the parts. The connections between the mandible and couplers were fixed with bushings and cap screws (Fig. 4.7).

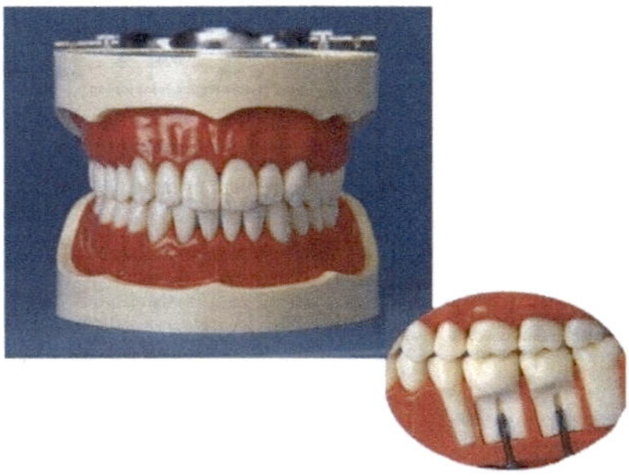

Fig. 4.6 Basic study model with replacement teeth

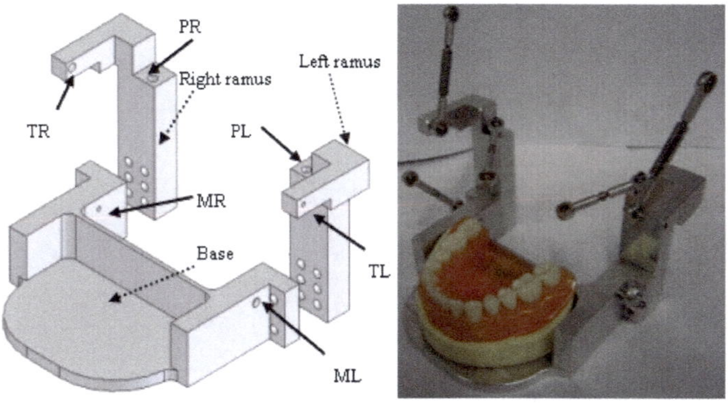

(a) (b)

Fig. 4.7 Assembly of the lower mechanical jaw, (a) SolidWorks model and (b) physical model

The lower part of the study model was fixed with screws at the base. Spatial requirements between the lower jaw and the bevel gearbox mounting unit of the right lateral pterygoid actuator, made a cut out on the base unit necessary (Fig.4.8).

The upper and lower jaws of the basic study model were reduced in depth, by approximately 10 mm in order to maintain the same relationship between the actuator attachment points on the mandible and the denture position (due to the position of the bevel gearbox mounting units for the lateral pterygoid). To stabilise the back molar teeth of both the upper and lower jaw, walls with a thickness of 2 mm were implemented in the mandible base unit and the upper jaw mounting unit, where the upper jaw was mounted. The upper jaw was fixed with one cap screw at the top and two countersink screws from behind. The upper jaw mounting unit was then connected to the middle mounting unit via a connecting unit (Fig. 4.9). All units were lined up, each with two dowels and were fixed with four cap screws.

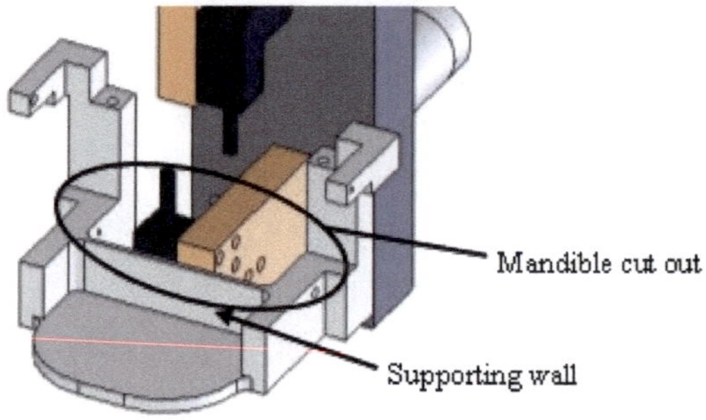

Fig. 4.8 Lower mechanical jaw with bevel gearbox mounting units for the lateral pterygoid muscle group

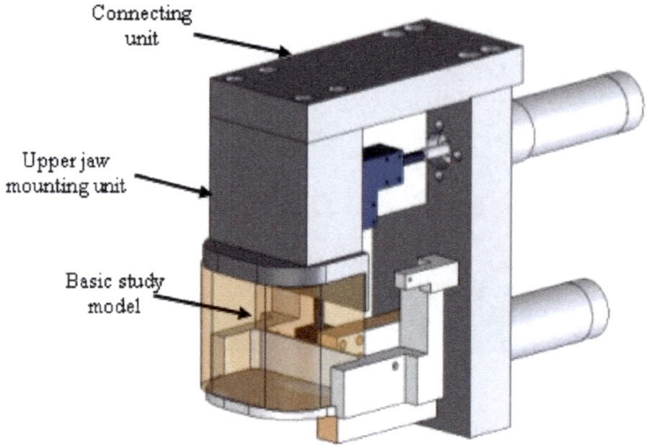

Fig. 4.9 The mounting system of the upper jaw

4.6 Motion Control

The constructed robot is shown in photographs (Fig. 4.10). Each actuation unit consists of a 60 W dc motor, a 66:1 gearbox, and a 500 count encoder. This gives rise to a constant speed torque of 86.2 mNm and a maximum speed of 121.2 RPM at the crank end. By using bevel gearboxes with a gear ratio of 1:1 in front of each motor unit, all six actuation units were oriented in the same direction. The crank was manufactured out of mild steel, whereas the coupler was made of two rod ends and a shaft. A test was conducted and revealed that the coupler could withstand a pulling force of 988N.

The crank and the coupler were connected in such a way that the coupler was able to rotate around its connecting shaft and could move freely in a frontal angle direction. The teeth were made out of melamine and were therefore strong enough

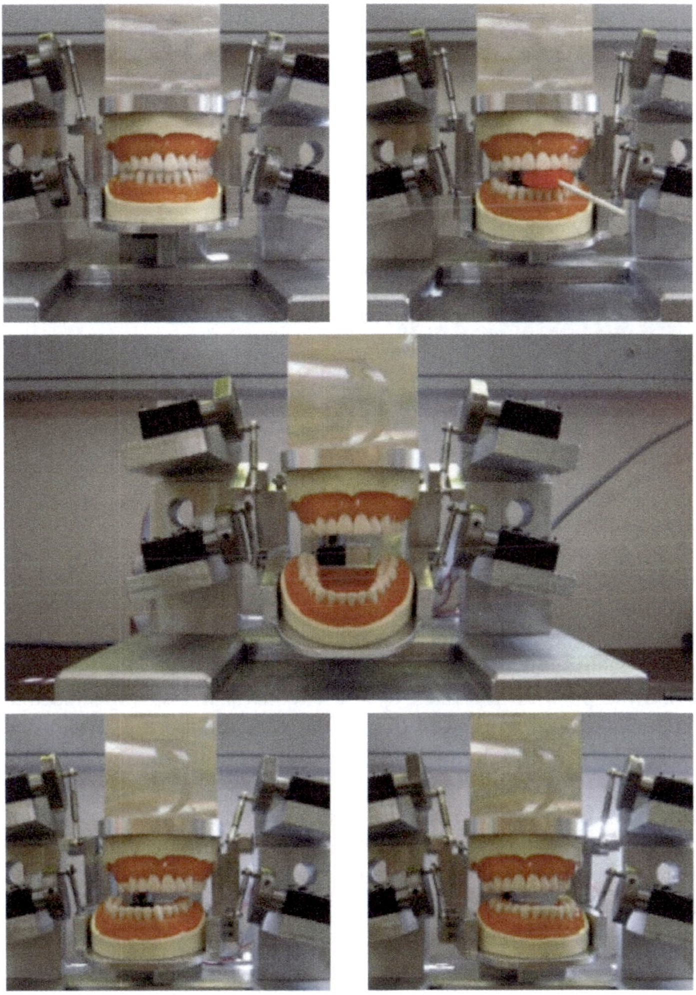

Fig. 4.10 Close-up of the physical robot at various configurations

to withstand chewing tests. During construction, the life-sized upper and lower jaws were reduced in depth, by approximately 10 mm in order to accommodate the bevel gearbox mounting units for the left/right pterygoid actuators.

The motion control system consisted of a six-axis motion control card (Galil DMC-1860), two amplifiers (each driving up to four motors), a power supplier, and a program (Galil DMC Smart Terminal), as shown in Fig. 4.11 The motion control card produced a control signal for each actuator that varied between -10V and 10V.

The low level motion control of the robot involved a trajectory-tracking proportional-integral-derivative (PID) control for each of the six actuators. In Galil, PID

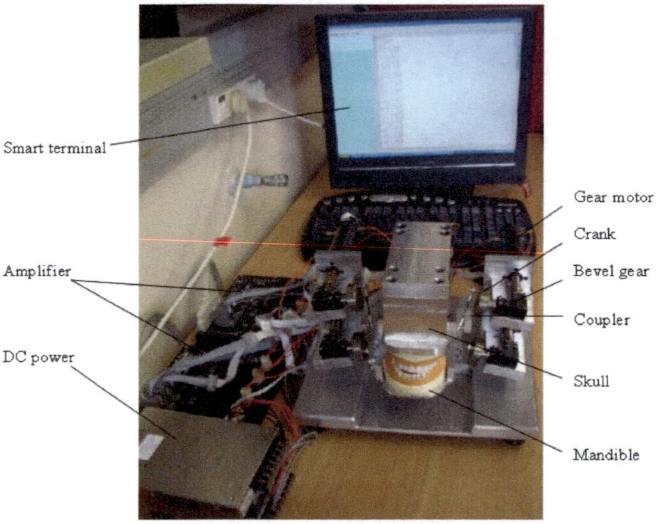

Fig. 4.11 The masticatory robot system

gains were tuned for each actuator, which act on differences between the commanded position and the position fed back by the encoders. It was expected that independent PID control would work well due to the gear reduction between the motor and the crank, which would greatly reduce the effects of dynamic coupling between cranks.

The trajectory data for each actuator can be specified in the form of a series of points versus time. This type of motion suits contour mode motion control in Galil and was implemented via the Galil Smart Terminal program. In order to program a motion, two arrays must be defined; one for the absolute position data, and the other for the difference between consecutive points. The absolute position data was able to be imported explicitly from the output of the inverse kinematics program in Matlab. The difference data array was then calculated using the terminal program. A constant time-interval between consecutive points was also defined. The contour mode works by moving the specified difference (in encoder counts) over a fixed time interval. Using this method, the commanded motion of all six actuators could be synchronized, giving rise to a coordinated movement of the robotic jaw.

4.7 Simulation in SolidWorks/CosmosMotion

Inverse kinematics for the robot was simulated in SolidWorks/CosmosMotion (Fig.4.12). The 3-D mandibular trajectory for simulation was recorded using a custom-made brace that had 6 markers on it [5]. The trace paths of this trajectory are shown in Fig. 4.13 and Fig. 4.14. A total of 4 'chews' were simulated. The time interval of the trajectory was 0.01 s.

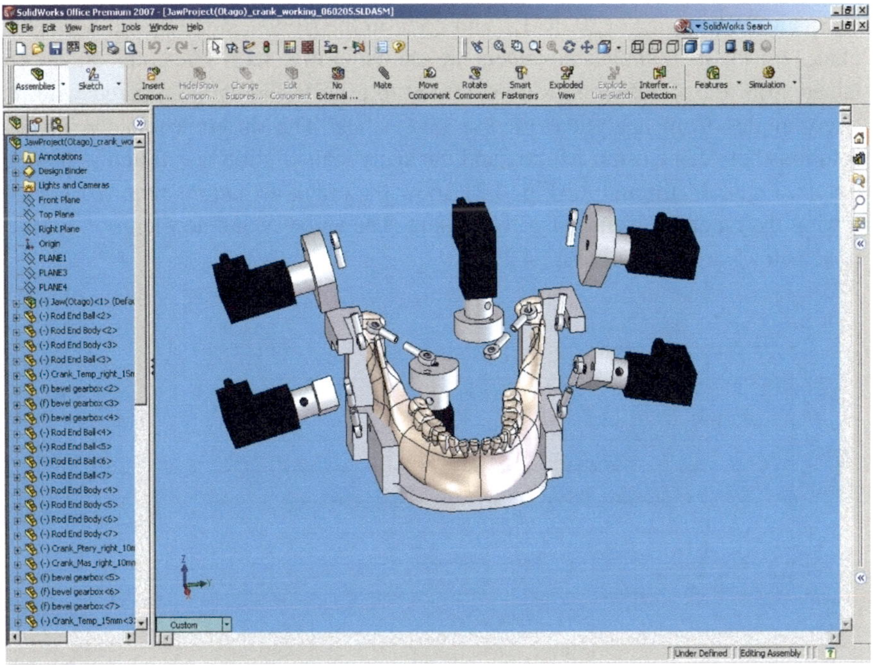

Fig. 4.12 The robot in SolidWorks/CosmosMotion

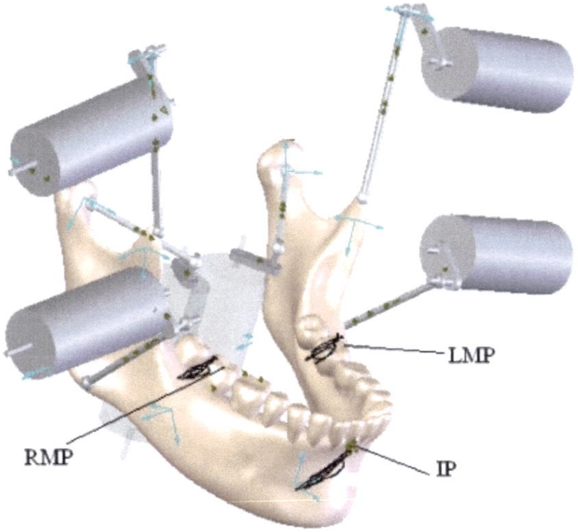

Fig. 4.13 Trace paths of a mandibular trajectory in 3-D space

Spatial trajectories of IP, RMP, and LMP in the frontal, sagittal, and horizontal planes for the mandibular movement are shown in Fig. 4.15. The crank angles obtained from the simulations were exported to a CSV file and compared with the crank angles from analytical inverse kinematics. The difference in the inverse solutions between CosmosMotion and the analytical solution was found to be negligible. Fig. 4.16 through to Fig. 4.21 show the crank actuations required for following the recorded mandibular movement. The velocity and acceleration profiles of actuator 5 are shown in Fig. 4.22.

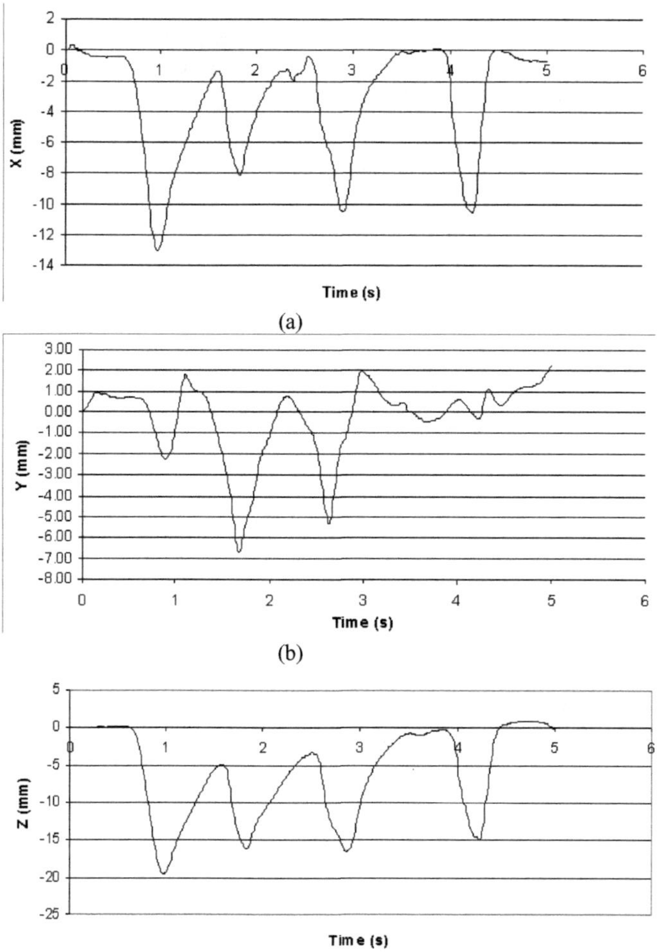

Fig. 4.14 Time history of IP trajectory, (a) x-axis, (b) y-axis and (c) z-axis

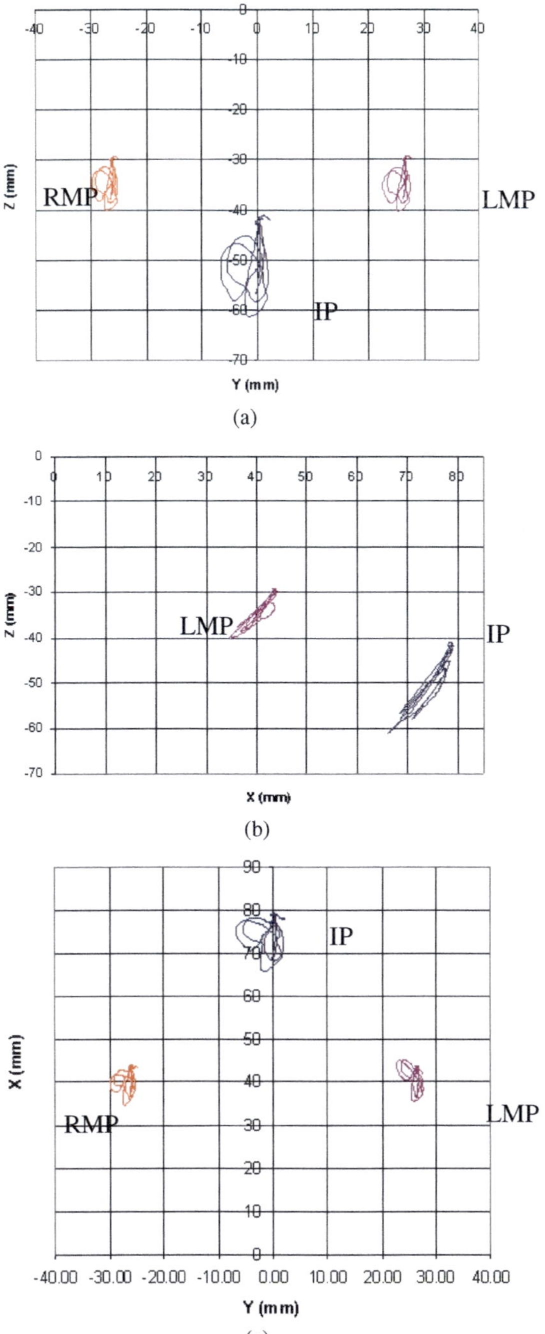

Fig. 4.15 IP, RMP, and LMP trajectories in (a) frontal, (b) sagittal and (c) horizontal plane

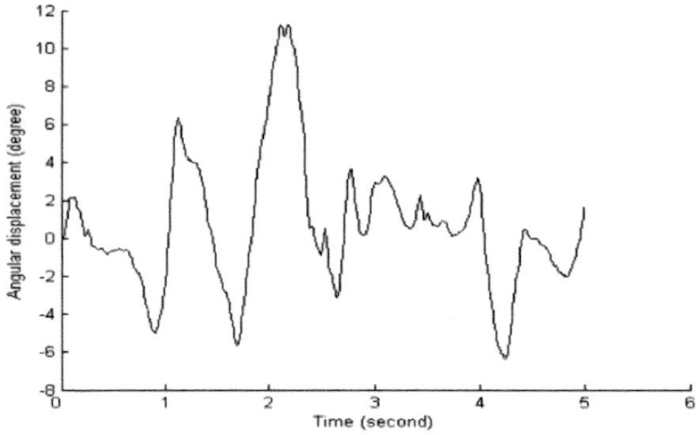

Fig. 4.16 The actuation of crank 1

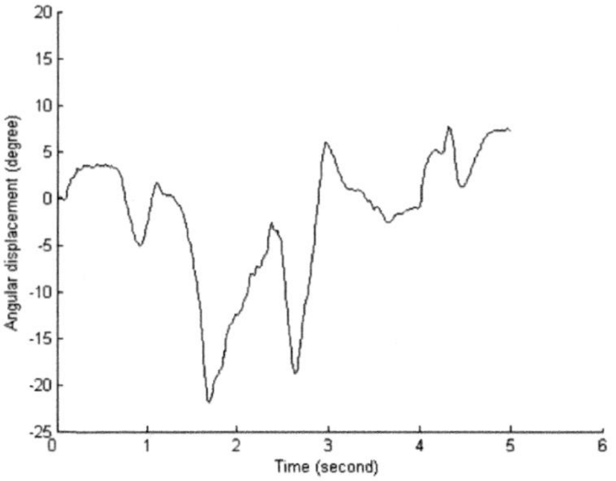

Fig. 4.17 The actuation of crank 2

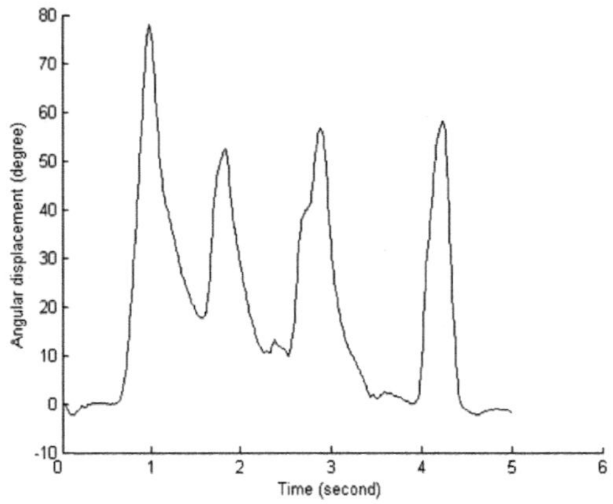

Fig. 4.18 The actuation of crank 3

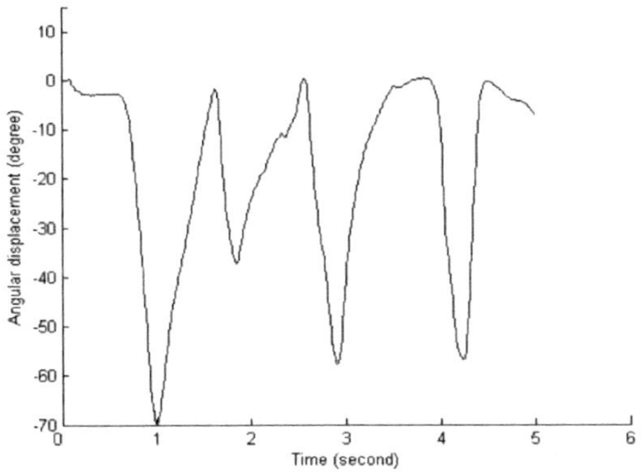

Fig. 4.19 The actuation of crank 4

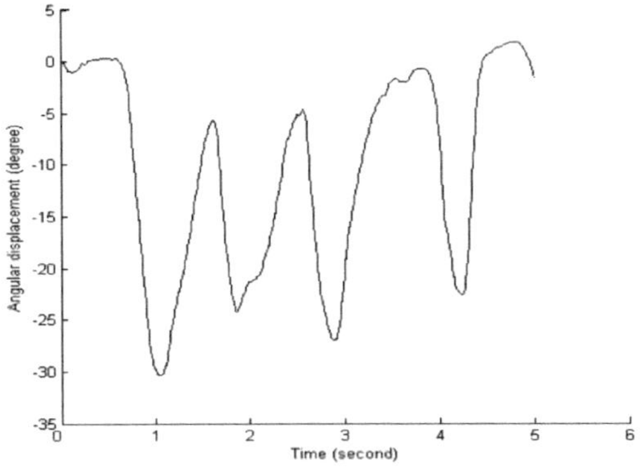

Fig. 4.20 The actuation of crank 5

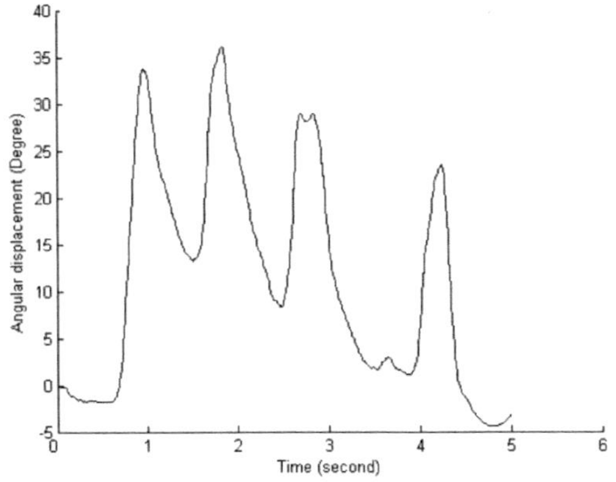

Fig. 4.21 The actuation of crank 6

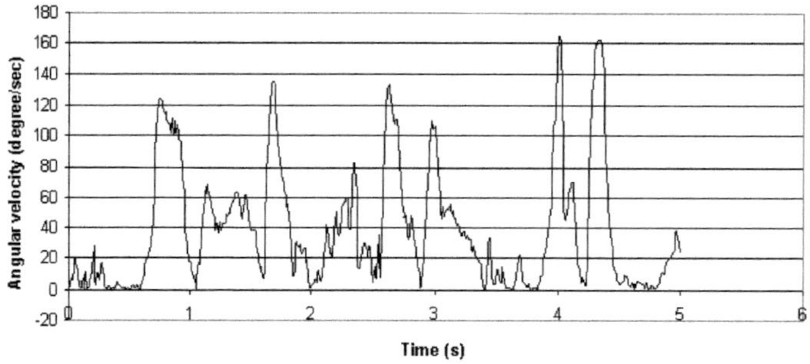

(a)

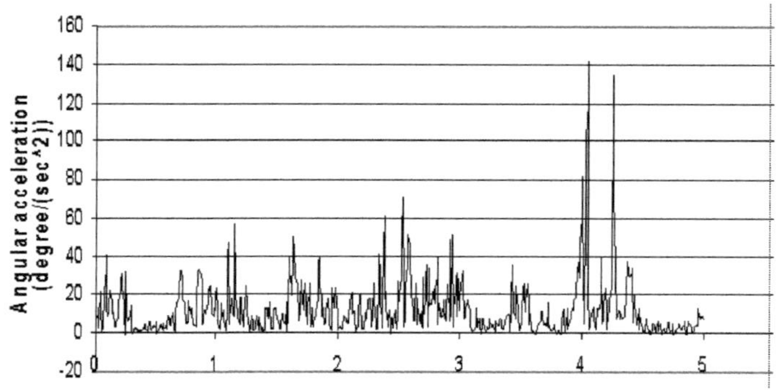

(b)

Fig. 4.22 Motion profiles of actuator 5, (a) velocity and (b) acceleration

4.8 Simulation in SimMechanics

The robot was also modeled in SimMechanics, as shown in Fig. 4.23. The model was developed using the following block components:

- Body, to represent a rigid body, whose pose is determined by a CS (coordinate system) and its mass properties by way of a CG (centre of gravity coordinate system).
- Ground, to represent an immobile link.
- Revolute joint, to represent a 1 DOF revolute joint between the crank and the ground.
- Spherical joint, to represent a 3 DOF ball-socket joint between the crank and the coupler, and the coupler and the mandible.
- Joint actuator, to specify the crank actuation either, under a generalised force, or a motion in terms of displacement, velocity and acceleration (see Fig. 4.24 for example).
- Joint sensor, to measure the motion of a joint or the reaction forces inside a joint (see Fig. 4.25 for example).
- Body sensor, to sense the motion of a body (see Fig. 4.26 for example).

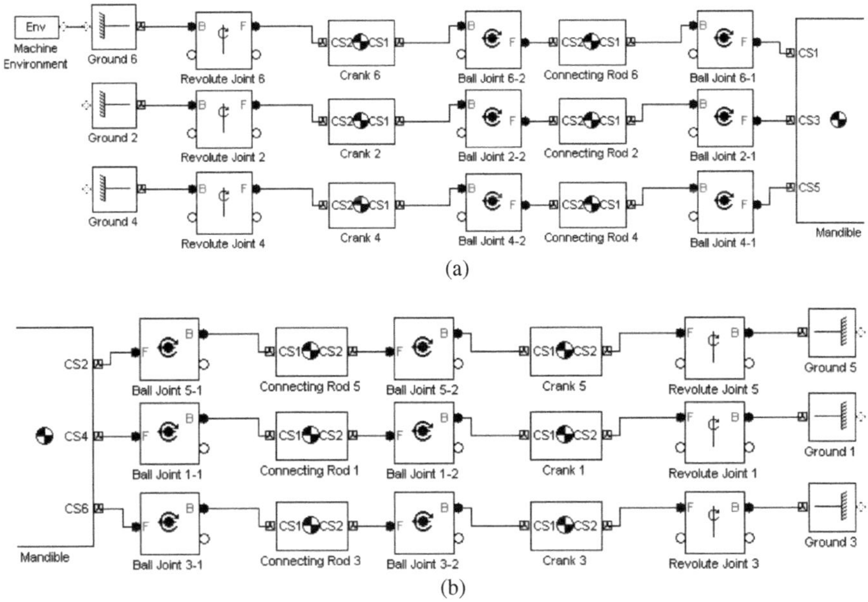

Fig. 4.23 The SimMechanics model of the robot, divided in two halves, (a) actuator 2, 4 and 6, and (b) actuator 1, 3 and 5

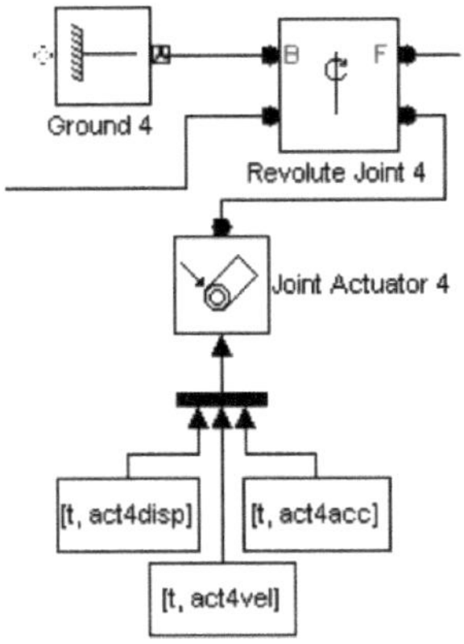

Fig. 4.24 The actuation input of actuator 4 for left masseter

All physical properties required for completing the model were taken from measurements obtained from the SolidWorks simulation model (Fig. 4.12). The robot's home position was taken to be when the mouth was fully shut. In the simulation, the actuations of the six cranks were given (as depicted in Fig. 4.27 through to Fig. 4.32) which were converted by use of inverse kinematics from the recorded mandibular movement. For this recording, the mouth remained shut for an initial period of approximately 7 seconds. The motion of the mandible, the torque required for the six actuators and the bearing forces inside the joints were required. Fig. 4.33 shows the animation model of the robot, in which the centres of gravity are represented by black circles and the CSs are represented as three-orthogonal vectors. Because no food was placed in the mouth, the simulated results are only valid for free jaw movement.

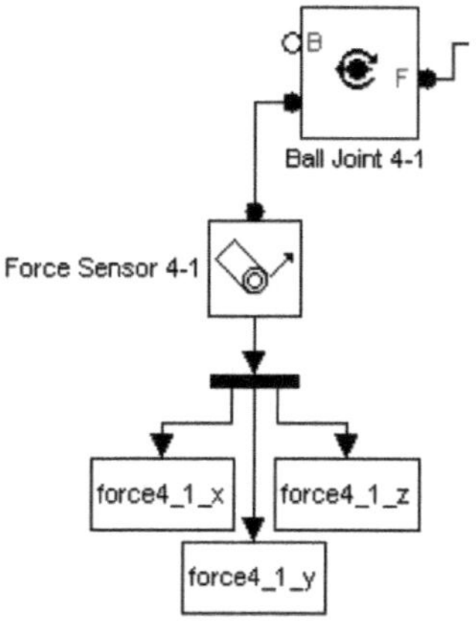

Fig. 4.25 Sensor for the measurement of the bearing forces of joint 4-1

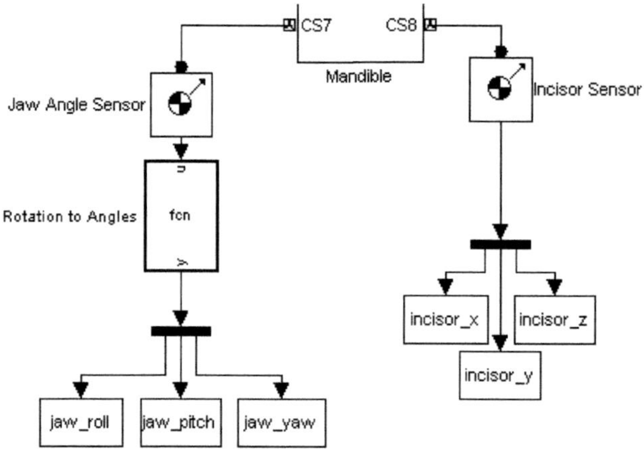

Fig. 4.26 Sensors for the measurement of the incisor trajectory and the jaw angles

The crank actuation was inputted by specifying a joint actuator block where the time-functions of the angular position, velocity and acceleration were those as shown in Fig. 4.27 to 4.32. By way of an example, Fig. 4.24 shows how actuator 4 was specified.

The crank torque required to drive each actuator is presented in Figs. 4.34 to 4.39. The torque predictions obtained had mean values for each actuator of 10.7, 39.8, 4.6, 4.6, 36.5 and 35.5 mNm for actuators 1 through 6, respectively.

Using the body sensor block (Fig. 4.26), the incisor trajectory and the angular trajectory of the mandible were plotted. The two body sensors were placed at the CG of the mandible and at the incisor, respectively. It should be noted that the roll, pitch and yaw angles were converted from the sensed rotation matrix, so that all the trajectories were defined in the world CS (the reference was carried over from the SolidWorks model) which is different from the skull CS (the reference was used to describe the robotic kinematics. Furthermore it is noted that the torque and joint forces calculated may not be very realistic due to the lack of the moments of inertia in the model.

The plots of the incisor trajectory (Figs. 4.40 to 4.42) show the amount of movement away from the rest position, which was for the first 7 seconds. The most movement occurred in the z-direction.

Figs. 4.43 to 4.45 show the angular trajectory of the mandible. The greatest angular movement of the mandible occurred in the pitch angle and the least in the roll angle. All the aforementioned findings mirrored the chewing movement recorded. Reaction forces within the joints were measured by attaching a joint sensor. Fig. 4.25 gives an example of this for spherical joint 4-1, and Figs. 4.46 to 4.48 show the x-, y- and z- force elements within the joint.

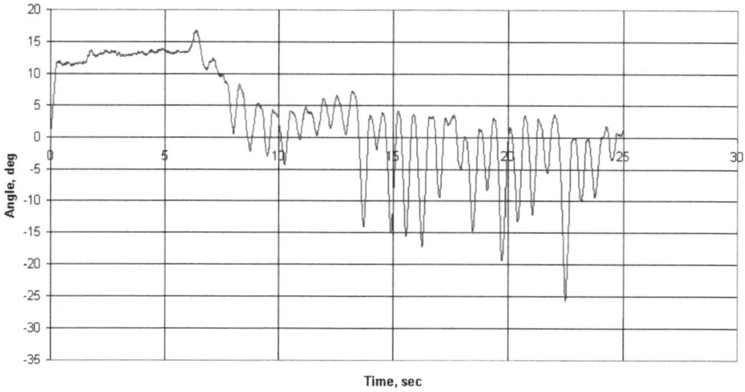

Fig. 4.27 The actuation of actuator 1 for right pterygoid

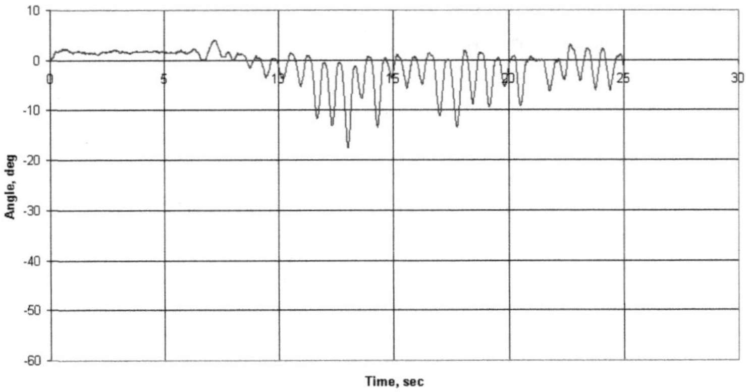

Fig. 4.28 The actuation of actuator 2 for left pterygoid

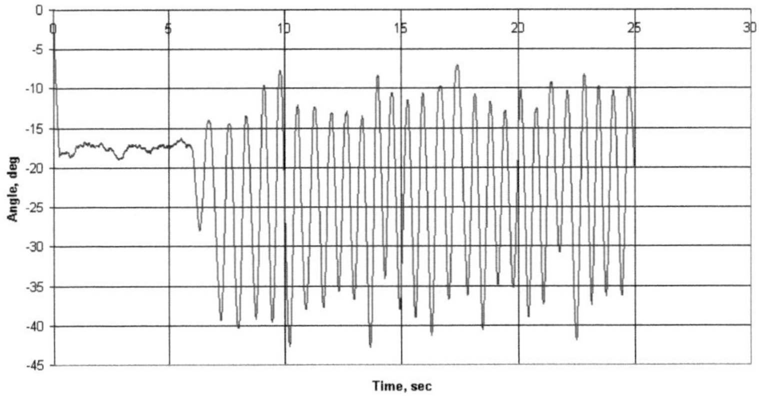

Fig. 4.29 The actuation of actuator 3 for right masseter

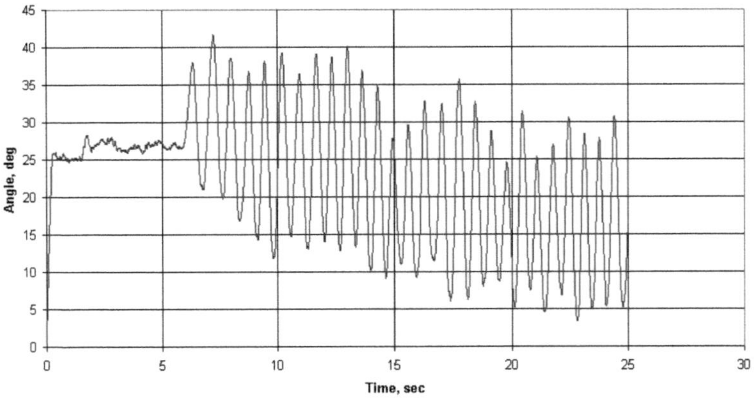

Fig. 4.30 The actuation of actuator 4 for left masseter

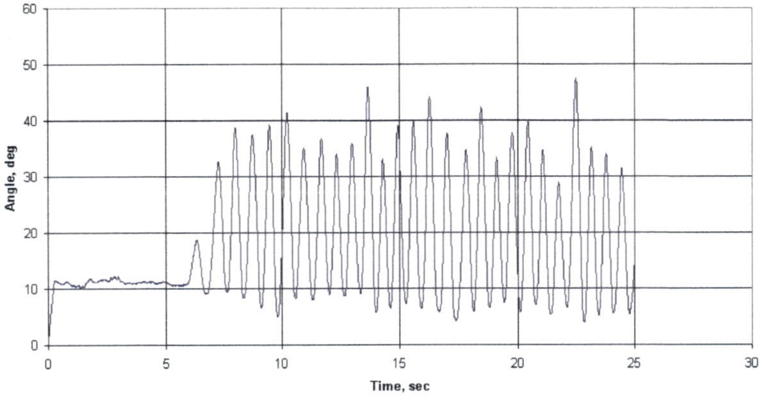

Fig. 4.31 The actuation of actuator 5 for right temporalis

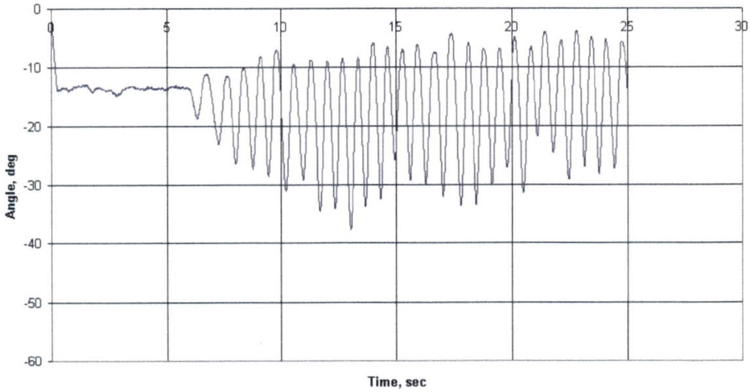

Fig. 4.32 The actuation of actuator 6 for left temporalis

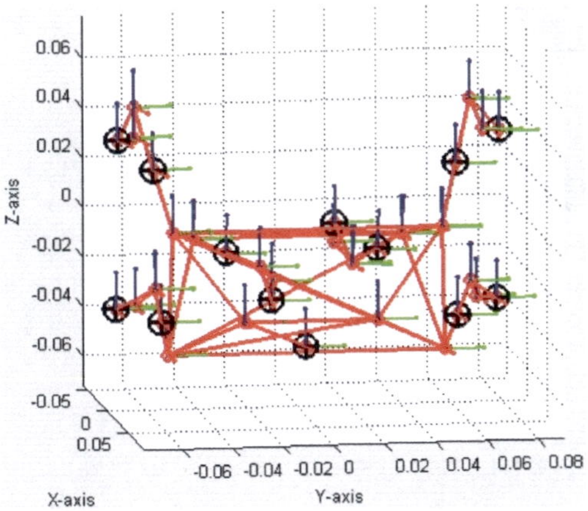

Fig. 4.33 Animation model of the robot in CosmosMotion

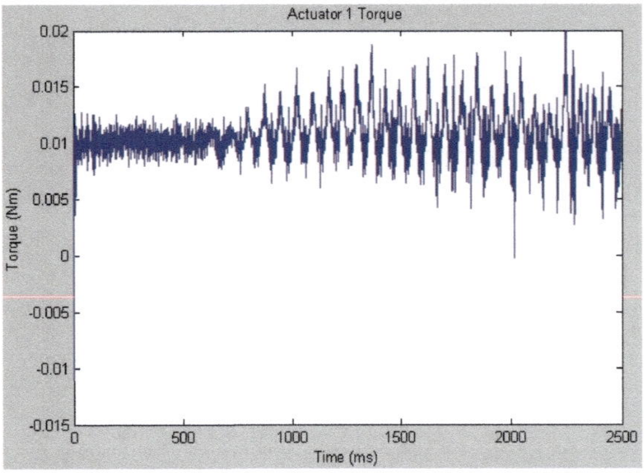

Fig. 4.34 The torque required to drive actuator 1 for right pterygoid

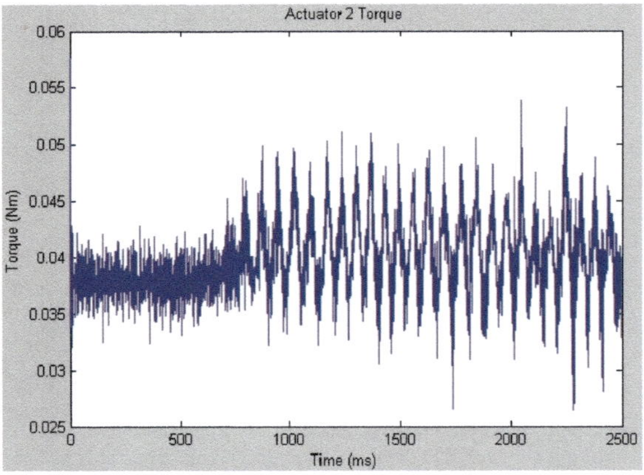

Fig. 4.35 The torque required to drive actuator 2 for left pterygoid

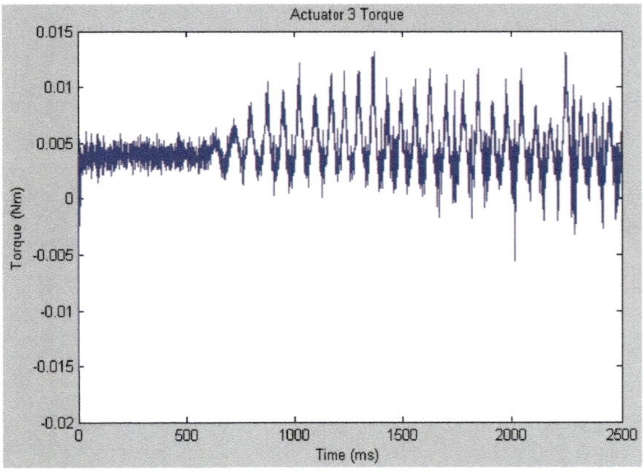

Fig. 4.36 The torque required to drive actuator 3 for right masseter

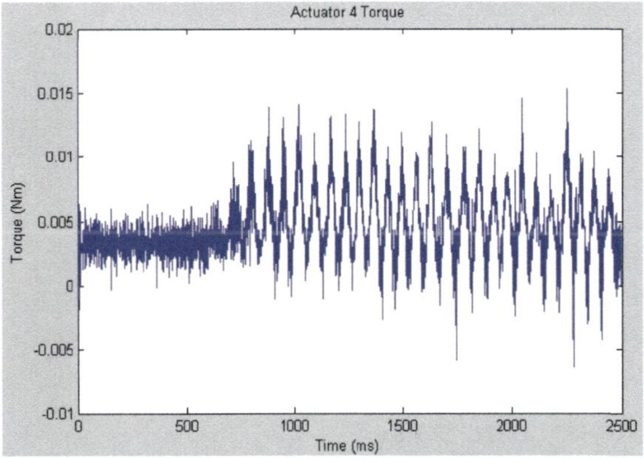

Fig. 4.37 The torque required to drive actuator 4 for left masseter

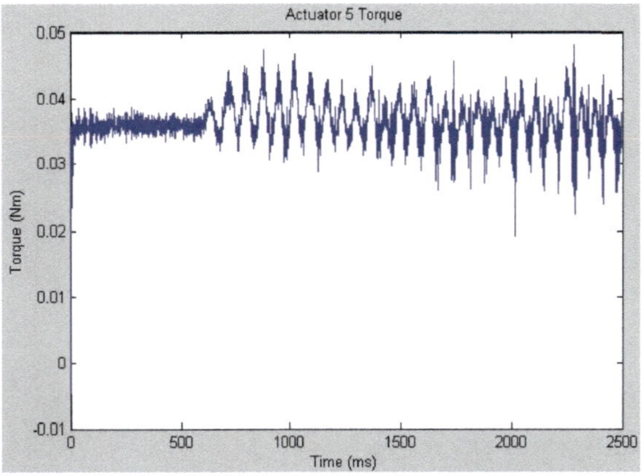

Fig. 4.38 The torque required to drive actuator 5 for right temporalis

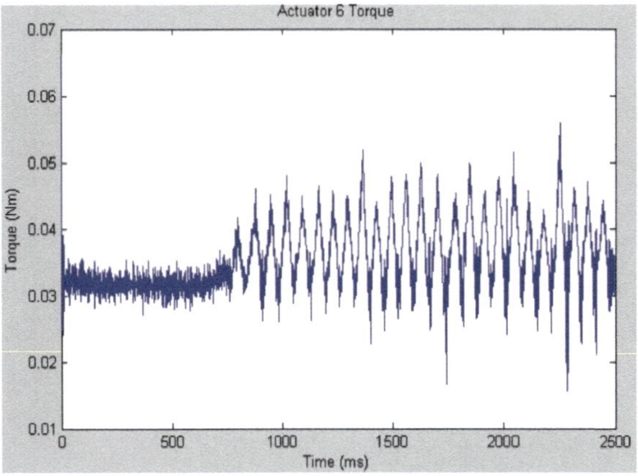

Fig. 4.39 The torque required to drive actuator 5 for left temporalis

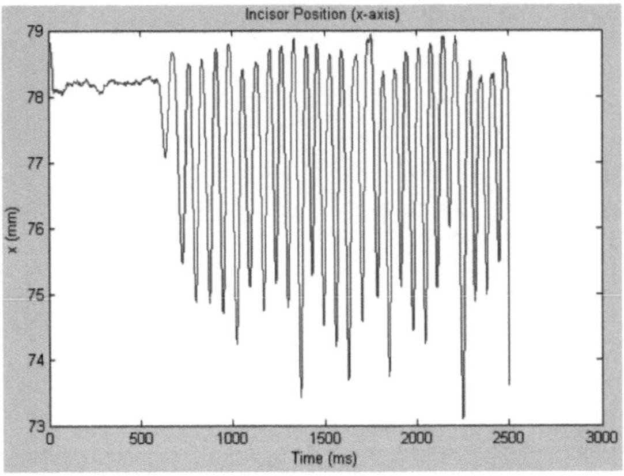

Fig. 4.40 The x-axis temporal trajectory of the incisor

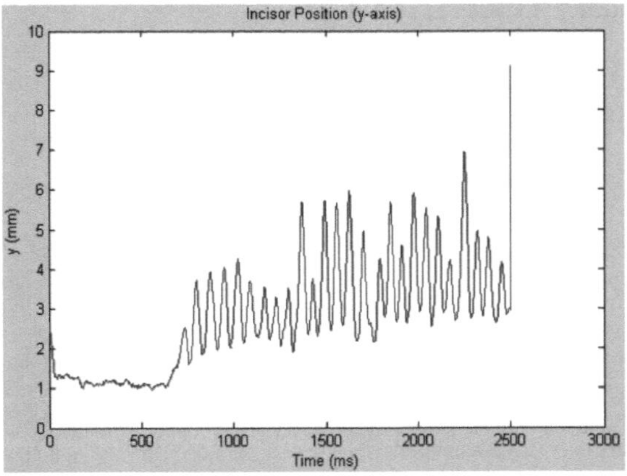

Fig. 4.41 The y-axis temporal trajectory of the incisor

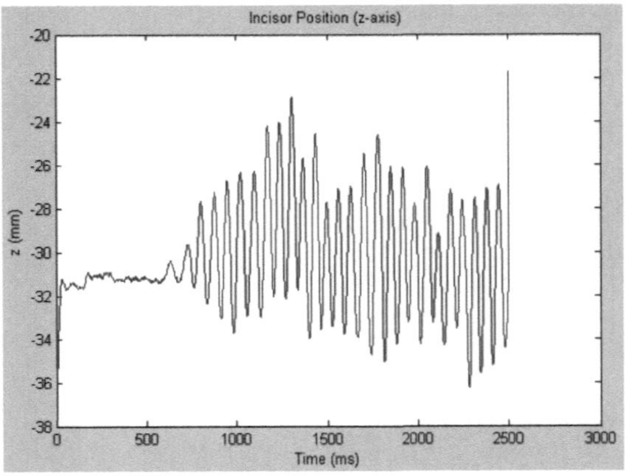

Fig. 4.42 The z-axis temporal trajectory of the incisor

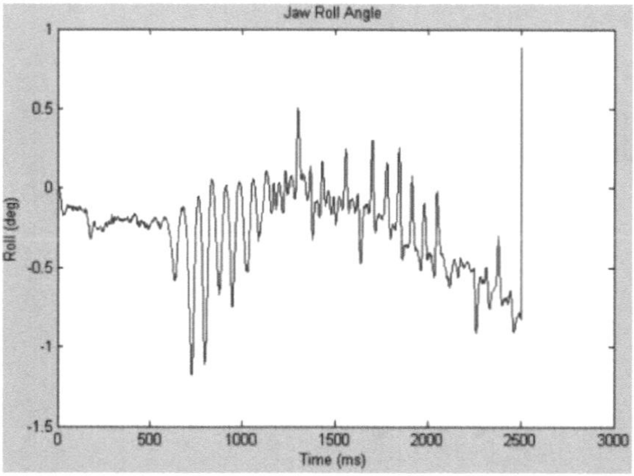

Fig. 4.43 Roll angle of the mandible

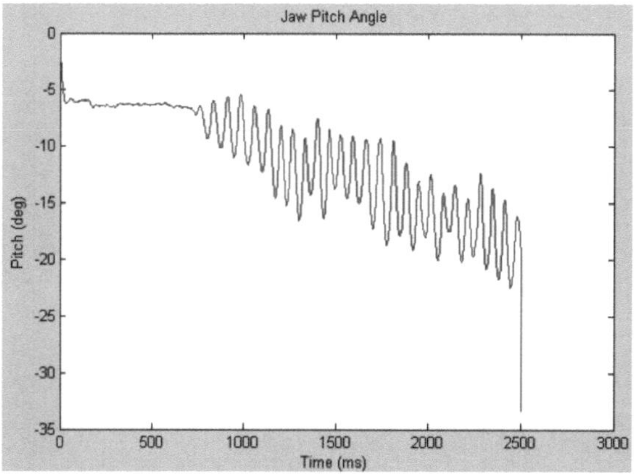

Fig. 4.44 Pitch angle of the mandible

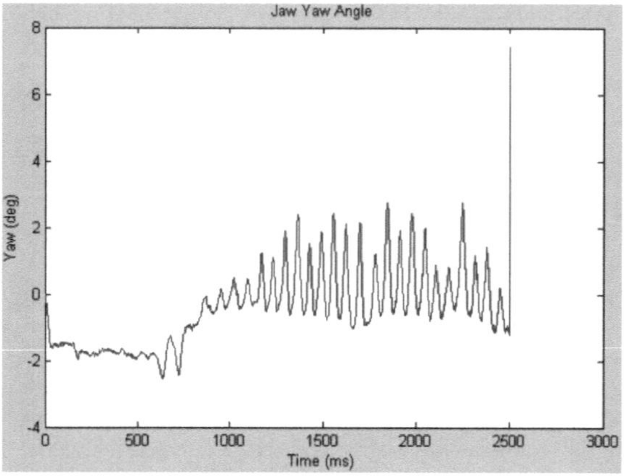

Fig. 4.45 Yaw angle of the mandible

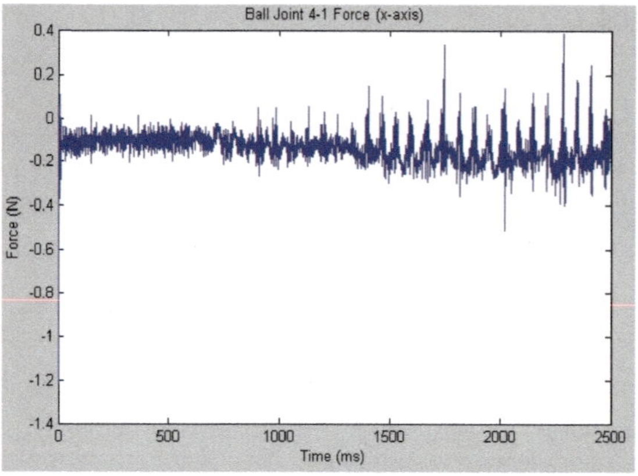

Fig. 4.46 The *x*-axis element of the bearing force of the joint 4-1

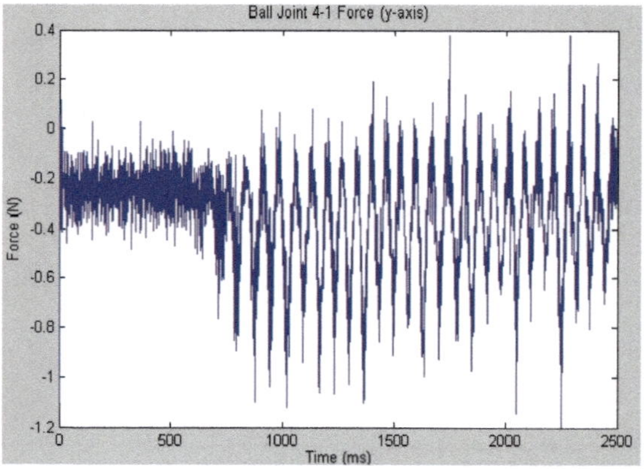

Fig. 4.47 The *y*-axis element of the bearing force of the joint 4-1

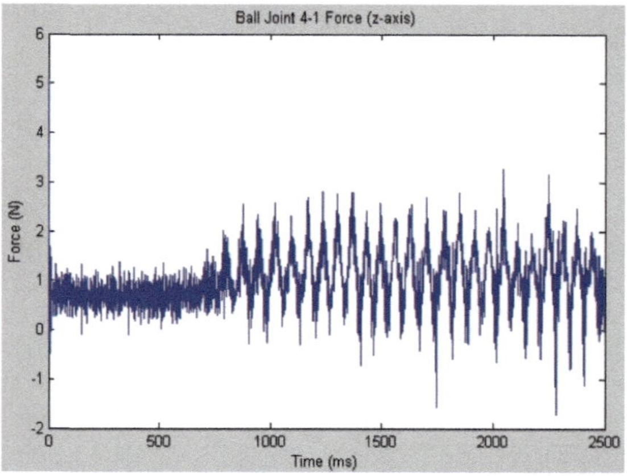

Fig. 4.48 The *z*-axis element of the bearing force of the joint 4-1

4.9 Initial Chewing Experiments

4.9.1 Masticatory Movements

A series of experiments were carried out was to test whether or not; the masticatory robot could follow a real-life human masticatory movement, if the robot could chew on foods of various texture properties, and if the robot could have a compliant ability in terms of giving way to an extremely hard food.

In the following experiments, the masticatory movement to be reproduced by the robot was specified by the trajectory of incisor point (IP) (or o_m in Fig. 4.2) in three coordinates (Fig. 4.49) and the roll-pitch-yaw angles of the mandible (Fig. 4.50), in the skull system $\{S\}$. The masticatory movement was modified from a masticatory measurement made for a human subject chewing a hard food [6]. The mastication lasted 45.32 s and was sampled every 10 ms.

Given the above mandibular movement, the inverse kinematics were solved to determine the individual trajectory of each of the six crank actuations required for the robot to follow the mandible movement. Because the Galil controller used could only implement a trajectory of a time interval of 2^n ms (where n is an integer between 1 and 8) the actual mandibular movement that was sampled every 10 ms needed to be interpolated before its implementation in Galil. A cubic polynomial was found to be capable of converting a 10-ms interval trajectory to an executable 8-ms interval trajectory without significant distortion in the motion profile (in terms of angular displacement, velocity and acceleration).

It can be seen from Fig. 4.49(b) that the subject chewed predominantly on the left side and performed only three peaks of the lateral excursion movement to the right. It can also be seen in Fig. 4.49(c) that for the first couple of chewing cycles, for the following few chewing cycles, and for the rest of chewing process, the maximum mouth opening was approximately 30, 25 and 20 mm, respectively, given that the incisor tooth rested initially at position z=-20 mm.

4.9.2 Robotic Chewing without Food

During the first experiment no food was involved and the robot was commanded to freely follow the defined masticatory movement (Fig. 4.49 and 4.50) for 5 seconds. All the motors were under contour mode PID control with K_p=8, K_d=290 and K_i=0.2 with a time interval of 8 ms. Motor speed and acceleration were set to be autoadaptive and the motor torque was set at a maximum torque of 60% (i.e., ±6 V). The PID control was implemented using a Galil controller [7].

Only representative results are presented within this section. Fig. 4.51 shows the actuation of the right pterygoid actuator. The occlusion took place at a crank angle of around zero and full opening of the mouth occurred at the valleys of the crank angle profile. As can be seen in Fig. 4.52, there is negligible error during the

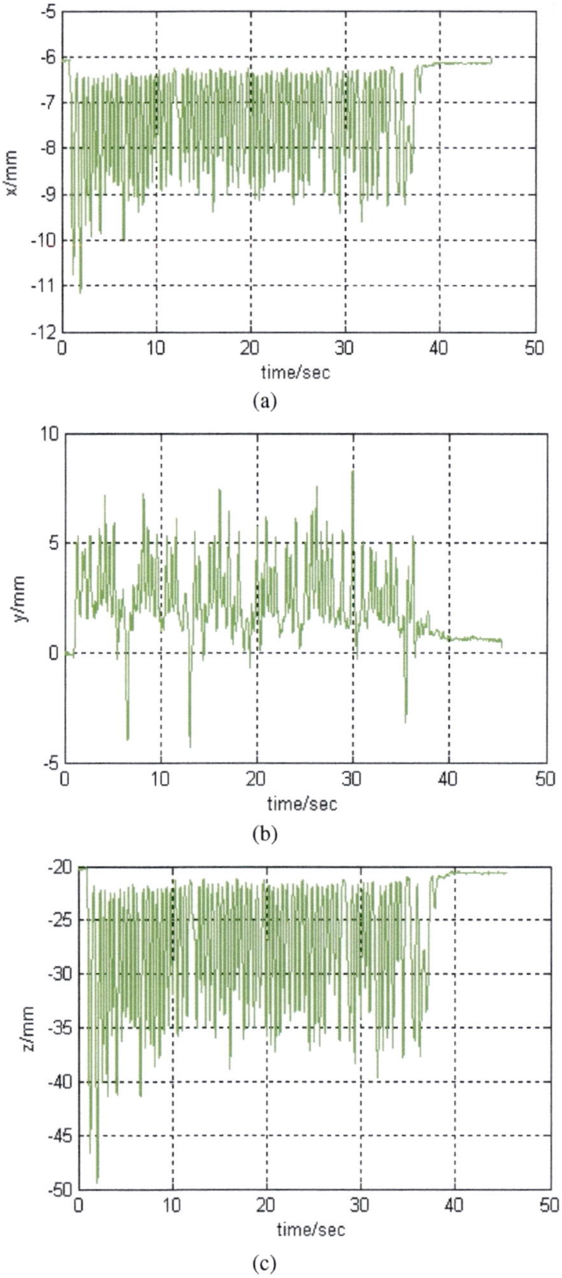

Fig. 4.49 Incisor's coordinates versus time, (a) *x*-axis, (b) *y*-axis and (c) *z*-axis

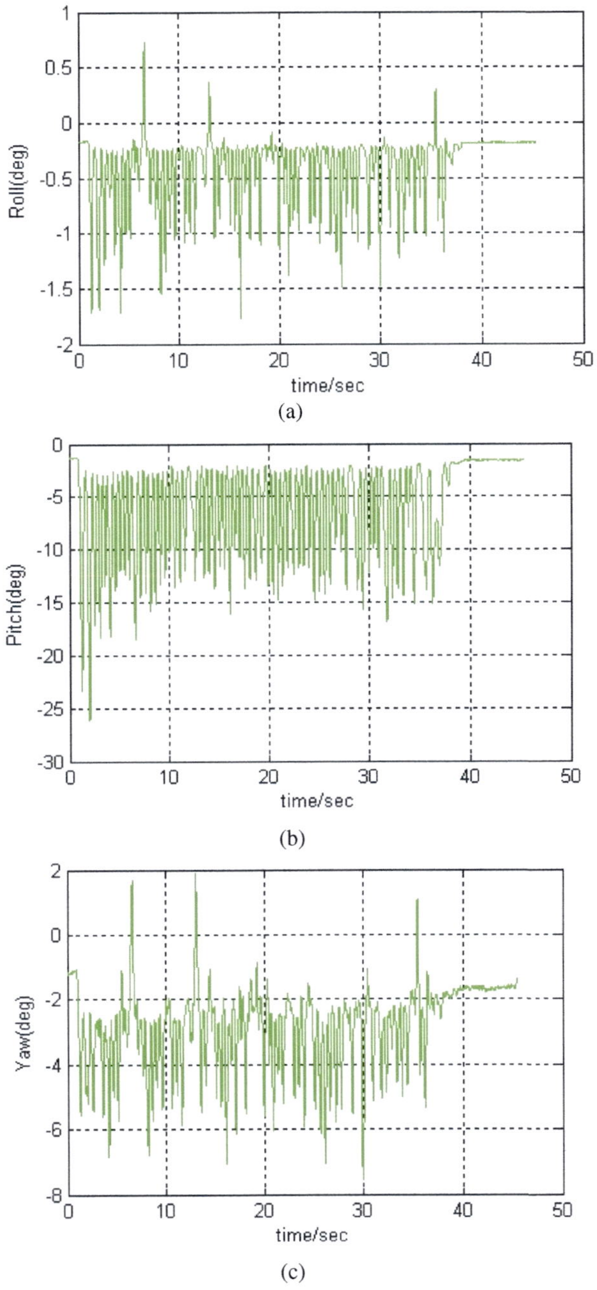

Fig. 4.50 Mandibular orientation versus time, (a) roll angle, (b) pitch angle, and (c) yaw angle

occlusal phase of chewing. This is the most important phase of the chewing process as it is supposedly involves interactions between food and teeth. The maximum error (of less than 1°) occurs during the transition from the opening to the closing phase of the mouth.

The motor torque (Fig. 4.51) varied at a frequency which was the same as that of the crank angle and reached a maximum at the point in time when the mouth transitioned from open to close. There was very little torque required during the opening and closing phases. These findings are self-explanatory given the fact that no food was involved during this particular chewing experiment. The other crank actuations behaved in correspondingly similar ways.

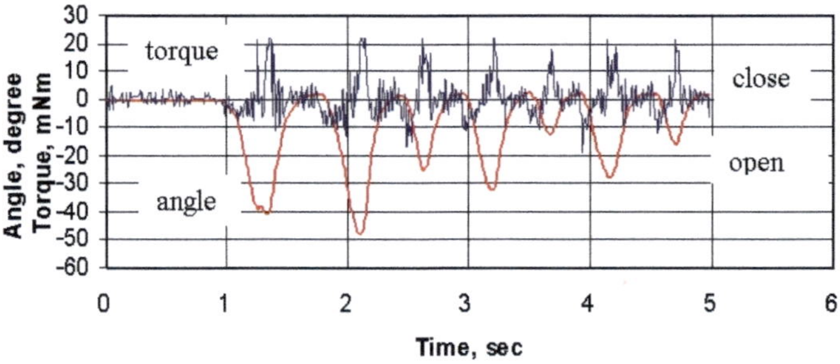

Fig. 4.51 Actuation of right pterygoid actuator in free chewing

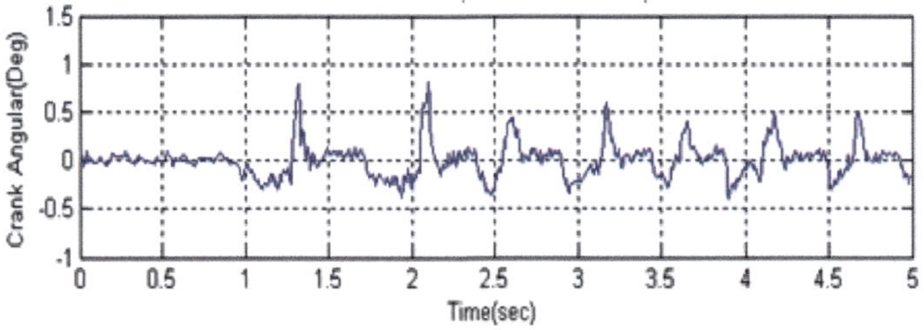

Fig. 4.52 Error between the commanded and actual right pterygoid actuation in free chewing

4.9.3 Robotic Chewing of Simulated Foods

In the following two experiments, foods were simulated by a 10-mm-thick aluminum plate for hard food and a 10-mm-thick foam plate for soft food. The masticatory robot was commanded to perform seven chewing cycles over the first 5 s (see the trajectories shown in Figs. 4.49 and 4.50); however, it was paused for 2.5 s in order that the simulated food could be inserted between the teeth when the mouth opened up during the first chewing cycle.

To avoid a large chewing force being exerted when the robot chewed on a hard plate (which would have resulted in the weakest part of the robot being damaged) the torque on the motor was limited. In the two experiments, the torque thresholds were set at the maximum torques for the free chewing referred to above (Table 4.3). The rest of the controller specifications were set the same as those used in the previous experiment.

4.9.3.1 Aluminum Plate as Food

For this experiment, an aluminum plate was inserted into the mouth and placed at the left side molars, which mirrored the left side chewing process specified in Section 4.8.1.

Table 4.3 Motor Torque Thresholds for Chewing Simulated Foods.

	Right Pterygoid	Left Pterygoid	Right Masseter	Left Masseter	Right Tempolaris	Left Tempolaris
Torque Limit	2.0V/10V	1.5V/10V	1.0V/10V	1.0V/10V	2.50V/10V	3.0V/10V

Fig. 4.53 shows the crank angle and torque of the right pterygoid actuator while the hard food was being chewed. Comparison between Figs. 4.51 and 4.53, reveals that there was no significant difference in either the crank angle or the torque between free chewing and chewing on hard food.

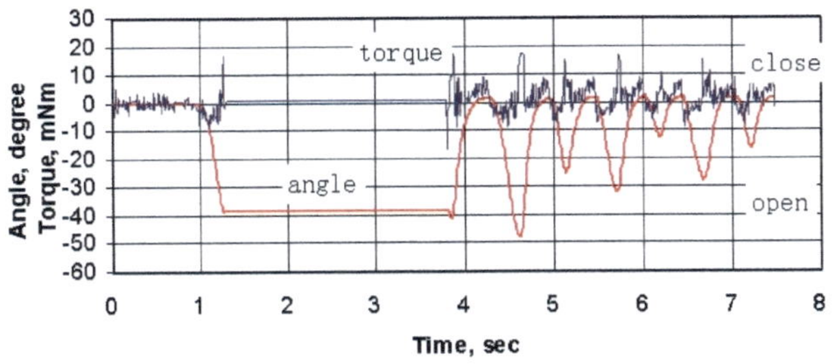

Fig. 4.53 Actuation of right pterygoid actuator while chewing on hard food

Data from the robot mechanism (Fig. 4.1), showed that the left and right pterygoid actuations were mainly responsible for the lateral movement of the jaw. When the robot chewed on the aluminum plate placed at the left, the jaw teeth slide on the plate surface and the left pterygoid actuator exhibited a larger torque (shown truncated in Fig. 4.54(a)) than that which occurred during free chewing (Fig. 4.54(b).

In Fig. 4.1 it can be seen that the right temporalis and right masseter actuations were mainly responsible for the lifting/lowering of the right side of the jaw. When the hard plate was inserted at the left molars, their actuations did not differ significantly from those for free chewing. Fig. 4.55 shows the crank angle and the motor torque of the right masseter actuator for hard food.

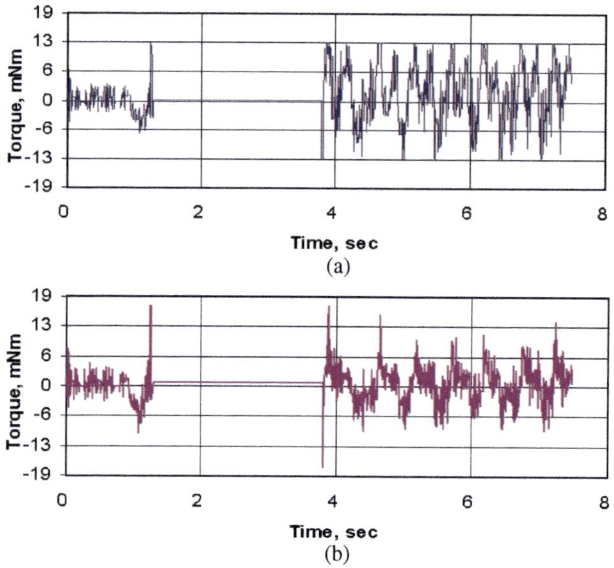

Fig. 4.54 Torque of left pterygoid actuator for free chewing and hard food, (a) chewing on hard food and (b) free chewing

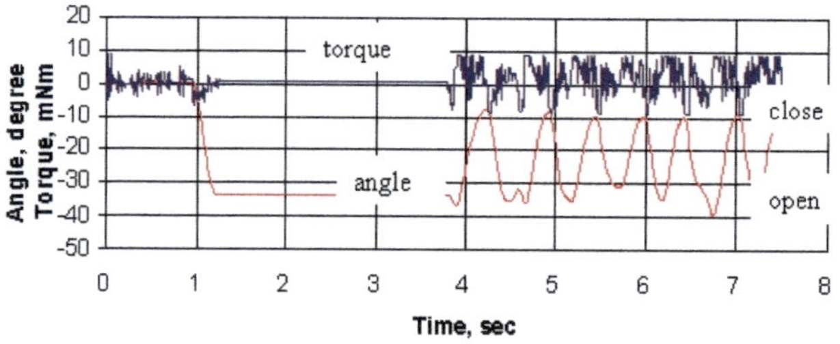

Fig. 4.55 Actuation of right masseter actuator for hard food

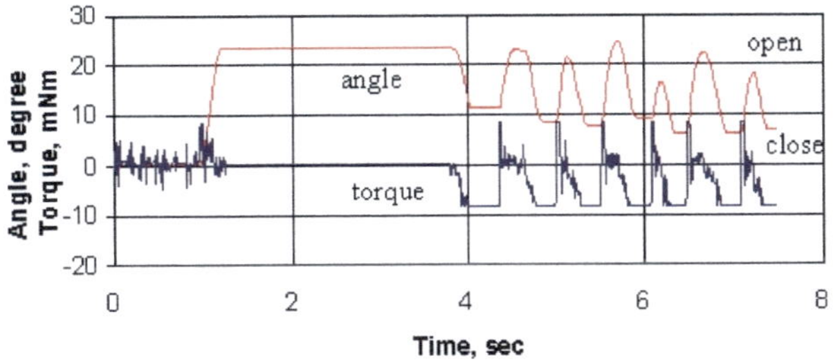

Fig. 4.56 Actuation of left masseter actuator for hard food

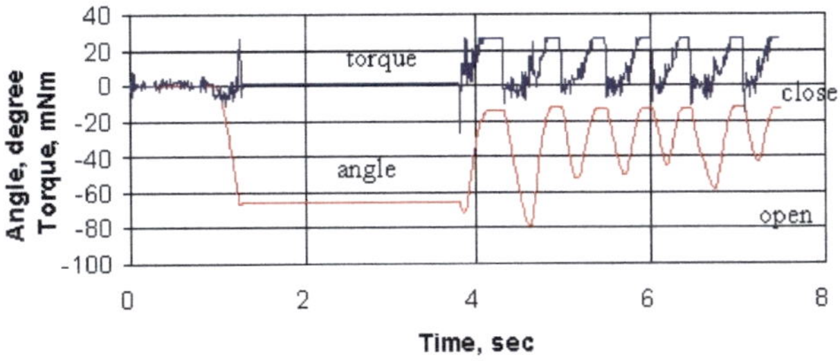

Fig. 4.57 Actuation of left temporalis actuator for hard food

Due to the hard plate having been inserted on the left molars, the left temporalis and left masseter actuations (responsible for the lifting of the left side of the jaw) differed greatly from those generated during free chewing. Because the plate is not compressible (with the forces possible using the actuators), the driving torques limited the motion, and the specified chewing trajectories of the two actuations could not be followed (as demonstrated in Figs. 4.56 and 4.57). It was found that the occlusal peaks of the chewing trajectory were truncated off. The actual trajectory involved an occlusal inaccuracy of between 5° and 10° at the crank shaft and the motor torque held at the threshold value during these periods.

This experiment showed that the robot motion was able to give way to the hard food by limiting the motor torque. This ability would thereby prevent the robot from being damaged when hard foods were being chewed.

4.9.3.2 Foam Plate as Food

In this experiment, a foam plate was inserted into the mouth in order that it would be chewed at both the left and right molars. The foam used was slightly rough in texture and could be broken into pieces when being chewed. Due to these properties, the lateral movement of the jaw was slightly frictional, as evidenced by the experimental results gathered from the left/right pterygoid actuations. However, none of the actuators behaved with any significant difference to the results observed whilst chewing without food.

4.10 Summary

A masticatory robotic model of crank actuation was presented, kinematic parameters having been derived according to the anatomy of its human jaw counterpart. The closed-form inverse kinematics was determined for joint actuations. The Jacobian matrices were derived, and their singularity was analyzed. The physical robot was presented and the mechanical and motion control system designs were discussed. Experimental measurements for chewing of simulated foods were conducted. The results of these experiments were analysed and compared in terms of both the crank actuations and driving torques required.

The robotic model was also simulated for an inverse kinematics problem in SolidWorks/CosmosMotion and for a forward dynamics problem in SimMechanics.

Further chewing experiments conducted using real foods are required in order to test the functionalities and control algorithms of the robot constructed in this research. Currently food retention is achieved simply by way of a plastic enclosure. To make the robot more human-like, a tongue, a cheek and mouth chamber need to be included. Furthermore, to enable the masticatory robot to perform more human-like chewing, it is suggested that the force contact transition control [8] be applied to achieve the mouth-opening motion (free motion). This would allow for a seamless change during the motion of mouth-closing (which involves food resistance) and finally to occlusal motion (which involves teeth contact).

References

1. van Eijden, T.M.G.J., et al.: Architecture of the human jaw-closing and jaw-opening muscles. Anat. Rec. 248, 464–474 (1997)
2. Craig, J.J.: Introduction to robotics, mechanics and control, 3rd edn. Prentice-Hall, Englewood Cliffs (2004)
3. Merlet, J.P.: Parallel Robots, 2nd edn. Springer, The Netherlands (2006)
4. Gosselin, C., Angeles, J.: The optimum kinematic design of a planar three-degree-of-freedom parallel manipulator. ASME J. Mech., Trans., Auto. Des. 110, 35–41 (1988)
5. Rohrle, O., et al.: From jaw tracking towards dynamic computer models of human mastication. In: IFBME Proceedings of 12th International Conference on Biomedical Engineering, Singapore (2006)
6. Peyron, M.A., et al.: Effects of increased hardness on jaw movement and muscle activity during chewing of visco-elastic model foods. Exp. Brain Res. 142, 41–51 (2002)
7. Torrance, J., et al.: Foster, Motion control of a chewing robot of 6 RSS parallel mechanism. In: Proc. of Int. Conf. Autonomous Robotics and Agents. Palmerston North, New Zealand (2006)
8. Xu, W.L., et al.: Contact transition control via joint acceleration feedback. IEEE Trans. Ind.Electron. 47, 250–159 (2000)

Chapter 5
Mastication Robot of a Crank-Slider Linkage*

Abstract. This chapter outlines the development of a linkage-based masticatory device for use in food evaluation that is capable of performing a range of standardised chewing trajectories. The robot design was required to match the trajectory of the first molar and the forces applied on the foods as measured during human chewing experiments performed in-vivo. A four-bar linkage was synthesized to achieve a lateral chewing trajectory of the molar. By adjustment of the ground link length, any trajectory between lateral (grinding) and vertical (crushing) chewing motions was possible. A six-bar crank-slider linkage was then designed in order to guide the movement of the mandible in a set orientation whilst maintaining the chewing trajectory produced by the four-bar linkage. The chewing device based on the six-bar linkage was constructed inclusive of: anatomically correct teeth for reducing the food particle size, a food retention device for collection of the food particles being chewed and a shock absorber for preventing the application of excessive chewing forces. The linkage chewing device was evaluated by way of kinematics and dynamics simulations together with a comparative analysis of the actual measurements of the trajectory and chewing force. The chewing velocity along the trajectory was profiled for both the occlusal and opening/closing phases of the chewing process and used for the set points for motion control. Variations to account for differences in chewing cycle velocity profiles were made to be adjustable through a graphical user interface developed in LabView. The masticatory device was validated against experimentation involving the chewing of example food systems and comparison of the resultant particle size within the bolus produced by the device and those produced by human subjects for the same food stuffs.

5.1 Why a Linkage Mechanism for Mastication

The human masticatory system is a complex system that comprises of an upper and a lower jaw, both of which have teeth. The system also includes a tongue,

* Part of this chapter is reprinted from Xu WL, Lewis D, Bronlund J and Morgenstern MP (2008) Mechanism, design and motion control of a linkage chewing device for food evaluation. Mech. Mach. Theor. 43:376-389, with permission from Elsevier.

W. Xu and J.E. Bronlund: Mastication Robots, SCI 290, pp. 129–162.
springerlink.com

cheek and saliva production capability. When mastication is performed, the lower jaw (mandible) is moved by muscles that are attached between it and the upper jaw [1]. The chewing movement begins with the mandible opening, thereby creating a space between the teeth located on the skull and the mandible. The tongue then relocates food particles that require chewing to the molars on one side of the mouth. The mandible then closes and breaks up these food particles. The particles then fall off the teeth back on to the tongue enabling repositioning during the next chewing cycle [2]. The opening movement of the mouth during the chewing cycle is approximately vertical [3]. The speed with which the mandible moves during the opening phase is initially slow and increases as the mouth opens. When the mouth starts to close, the mandible moves laterally outward, initially it closes quickly as it comes back towards the teeth, and then slows for occlusion.

In the frontal plane, the trajectories of the teeth during chewing vary substantially for different food types, but, are very close to a straight line in the sagittal plane [4]. This line may vary from a vertical line, where the teeth come together at a 0° angle, and a line where the teeth come together at a 30° angle. The trajectories used to chew different food particles differ depending upon both the shape and the texture of the food particles, thus generating a different chewing action for different foods. If a vertical chewing motion is used, the cusps of the teeth are used in order to fracture food particles. In comparison, when a more lateral chewing motion is employed, the sharp edges of the teeth are used and function as blades and allow food particles to be cut up into pieces [5].

As food properties affect the chewing trajectories, a considerable amount of work has been done to determine chewing movements in food sciences [3, 4, 5]. Measurements of these various chewing movements have been made continuously over the masticatory process, and include the following parameters: frequency, length of chewing time, tracking of jaw movement, force distribution, application of compression and shear forces on the food together with the particle size and structure of the bolus just prior to swallowing. Results vary between subjects (e.g. due to differences in jaw geometry, teeth shape and sensitivity to pain) and across various food textures (e.g. elasticity, hardness, adhesion especially to dentures etc.).

There are a variety of instruments or devices available for evaluating food properties. However, such devices usually use a simple straight motion (mostly food compression) and are not capable of simulating the entire suite of complex functions and movements involved during human mastication. Since the early 1990s, there have been various attempts at developing masticatory robots for food texture assessment (as reviewed in Chapter 1). While robotic chewing devices that possess multiple DOFs are able to reproduce chewing behaviour in a 3D space, a single DOF linkage device for chewing is the objective in this Chapter. A linkage device is simple in both structure and motion control and is more reliable than more complex systems in terms of its operation. The presence of a straight line trajectory in the sagittal plane presents the opportunity to approximate human chewing motion in 2D using a simple linkage device.

5.2 Design Specifications

Each mandibular tooth has its own trajectory during chewing. A typical trajectory can be defined by firstly, vertical and lateral displacements, secondly, the opening (exit) and closing (entry) angles (as illustrated in Fig. 5.1) and thirdly, the time taken to complete each cycle [2]. The trajectory of the first molar is simply a vertically compressed version of the incisor trajectories. However, the entry and exit angles, both to and from occlusion, are not greatly different. The incisor trajectory can be measured and vary between lateral chewing (grinding) and vertical chewing (crunching), depending on the type of food being chewed. Due to the fact that chewing is performed on the molar teeth, the chewing device must follow the trajectories of the molar teeth. As no actual data for molar trajectories during chewing is available, they were estimated by simulation of the trajectories of the incisor [6], as given in Table 5.1.

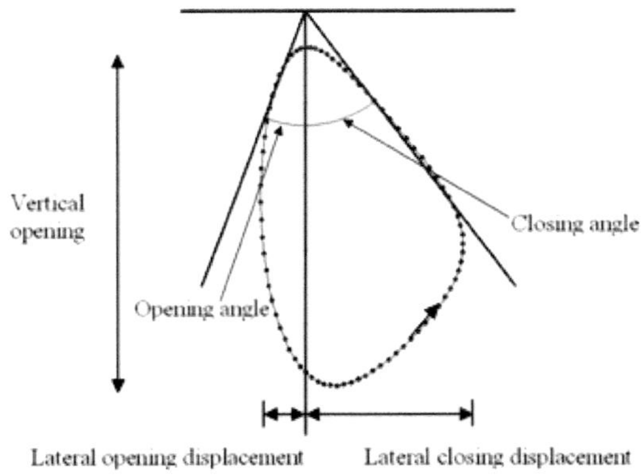

Fig. 5.1 A tooth trajectory and its defining parameters

Table 5.1 Values of the trajectory parameters for lateral and vertical chewing

	Lateral chewing	Vertical chewing
Closing angle (°)	46.6	72.5
Opening angle (°)	113.1	78
Total angle (°)	66.5	5.5
Vertical opening (mm)	14.6	15.1
Lateral displacement in opening (mm)	1.1	0.4
Lateral displacement in closing (mm)	4.0	3.1
Cycle time (s)	0.77	0.7
Occlusal time (s)	0.12	0.16
Opening/closing time (s)	0.65	0.54

In order to evaluate different foods, the chewing device required the ability to achieve any trajectory between the extreme lateral and vertical trajectories. The device was also required to meet both cycle and occlusal times. Therefore, the linkage for chewing was specified by the set of parameters given in Table 5.1.

Furthermore, the forces applied to the teeth vary according to the type of food being chewed. The force applied to a single tooth is also different to that of the total force between all the contacting teeth during chewing. On foods such as biscuits, carrots and cooked meats, forces range between 70 and 150 N on a single tooth [7]. Thus, the chewing force that the linkage could apply on food samples was specified to be 150 N at maximum.

5.3 Basic Linkage Mechanism for an Incisor Trajectory

A four-bar linkage (Fig. 5.2) is a relatively simple mechanical mechanism. Point 'P' on the coupler can trace a 2D trajectory. The kinematic parameters for this linkage include crank link '*a*', coupler link '*b*', follower link '*c*' and ground link '*d*', as well as an angle of 'γ' and distance '*BP*' for the coupler point '*P*'. A four-bar linkage, in its standard form, can only perform one set trajectory. In the event that a range of trajectories must be reproduced, the ground link must be manually adjusted.

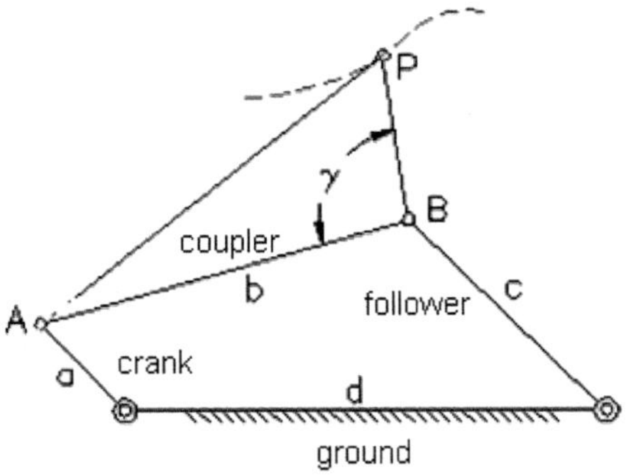

Fig. 5.2 Kinematic parameters for a four-bar linkage

The Cedarville engineering atlas [8] was used to find a number of suitable trajectories that have entry and exit angles which closely match a lateral chewing cycle (as specified in the previous section). When a trajectory was found that matched the desired occlusal angles of a lateral chewing motion, it was marked down as a possible solution. As vertical chewing motions were also required, this was achieved by further altering the ground link length. Fig. 5.3 shows the final specifications chosen for a lateral chewing trajectory where the link parameters are shown at the top with varying '*BP*' length.

After the occlusal angles were examined, the final design specification values chosen (Fig.5.3) were as follows:

Crank (link '*a*') = 1,
Follower (link '*c*') = 3,

Ground (link '*d*') = 3.8 – 5 (to achieve horizontal and vertical chewing motions), Coupler (link '*b*') = 3.5,
Coupler (distance '*BP*') = 3, and
Coupler (angle '*γ*') =60 degrees.

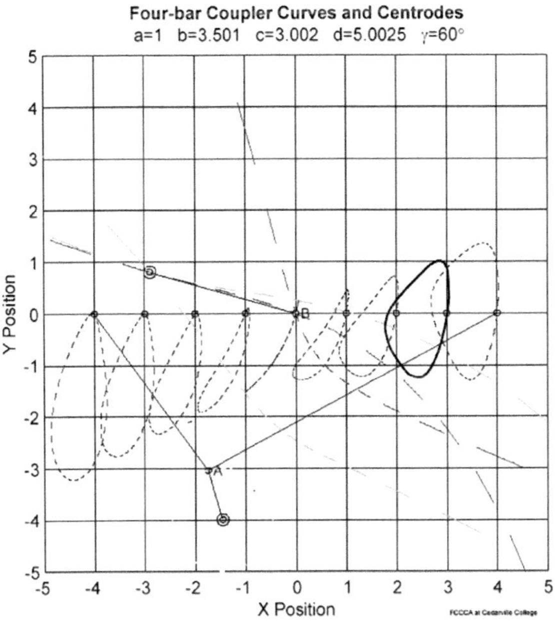

Fig. 5.3 Candidate trajectories of the four-bar linkage

These values are ratios of the link lengths relative to the Crank (a). The actual links can be of any length so long as the ratios are obeyed. In order that the actual linkage chewing device was as compact as possible, the smallest feasible physical crank was chosen, at 10mm in length when its pivotal and joint bearings were taken into account.

The chewing trajectories produced by the linkages were compared with measurements taken during real human chewing cycles (as given in Table 5.2). It can be seen that in terms of the occlusal angles, the linkage achieved a close match with the lateral chewing trajectory while still having reasonable trajectories in relation to vertical chewing. However, the linkage had a larger vertical opening and greater lateral displacements. These differences were acceptable as these particular aspects of the chewing trajectory do not impact upon the food breakdown process, so long as they are sufficient to clear the food between chewing cycles. The motor can be sped up during this part of the cycle to ensure representative human masticatory behaviour is simulated.

Table 5.2 Comparison of trajectories between linkage and human chewing.

	Lateral chewing		Vertical chewing	
	Human	Linkage	Human	Linkage
Closing angle (°)	46.6	45	72.5	99
Opening angle (°)	113.1	112	78	105
Total angle (°)	66.5	67	5.5	6
Vertical opening (mm)	14.6	23	15.1	34
Lateral displacement in opening (mm)	1.1	3	0.4	3
Lateral displacement in closing (mm)	4.0	9	3.1	0

Fig. 5.4 depicts three trajectories of the linkage at ground lengths of 38, 44 and 50 mm. It can be seen that the occlusal position shifts, the vertical opening displacement increases and the lateral displacement decreases, as the length of the ground link reduces. This confirms that the linkage is capable of reproducing various chewing trajectories.

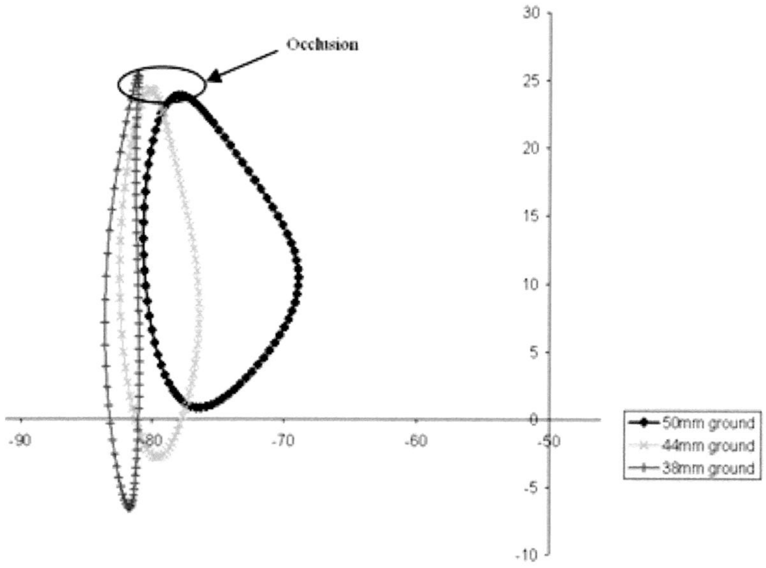

Fig. 5.4 Three chewing trajectories produced by an adjustable four-bar linkage

The occlusal position of the teeth in the chewing device can be adjusted by varying the distance between the upper and lower teeth.

The vertical and horizontal trajectories can be viewed as a function of the crank angle (as shown in Fig. 5.5 and Fig. 5.6). Fig. 5.5 shows that the actuating point on the coupler reached maximum displacement at approximately 150 degrees when the ground link was 50mm. While Fig. 5.6 shows that there is a maximum horizontal movement of approximately 11 mm when the ground link is 50 mm. This gives a good indication of how the mechanism responded in both the vertical and horizontal directions when the crank was driven.

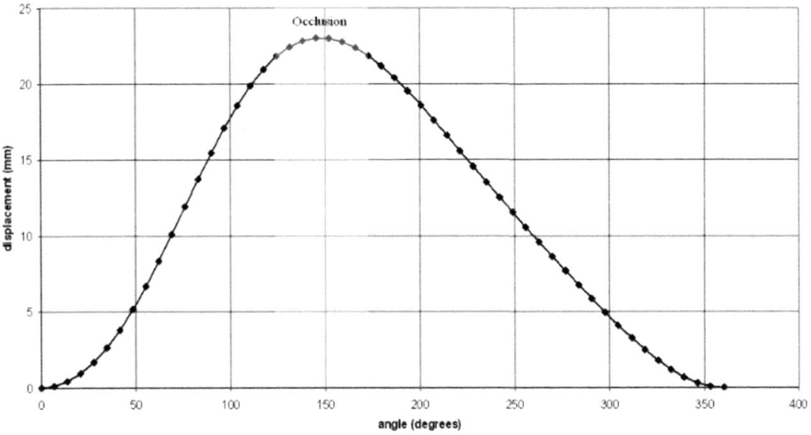

Fig. 5.5 Vertical trajectory as a function of crank angle, with a 50 mm ground

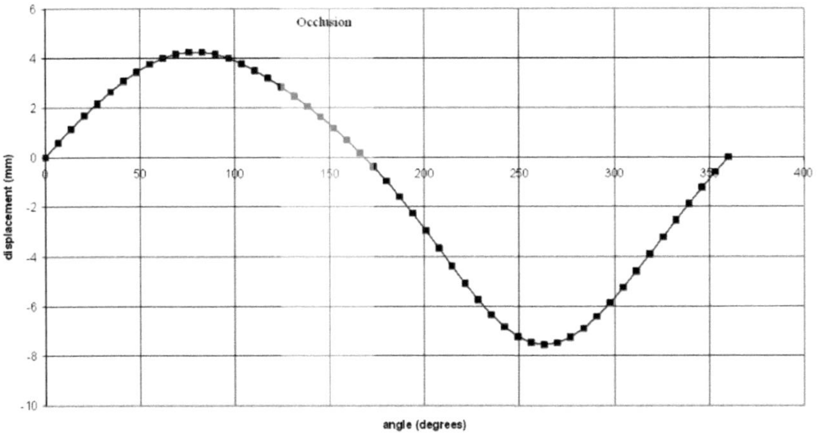

Fig. 5.6 Horizontal trajectory as a function of crank angle with a 50 mm ground

The transmission angles of the mechanism were then checked to ensure that they were not less than 30° or greater than 150°. This is due to the fact that the linkage would be inefficient if the transmission angles were outside this range. Some example transmission angles for the maximum and minimum ground lengths can be seen in Fig. 5.7 and Fig. 5.8. These figures show that the transmission angle stayed well within the required range. The occlusal position of the teeth on these plots occurred at time 0.34 s. At this time, the transmission angles were shown to be between approximately 72° and 95°. Therefore the linkage is most efficient at angles very close to 90°. It also stayed well clear of dead positions (i.e. 0° and 180°).

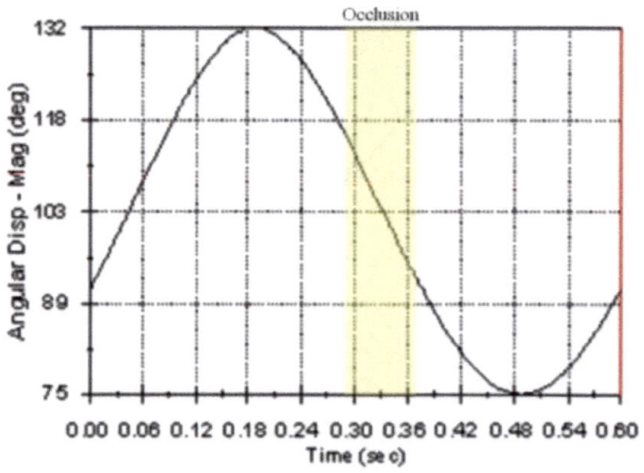

Fig. 5.7 Transmission angle when the ground length was 50 mm

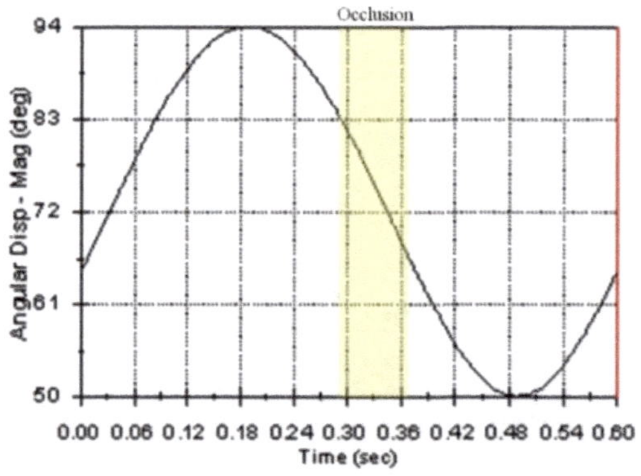

Fig. 5.8 Transmission angle when the ground length was 38 mm

The design of the mechanical device could commence once the link lengths of the four-bar linkage were determined. The basic designs of the crank, coupler and follower were straightforward so long as the joint points of the links matched the lengths determined. The ground link needed to have a 12 mm adjustment that effectively changed the link length from 38 to 50 mm. Fig. 5.9 shows the final four-bar linkage constructed where the adjustable ground link was achieved using a threaded rod to move a block when the rod was turned.

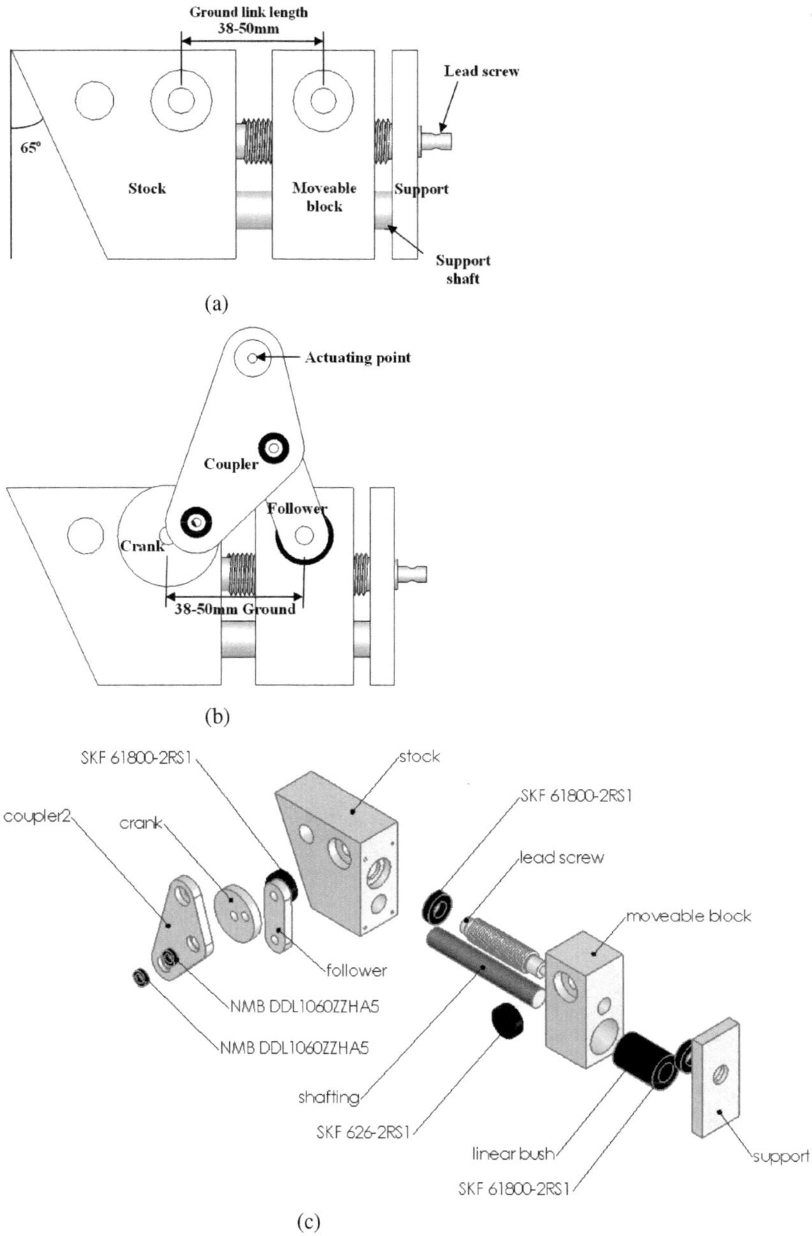

Fig. 5.9 Construction of the adjustable four-bar linkage, (a) adjustable ground link, (b) adjustable four-bar linkage and (c) major component design

5.4 Six-Bar Linkage Mechanism for Jaw Movement

5.4.1 The Six-Bar Linkage

As discussed previously, the adjustable four-bar linkage was capable of producing the required trajectories. However, it could not keep the mandible or the teeth in the correct orientation over the entire trajectory. A simple way to resolve this issue was to add another two links (slider and guide in Fig. 5.10 or link 5 and 6 in Fig. 5.11) to the four-bar linkage, thus making it a six-bar linkage. The two additional links were connected by a sliding joint between them. Link 5 was attached

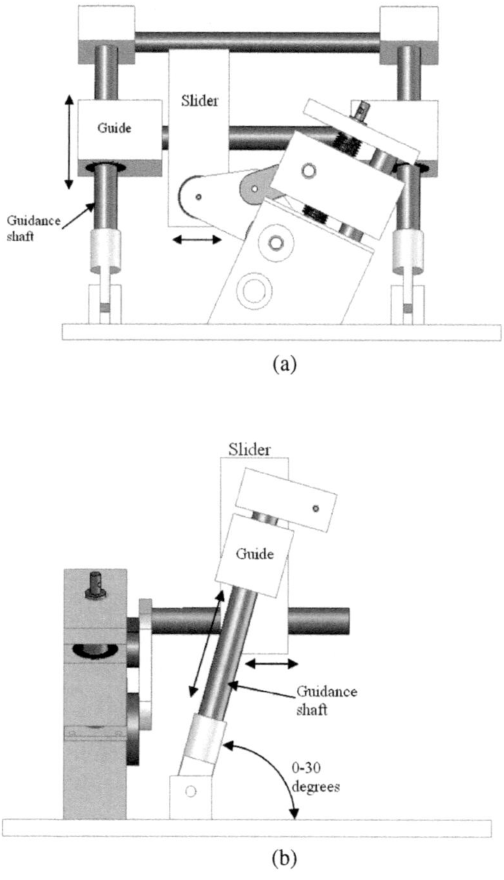

Fig. 5.10 The six-bar linkage

to the coupler by a revolute joint and link 6 onto the ground by another sliding joint. The set of teeth was mounted atop link 5, which is forced to move in a plane constrained by the two sliding joints. To produce the chewing trajectories in the sagittal plane with an angle ranging between 0° and 30° relative to the horizontal plane, the base of the six-bar linkage had to be tilted manually. In order to have balanced dynamics, thereby reducing vibration, the chewing device was constructed symmetrically by placing two identical four-bar linkages on each side of the device (Fig. 5.11). The cranks of the two linkages were mounted on a single shaft driven by a single motor via a spur-gear train.

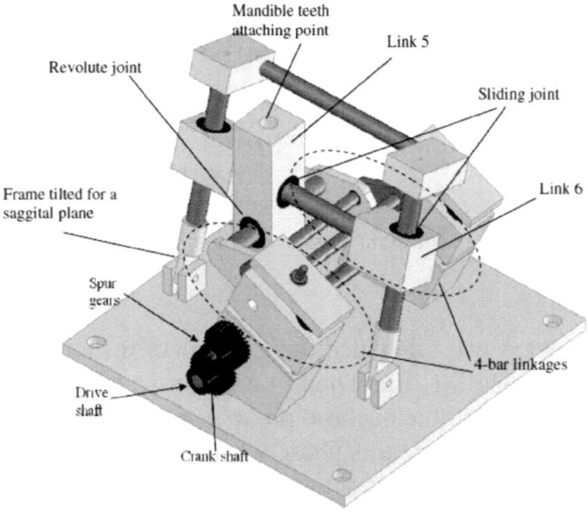

Fig. 5.11 Design of a six-bar crank-slider linkage

5.4.2 Motion Planning

While the linkage could trace a desired chewing trajectory, the velocity along the trajectory still needed to be profiled. In a chewing cycle (Fig. 5.12), the molar teeth move at a constant velocity during the occlusion phase. During opening the jaw will speed up from this occlusal velocity and then decelerate back to the occlusal velocity in order to complete the cycle in the specified period. The occlusion starting and ending positions were defined at the points where the molar teeth reach a horizontal line of 0.5 mm from the maximum intercuspal position (i.e. just prior to teeth-teeth contact) [6].

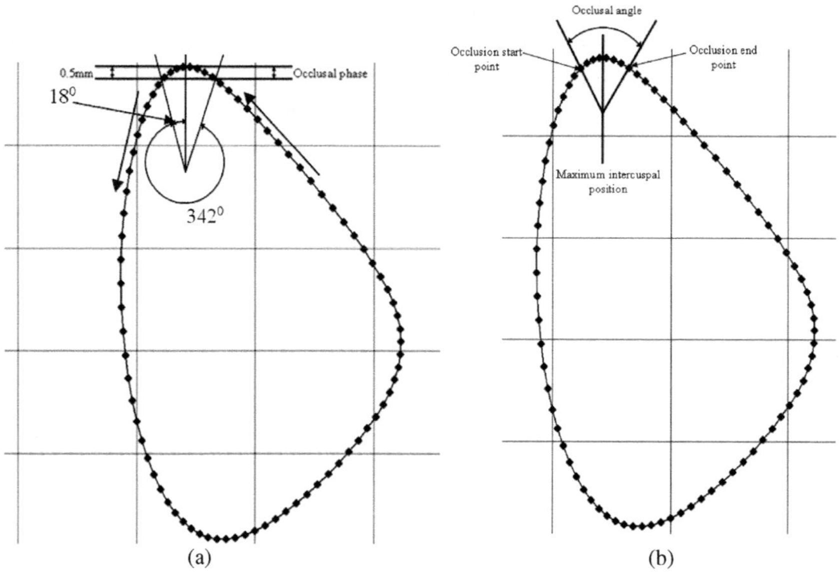

Fig. 5.12 Definition of the occlusal phase, (a) start/end crank angles and (b) notions for trajectory

The linkage mechanism constructed was simulated in SolidWorks for a lateral chewing trajectory as specified in Table 5.1. Each point on the trajectory corresponded to the crank shaft rotating 4.5° and the occlusal phase of a 36° turn of the crank shaft was found. As the time to complete this phase was specified as 0.12 s (Table 5.1), the occlusal velocity of the crank was set at 300°/sec. The start angle and final angle of the crank were found to be 18° and 342°, respectively (Fig. 5.12a). Taking into consideration a 1:42 gear reduction between the motor and the crank, the occlusal velocity, start angle and final angle of the motor shaft were 12600°/s, 756° and 14364°, respectively. The time taken to open and close the mouth was specified at 0.

Based on these values, a cubic trajectory of the motor shaft was found [9] as follows:

$$\theta(t) = 756 + 12600t + 384711t^2 - 39457t^3 \quad (5.1)$$

$$\dot{\theta}(t) = 12600 + 76942t - 118371t^2 \quad (5.2)$$

$$\ddot{\theta}(t) = 76942 - 236742t \quad (5.3)$$

The planned trajectory described above reproduces the specified lateral chewing motion specified in Table 5.1. As the chewing device was intended to perform a variety of trajectories between lateral and vertical chewing, the trajectories (other than that for lateral chewing) will vary. Any actual planned trajectory is determined by occlusal angle, occlusal time and opening/closing time (Fig. 5.12b), which can be set up in the motion control GUI (see Section 5.8.3).

5.4.3 Motor Selection

After the materials for production of the linkage parts had been specified the linkage mechanism could be simulated in SolidWorks in order to determine the driving torque required at the crank shaft when the maximum force of 150 N was applied vertically at the coupler point '*P*'. Fig. 5.13 shows a crank torque versus time plot with the crank running at a speed of 300°/s. The maximum torque the crank requires to allow the chewing device to run was 2.6 Nm. Consequently, the output of the geared motor had to be able to produce a minimum torque of 2.6 Nm at 300°/s (or 78 rpm) to achieve the desired chewing force of 150 N. In addition, the required acceleration was estimated at 190 °/s^2 [10].

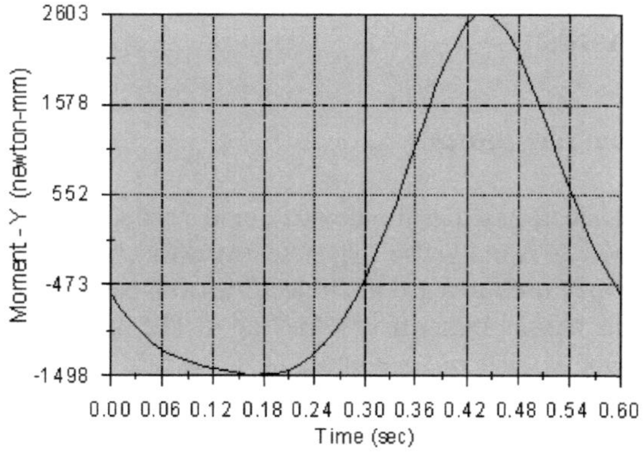

Fig. 5.13 The torque required at the crank shaft

Table 5.3 Comparison of required and achievable specifications at the crank

Parameter	Required	Estimated capability
Speed (rpm)	100	190
Torque (N m)	2.6	3.82
Acceleration (°/s^2)	190	414

A brushless DC motor was selected that could deliver 6.0 Nm of torque continuously in addition to 7.5 Nm of torque for short periods. With a gear ratio of 1.57, the torque, speed and acceleration at the crank were 3.82 Nm, 190 rpm and

414 °/s^2, respectively. This met the required performance of the crank (Table 5.3). Fig. 5.14 shows the motor and the two stages of spur gears arranged so as to drive the two cranks of the chewing device.

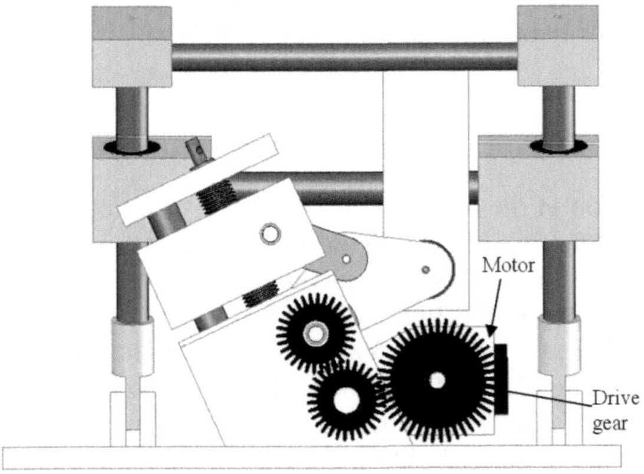

Fig. 5.14 The location of the motor and the drive line

5.5 Analysis of the Mechanism

5.5.1 Trajectory and Force Evaluation

A simple way to compare the trajectories that the device could produce and the desired chewing trajectories, was to simply use a pen to trace the trajectories achieved and overlay them. The pen used to trace the trajectories was modified so that a spring was used to push the nib out rather than retracting it. The pen was securely attached to the slider of the six-bar linkage and a piece of card was setup vertically, enabling the nib to remain in contact with the card throughout the complete trajectory. It was found from numerous measurements that the trajectories produced by the chewing device were close to the desired ones during the occlusion phase but slightly different in the opening/closing phases, as illustrated in Fig. 5.15. This minor difference was due to the play in the joints of the linkage. Because it occurred only during the opening/closing free motion phase of a cycle it did not significantly effect the occlusal phase of interest when chewing foods.

Force testing was performed by the use of a load cell. This idea involved having the teeth attaching point (link 5) pressing down on the load cell in order to measure the force applied. The chewing device was set to run continuously and the force that the linkage could apply was measured. The results revealed that the chewing device could comfortably apply the maximum desired chewing force of 150 N and would stall at approximately 260 N.

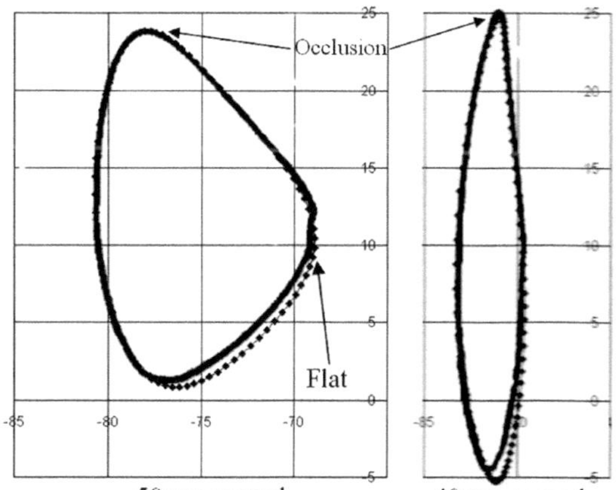

50mm ground (a) 40mm ground (b)

Fig. 5.15 The overlay of the achieved trajectory and desired trajectories (solid line = actual, dotted line = desired), (a) later chewing and (b) vertical chewing

5.5.2 Stress and Deformation

The six-bar linkage was simulated in CosmosWorks to test if the device could withstand the forces that will be applied. A stress analysis was performed at the occlusal position with a force of 150 N applied at the teeth attachment point. Results showed that no excessively large stresses were produced. The largest stress was concentrated in the linear bush of link 5 (Fig. 5.16). This was due to the linear bush being a relatively sharp edge within the structure. The design could account for this level of applied stress.

The deformation of the device was also analysed in order to examine the buckling that occurred when the 150-N load was applied at the teeth attachment point. The results show that the deformation is 0.004 mm at the point where the linear bush met the shaft that it slid upon (Fig. 5.17). This was considered to be negligible as this amount of buckling in the shaft would not cause the linear bush to jam.

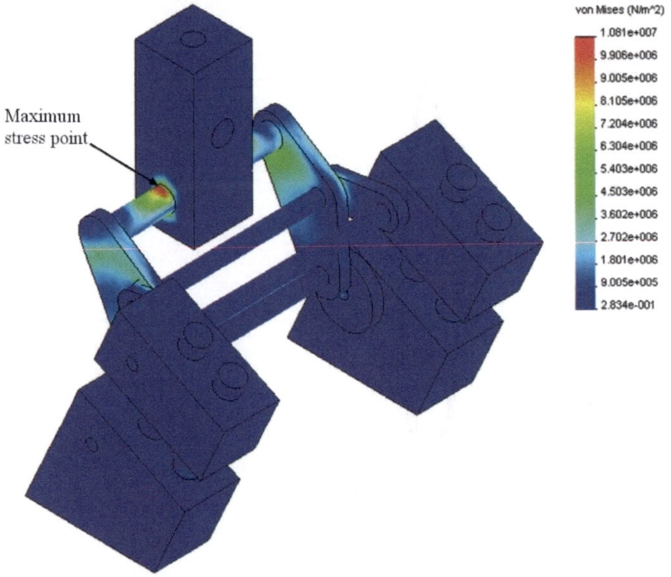

Fig. 5.16 Stress analysis of the linkage

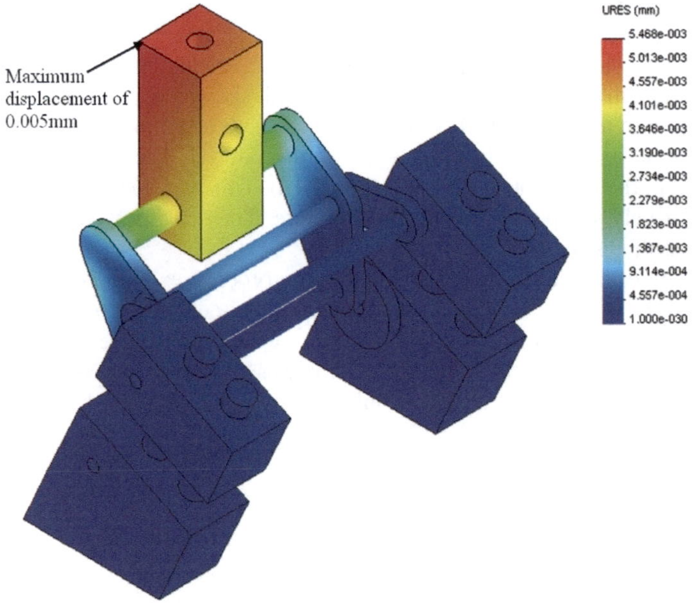

Fig. 5.17 Deformation analysis of the linkage

5.6 Mechanical Design of the Mechanism

5.6.1 Limiting the Chewing Force

The maximum chewing force the chewing machine could apply to the teeth had to be limited to 150 N, according the design requirements. It was achieved by way of a preloaded spring system, designed as shown in Fig. 5.18. The system consisted of five parts (Fig. 5.18b): 1) a hard stop, 2) guidance pins, 3) a pre-tightening plate, 4) a spring, and 5) an aluminum sleeve. The slot together with the guidance pins in the hard stop, allowed the spring to be compressed to 12 mm, and also provided a high damping function which largely reduced the oscillation of the spring, which may have affected the chewing force profile. The pre-loading plate was used to adjust the initial spring compression in order to provide variable initial force. Thus, the system was adaptable to the chewing force applied during occlusion as required for chewing different foods. The selected spring had the following properties: outer diameter 14.5 mm, free length 71 mm, minimum work length 33 mm, rate 6.69 N/mm, and maximum load 254 N.

5.6.2 The Enclosure

The chewing device enclosure refers to the housing that holds all the sub systems together. The device was originally chosen to be an inverted design. This was due to the fact that the original idea to retain food, involved the use of a bowl with teeth embedded in it. This used gravity to ensure that the food was capable of being guided back onto the teeth by the profile of the bowl. The human masticatory system involves moving the lower set of teeth while the top set of teeth remains fixed. However, it is not important which set of teeth moves, so long as there is the correct relative motion. Therefore the upper set of teeth was chosen to be actuated in the chewing device. This meant that the lower teeth would have to be located in the bowl in order to allow the use of gravity, hence the inverted design.

Because the chewing device had to be inverted, the six-bar linkage needed to be located above the teeth. The majority of the weight would therefore be located near the top of the device. This meant that the enclosure needed to be designed so that it was stable and hence the overall height of the device had to be as low as possible.

As the dimensions of the six-bar linkage mechanism and the clearance allowed for the teeth could not be reduced in size, only the position of the power supply and control card could be used to alter the centre of gravity. The two main options for the location of the power supply and control card were having them located either on top of or below the device. It was decided that they should be placed above the six-bar linkage mechanism due to the fact that the combined weight of these component parts was less than the six-bar linkage. This therefore meant that the device had a lower centre of gravity than it would have if the parts had been placed below the device. Fig. 5.19 shows the housing of the chewing device.

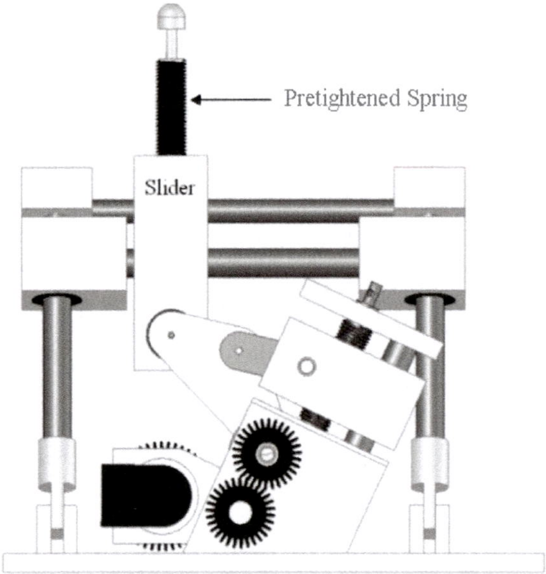

(a)

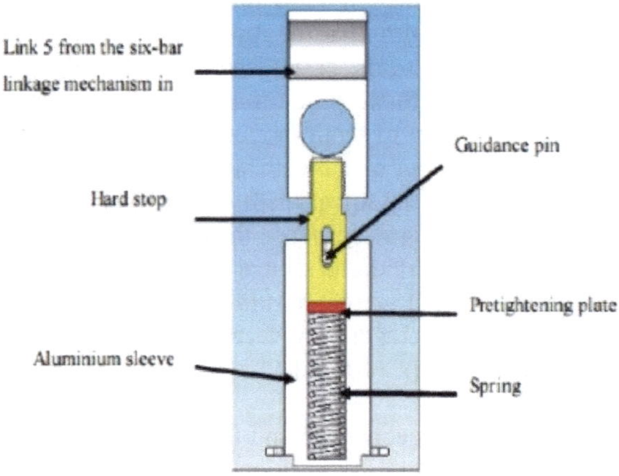

(b)

Fig. 5.18 Pre-loaded spring system for force limiting, (a) the placement and (b) the parts

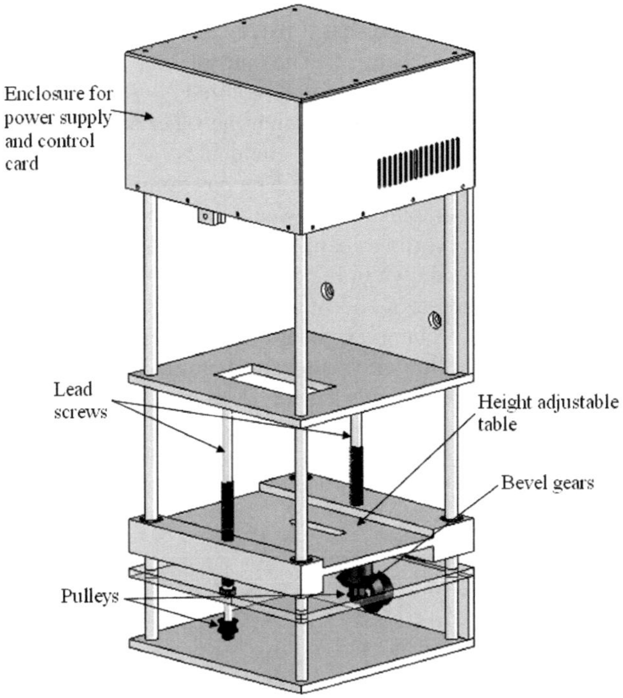

Fig. 5.19 The enclosure of the chewing device

The device also needed a moveable table in order to allow the height of the teeth to be adjusted. This was due to the fact that the height at which the teeth occlude changed when the ground link of the four-bar linkage was adjusted. It was decided that this would be achieved by use of a screw thread device that moved a platform up and down with the aid of linear bushes which removed any rotational movement. Two lead screws were used to move the table up and down. These lead screws were kept in time with each other by way of a toothed belt and matching pulleys. The belt was kept tensioned by the use of a third pulley. Linear bushes were used in the four corners of the table that slid on shafts. This stopped the table from twisting when force was applied to it. These shafts were also used to position all parts in the correct orientation and at the correct height. The lead screws were driven via a handle that when turned drove a set of bevel gears that connected both the lead screws and the handle. The enclosure also had a case positioned on top of it enclosing the power supply and motor control card.

5.6.3 Teeth Locking

The mechanisms holding the teeth in position were designed so that the operator could easily remove the teeth for cleaning and in order to gather the food sample that had been chewed. As teeth could have different occlusal positions depending upon the setting of the ground link of the four-bar linkage, the teeth had to be positioned every time the settings were changed. This meant that a method of quickly adjusting the occlusal position of the teeth was required.

Only one set of teeth needed to be adjusted to match the other set. Therefore, the teeth that were attached to the six-bar linkage could have a fixed position whilst the bottom teeth position was made adjustable. The positioning and locking system of the lower set of teeth was simple and allowed the operator to: slide the teeth into the locking mechanism, position them correctly and then lock the mechanism in place. This was done as shown in Fig.5.20.

The plate that had the teeth attached to it slid into an enclosure. The enclosure could then be locked in place by two clamping levers that tightened themselves onto bolts which in turn clamped the top enclosure and bottom plates together. This system allowed the lower teeth locking mechanism to be positioned in the entire operating range and was achieved by having slots milled in the table that it sat upon. These slots accommodated the clamping bolts and allowed them to move by the desired amount.

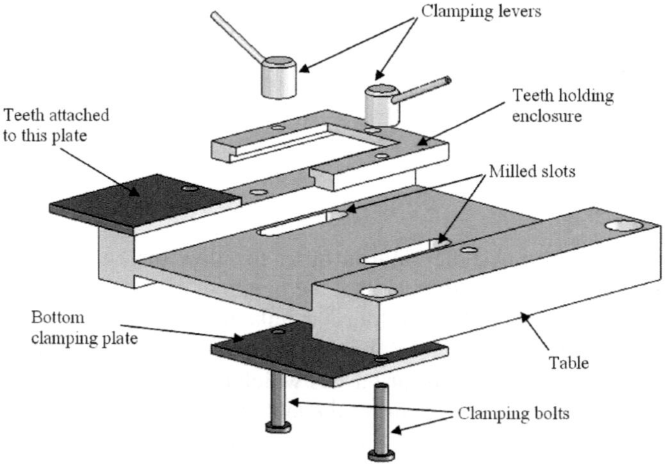

Fig. 5.20 Exploded view of lower teeth locking mechanism

Once this enclosure was clamped into place, the plate with the teeth on it could then be slid into position. The teeth plate was held in place by the flanges of the locking enclosure together with a locating pin. This allowed the teeth to be removed and inserted easily without having to re-align the upper and lower teeth every time a food sample was inserted. The location of the linkage and the teeth locking mechanism is shown in Fig. 5.21.

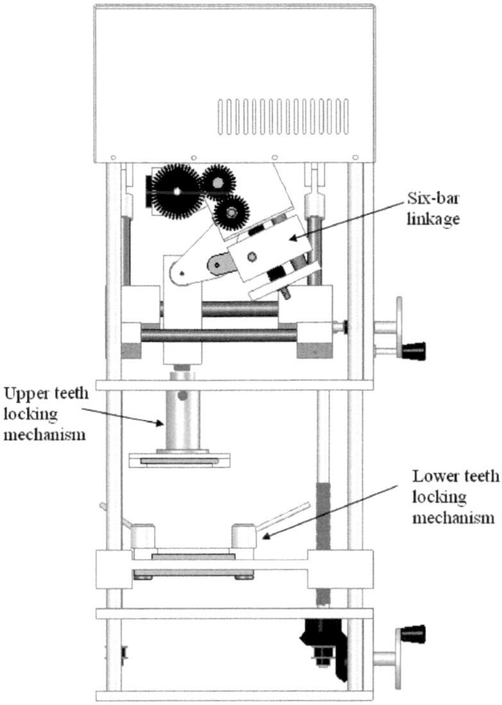

Fig. 5.21 The location of the six-bar linkage and the teeth locking mechanisms

5.6.4 The Teeth and Food Retention System

In the human mastication system only the pre-molars and molars are used for mastication. Therefore these are the only teeth that need to be included in the chewing device. Mastication also only occurs on one side of the mouth at a time, therefore only one set of pre-molars and molars were included in the chewing device. The teeth used were taken from a dental study model made of melamine polymer, and were anatomically correct.

The system employed to retain the food sample and keep it located on the teeth was a rather simple design. It used silicon flaps around the lower set of teeth in order to retain the food as shown in Fig. 5.22. This simulated the retention function of the tongue and cheek. The silicon material used was food safe and therefore did not adversely affect the food sample. Although this system retained the food sample with only a minimal amount of particles falling out of the enclosure, a lot of the food sample got stuck on the top set of teeth.

In the human mastication system, the tongue and cheek keep the food retained within the mouth and positioned upon the teeth. They also reposition the food particles between the occlusal phases of each chewing cycle so that the larger particles get broken up by the pre-molars and the smaller particles get ground down by

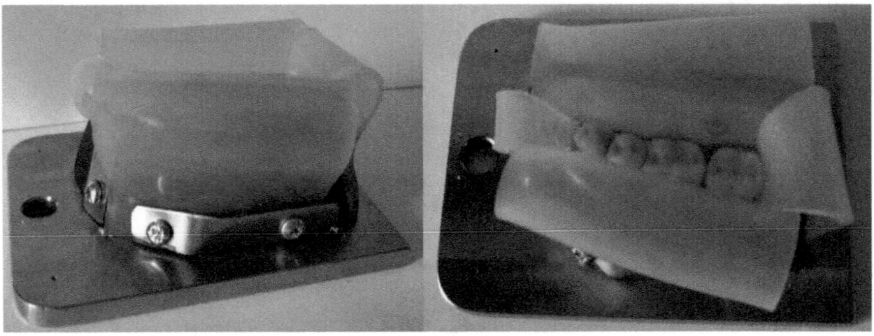

Fig. 5.22 The food retention system (note: the teeth shown are the early version made of melamine plastic)

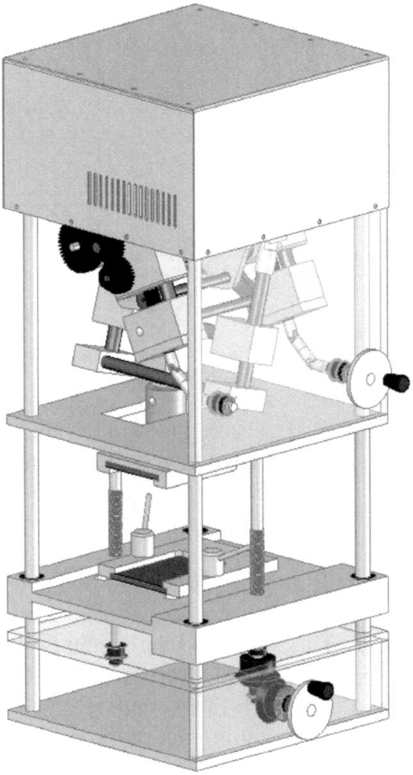

Fig. 5.23 The final assembled SolidWorks model of the chewing device

the molars. This repositioning function performed by the tongue is very complex and is therefore difficult to simulate by way of an automated system. Therefore, it was decided that the repositioning function be performed by the human operator, for the time being. This was achieved by running the device for a set amount of cycles with the food sample on the pre-molars, stopping it and repositioning the food on the molars and then running the device again.

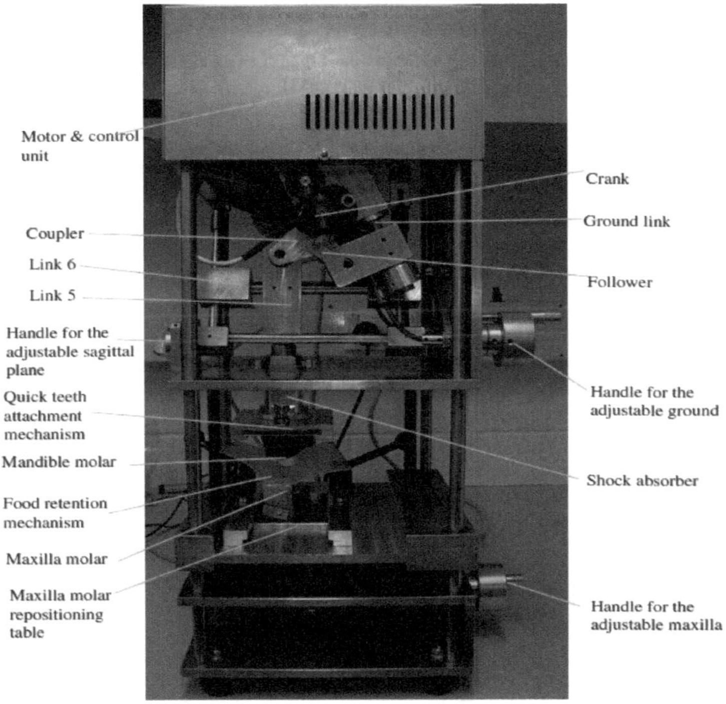

Fig. 5.24 The constructed chewing device

However, the repositioning function that the tongue and cheek perform is a very important part of the human mastication system. This is due to the fact that the process enables food particles to be placed upon the correct teeth at the beginning of every chewing cycle. This ensures that every particle is sufficiently broken down. In humans, much larger portions of food are masticated than the relatively small food samples that can fit on the occlusal surfaces of the teeth in the robot. The tongue and cheek manage the movement of this larger sample across the functional surfaces of the teeth. These functions were not achievable by the operator and therefore could be the focus of future work. Both the model of and the actual constructed chewing device are shown in Fig. 5.23 and Fig. 5.24, respectively. The model shown in Fig. 5.24 was ready to be instrumented with sensors and control system.

5.7 Measurement of the Chewing Force

In order to understand the actual chewing force applied to the food for the purpose of texture analysis, real-time force measurement was required. In selection of the force sensor four factors were taken into account: 1) the force sensing range, 2) dimensional requirements, 3) sensing accuracy, and 4) data acquisition (DAQ) card, supported by LabView.

An ATI Mini40 force sensor was chosen and was installed beneath the lower teeth. A NI PCI-6036E DAQ card was used for acquisition of the sensing data that involved three force and three moment elements. The real time data recording function was programmed into LabView, as part of the control and signal processing program of the device (Fig. 5.25a).

Due to the excessive amount of calculations required to calibrate 6-axis sensed force data, the force sensor could not be used for any real-time purpose. Post-processing would have been required in order to convert measurement data into force data that had any practical meaning (Fig.5.25b). The encoder data was also processed in order to obtain the spatial coordinates of the upper teeth. This data was used in conjunction with the measured chewing force to construct force-time and position-time curves to characterise the chewing process.

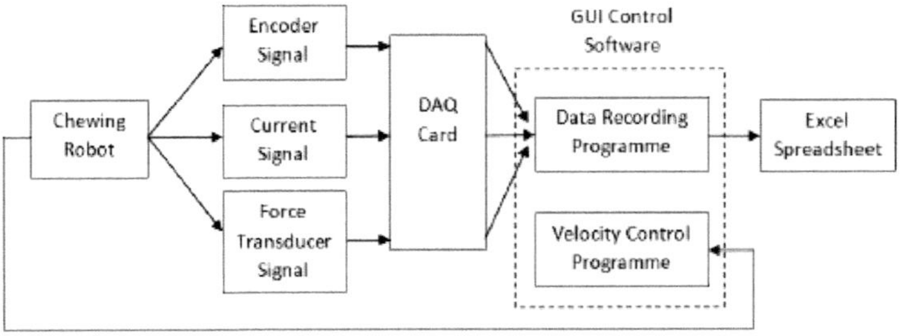

(a)

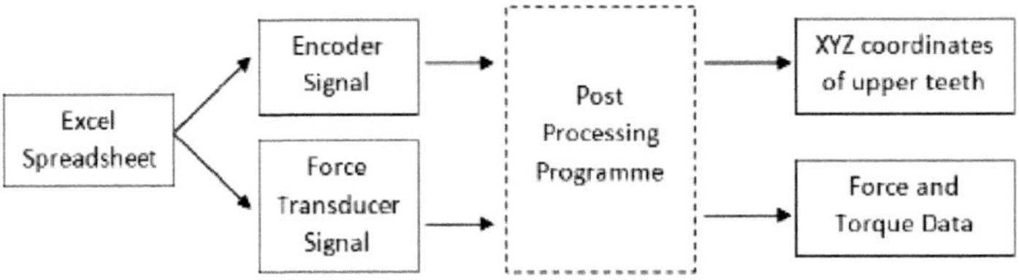

(b)

Fig. 5.25 Signal sensing system, (a) real time measurement and (b) offline processing

The force measured is in the sensor frame, $\{S\}$, and must be converted into the frame at the molar, $\{T\}$, for actual chewing force to be determined:

$$^{T}F_{T} = {}^{T}_{S}T_{f} \bullet {}^{S}F_{S} \tag{5.4}$$

in which ${}^{S}F_{S}$ is the measured force in $\{S\}$ and ${}^{T}F_{T}$ is the chewing force in $\{T\}$; they are related to each other by:

$$ {}^{T}_{S}T_{f} = \begin{bmatrix} {}^{T}_{S}R & 0 \\ {}^{T}P_{SORG} \times {}^{T}_{S}R & {}^{T}_{S}R \end{bmatrix} \tag{5.5} $$

According to the design, the force sensor frame and the molar frame are identical in direction, with an offset in their origin, i.e.:

$$ {}^{T}_{S}R = \begin{bmatrix} 1 & 0 & 0 \\ 0 & 1 & 0 \\ 0 & 0 & 1 \end{bmatrix} \tag{5.6} $$

$$ {}^{T}P_{SORG} = \begin{bmatrix} 0 \\ -0.0005 \\ -0.04 \end{bmatrix} \tag{5.7} $$

Consequently, Eq. (5.5) can be expressed explicitly as:

$$ {}^{T}_{S}T_{f} = \begin{bmatrix} 1 & 0 & 0 & 0 & 0 & 0 \\ 0 & 1 & 0 & 0 & 0 & 0 \\ 0 & 0 & 1 & 0 & 0 & 0 \\ 0 & 0.04 & -0.0005 & 1 & 0 & 0 \\ -0.04 & 0 & 0 & 0 & 1 & 0 \\ 0.0005 & 0 & 0 & 0 & 0 & 1 \end{bmatrix} \tag{5.8} $$

5.8 The Control System

Because the chewing cycle time and occlusal time vary between people and the different foods being chewed, the motor velocity had to be controlled. This control was implemented in LabView to allow flexibility in the design of subsequent prototypes. LabView was suitable for the control of the chewing device as it was only be used as a supervisory controller. The control of the motor itself had to be carried out by a dedicated controller in the form of a micro-controller or control card.

5.8.1 Computer/Motor Interface

As the device was to be controlled from a LabView program it made sense to choose a dedicated control card that could easily be interfaced with LabView. After investigating a few different options, a Maxon control card (DES 50/5) was selected. This worked seamlessly with the Maxon EC 32 motor used to actuate the device. The control card required an encoder with a line driver to run. Fig. 5.26 shows the line driver circuit built around a DS26LS31CM surface mount line driver chip. The communication between the control card and the computer based software was established via a RS-232 serial interface. This way, the software had to be adapted to access the appropriate transmit and receive registers.

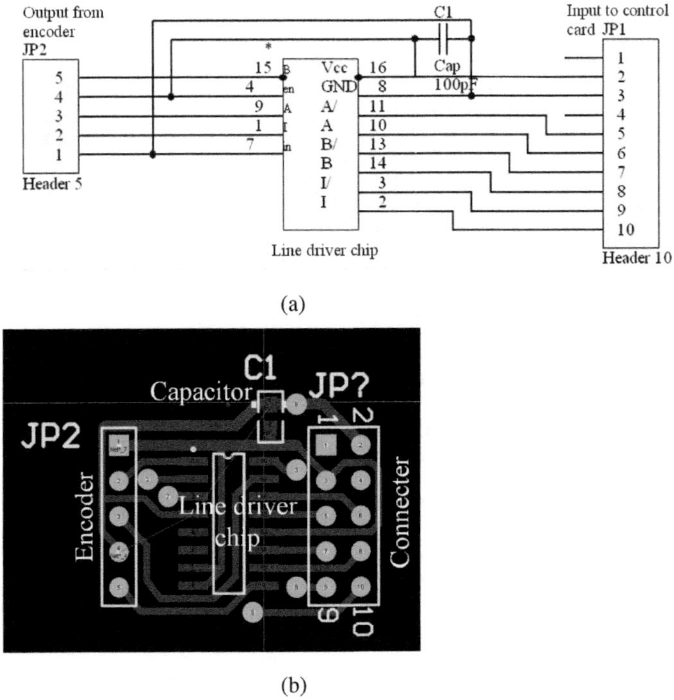

Fig. 5.26 The line driver circuit with the line driver chip in the middle (a) circuit diagram and (b) printed circuit board

5.8.2 *Hall Effect Sensor*

To be able to set the chewing device to the lower position where the teeth occlude, a reference point on the crank that corresponded to this lower position had to be known. The control card was unable to access the pulses that the encoder provided to determine location and only the speed of the motor in RPM could be measured. Therefore, the positioning of the motor had to be measured another way. To enable the lower and top positions to be sensed, a system that consisted of Hall Effect sensors and a small magnet was designed. The magnet was mounted on one of the gears. This system is very robust and simple, in comparison to counting the encoder pulses.

The pulses that the Hall Effect sensors provided were easily detected as the gear that the magnet was mounted on turned at a maximum speed of 100 RPM. This corresponded to one pulse every 0.6 s on each sensor. Fig. 5.27 shows the Hall Effect sensing circuit and board, where the board was fabricated in a ring shape so that the gear was able to be positioned inside it. The circuit consisted of two Hall Effect sensors with a resistor connecting the output of the sensor to the five Volt VCC rail, a couple of decoupling capacitors and a couple of connectors (for the power and signals). The board provided a +0 V output (when the magnet on the gear was positioned at the sensor locations) or a +5 V output. This signal was passed onto the control program via the parallel port.

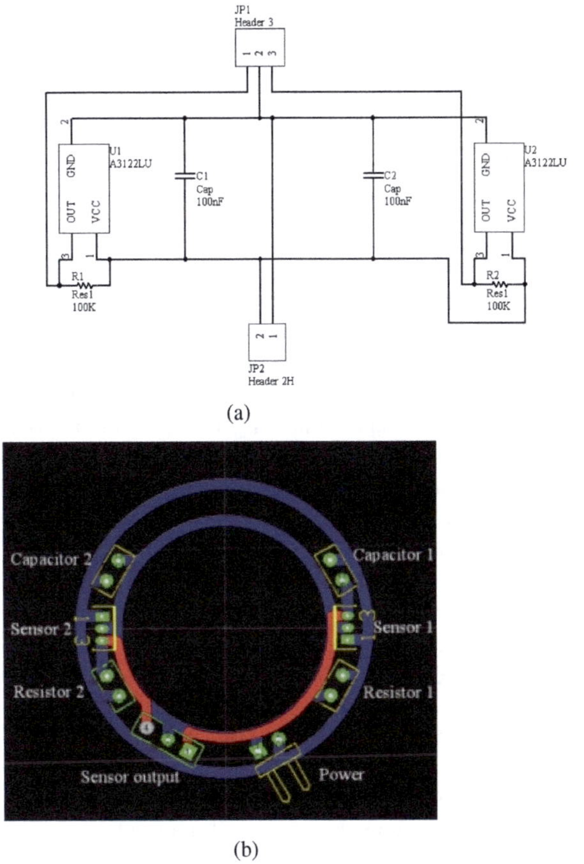

Fig. 5. 27 The Hall Effect sensor, (a) circuit and (b) board

5.8.3 The Software Functions

For the chewing device to function in a satisfactory human-like manner, it was important to include all the functions that would be necessary for simple operation. Fig. 5.28 shows the functions in the GUI as follows:

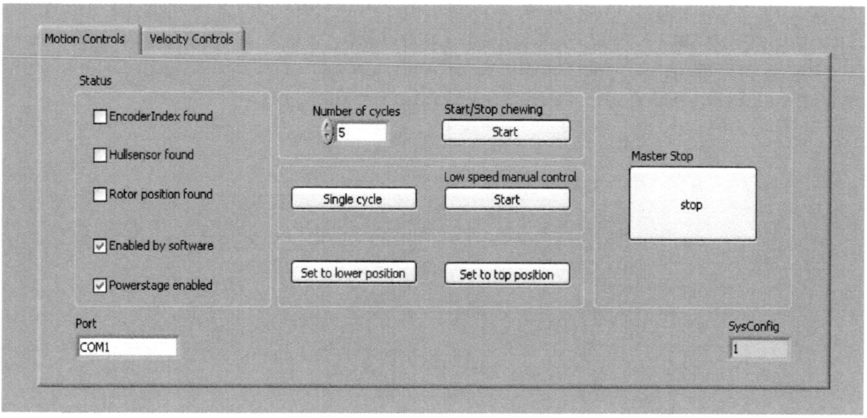

Fig. 5. 28 The software functions in the GUI

- *Set to lower position* – this function set the chewing device to the occlusal position so that the teeth could be aligned when setting up the device.
- *Set to top position* – this function set the chewing device to the position where the teeth were the maximum distance apart enabling a food sample to be placed in it. This can be thought of as the mouth being open.
- *Start chewing* – this function was used to make the device chew by making the device follow a velocity profile that matched that of a human chewing velocity profile. The number of chews was also able to be set.
- *Single cycle* – this function was used as a quick select 'one chew' button and would only perform one chewing cycle.
- *Low speed manual control* – the user was able to use this control to make the chewing device move at a slow speed in order to check the occlusion once the teeth have first been set up.
- *Master stop* – this was used as a safety feature that allowed the user to stop it at any time during any function. It should be noted that this function stopped only the execution of those motion programs, by setting the velocity of the motor to zero. To enhance safety, an emergency button is still required.

5.8.4 Chewing Trajectory Setting

The 'start chewing' function was the most complicated to develop. It involved making the chewing device perform the number of chewing cycles specified by the user and to follow a velocity profile (refer to Section 5.4.2). The velocity profile that the chewing device had to follow did not need to be exactly the same as that of a human in all cases. It only needed to match the occlusion phase velocity of a human and return to the occlusal position in the same time as that of a human. The operation of the 'start chewing' function operates by firstly moving the actuator, at the occlusal velocity, until the 'lower position' is reached – this being the middle point of occlusion. It then moves for a specified amount of time to reach the position where occlusion ends (as shown in Fig. 5.29a). The chewing device then speeds up from occlusal velocity to a maximum velocity and returns to occlusal velocity in a specified time (as shown in Fig. 5.29b). This allows the motor to be turned at the correct angle in order to return it to the start position. Therefore, the chewing device actuator returns back to the position where occlusion first occurs in the correct amount of time. This process is repeated for the number of times that the user specifies. The actuator then returns to the 'top position' so that the food sample can be removed (as shown in Fig. 5.29c).

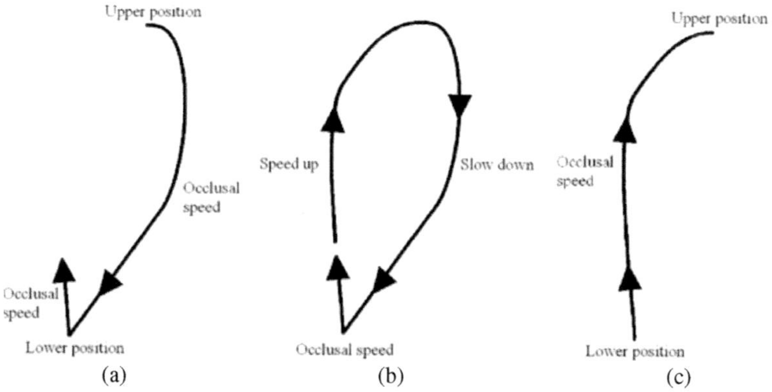

Fig. 5. 29 The general velocity profile of the chewing device, (a) closing, (b) occlusal and (c) opening

The operator may command the chewing device to chew foods differently. To this end, the device must allow for different velocity profiles. This can be achieved by changing the parameters for the velocity profiles. These control parameters include (Fig. 5.30):

- *Occlusal time* – this specifies the time that the occlusal phase is to take.
- *Opening/closing time* – this specifies the time that it takes to return to the start of the occlusal phase.
- *Occlusal angle* – this specifies the point where the occlusion begins and ends.

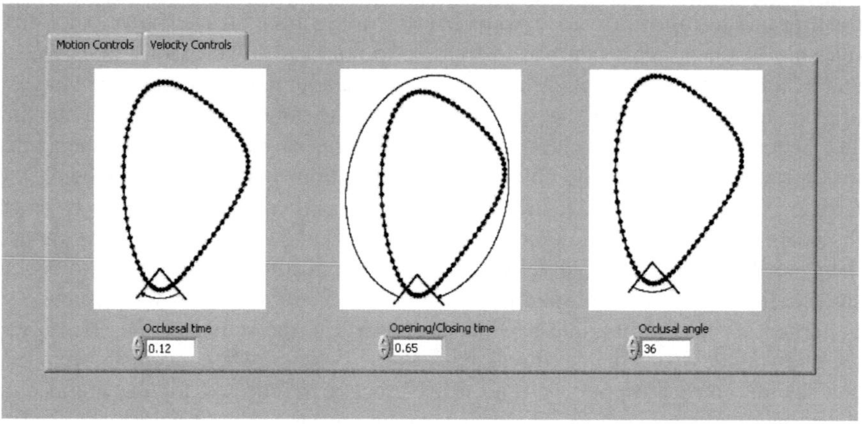

Fig. 5. 30 The velocity control GUI

5.9 Chewing Experiments

5.9.1 Operation of the Chewing Robot

The chewing device is designed to reproduce the trajectory of the molar teeth during human chewing cycles. It employs anatomically correct teeth geometries with accurate occlusion. The trajectory of the jaw can be adjusted to give a range of vertical and lateral movements so that different foods can be processed appropriately. The speed of the jaw can also be adjusted to simulate the motions observed during human chewing.

A set of experiments were carried out to examine the reproducibility and sensitivity of the functional parameters of the chewing robot. This included the number of chewing cycles, spring pre-tightening force (which refers to the potential force from the pre-tightened spring), shearing angle in the sagittal plane and length of the ground link required to adjust the shearing angle in the frontal plane. In the chewing experiments on peanuts, one whole peanut (1.0 ± 0.01 g) was put into the chewing robot for 9, 18 and 27 chewing cycles. The spring pre-tightening force was adjusted to 35 and 70 N. The length of the ground link was varied from 38 mm (vertical chewing) to 50 mm (lateral chewing). The chewing velocity of the robot was controlled in such a way as to match a human chewing velocity profile in accordance with human measurements [6]. 3D force profiles were recorded dynamically for the entire chewing cycle. Food particles were manually repositioned on the occlusal surface every three cycles in order to prevent particles from sticking to the teeth. Artificial saliva was not added in the experiment. The particle size distribution of the peanut bolus from a human was compared with the one from the chewing robot. Each experiment was replicated five times.

Particle size distribution of chewed peanuts was then measured by image analysis. The chewed food samples were collected in a Petri dish. Ethanol was added to eliminate the agglomeration of chewed peanuts without affecting the size of particles. ImageJ software (Version 1.40g) was used to analyse the particle area sizes of the samples as captured by an Epson scanner (Epson Perfection 3490 Photo) with a 1200 dpi resolution. The assumption that all particles were spherical was made and the diameters of the particles are obtained. The area percentages were calculated for particle bins with cut off diameters at 0.125, 0.177, 0.25, 0.354, 0.5, 0.707, 1, 1.414, 2, 2.828 and 4mm.

5.9.2 *Results and Discussion*[1]

5.9.2.1 Chewing Force Profiles

Chewing forces mainly depend upon the texture of the food. The chewing robot was able to generate various force profiles by adjustment to the pre-tightening force of the spring. Fig. 5.31 shows the difference of resultant chewing force from three dimensional forces on a 3mm thick hard plastic with different spring pre-tightening force. As shown by the green solid line in Fig. 5.31, the distance between the upper and lower teeth during mastication remained at 3mm. This indicated that there was no deformation of the plastic during chewing. When the upper teeth started contacting the plastic, the spring installed behind the upper teeth began to be compressed and generated compression forces onto the plastic. When the spring was pre-tightened before being compressed, the force applied from the spring increased by the pre-tightening force. For example in Fig. 5.31, by adjusting the spring pre-tightening force from 0 to 45 N, the maximum chewing force for each cycle was increased by 45 N.

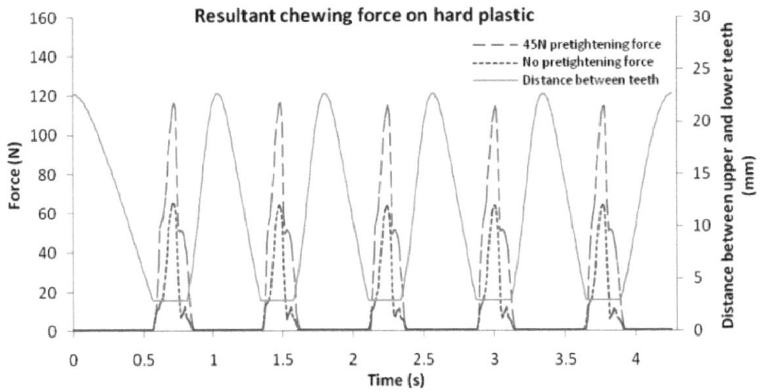

Fig. 5.31 Changes to resultant chewing forces with different pre-tightening forces throughout the chewing process of 3mm thick hard plastic

[1] More results refer to Section 8.3.4.

The chewing force profile results from the experiment involving the pre-tightened spring force of 35 N and maximum chewing angles in both sagittal plane and frontal planes, was plotted.

Since the food was repositioned every three chewing cycles, the force profiles revealed a decreasing trend over a series of three chewing cycles. The irregular surfaces of the teeth along the path of forward movement in the sagittal plane resulted in component forces along all axes. The directions of the resultant force (in the x-y-z plane) during the second cycle of the trial can be seen in Fig. 5.32. As the upper teeth moved towards and compressed the food, the chewing force increased and reached the largest value at complete occlusion. The differences in directions between positional change and force propagation illustrate the importance of teeth shape on mastication mechanisms.

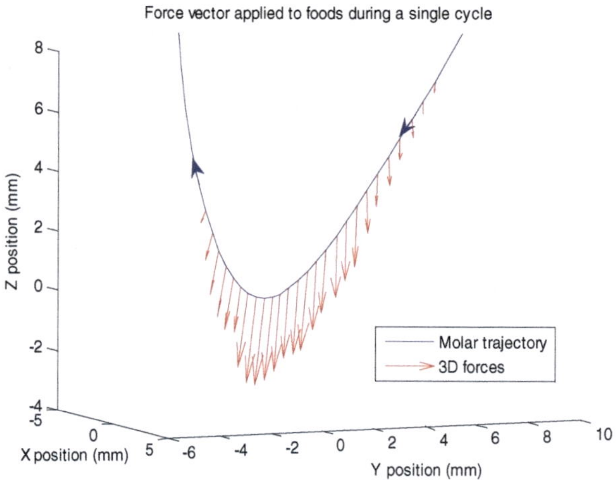

Fig. 5. 32 Resultant force vectors projected on the chewing trajectory of one cycle on one peanut

5.9.2.2 Motor Torque Profiles

The motor torque profile during the mastication process of a 10x5x15 mm piece of apricot pie over three chewing cycles was produced in Fig. 5.33. In general, an increase in motor torque resulted in an increase in the resultant chewing force. Peak force as well as motor torque during were largest during the first chewing cycle this being related to the hardness of apricot pie. As chewing continued, motor torque decreased as well as the resultant force, which indicated that the hardness of the food was reduced due to changes in texture during mastication.

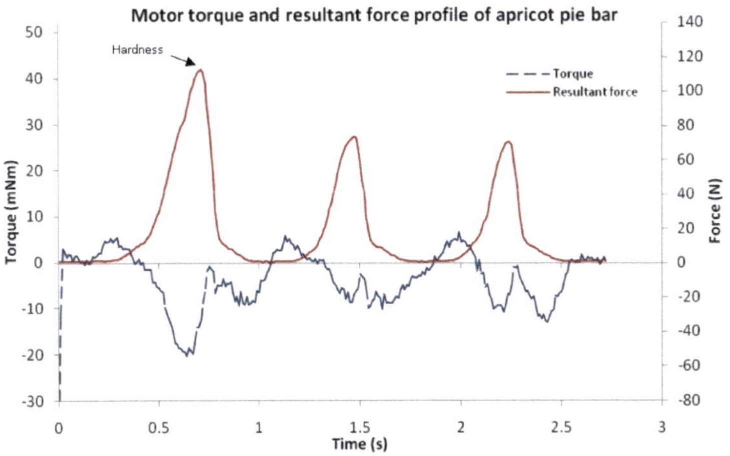

Fig. 5. 33 Motor torque and resultant force profile throughout the chewing process of apricot pie for three chewing cycles

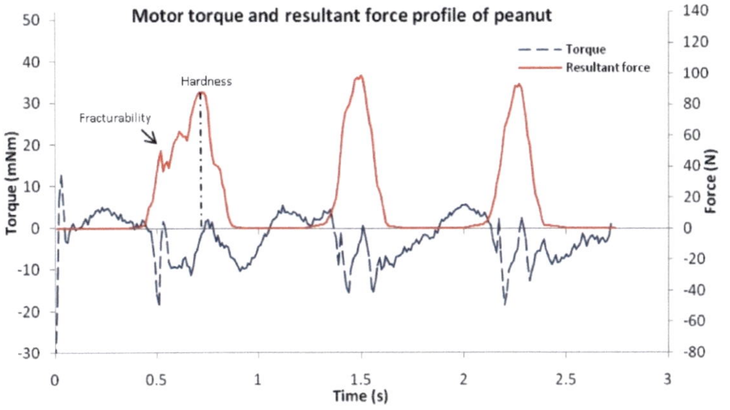

Fig. 5.34 Motor torque and resultant force profile throughout the chewing process of a peanut over three chewing cycles

The motor torque profile during the chewing process of a one gram peanut is presented in Fig. 5.35. In the first cycle, fracturability of the peanut can be observed by the force of the significant break, at which point peak motor torque also occurred. A possible explanation for this occurrence may be that motor torque is related to food texture. More experimentation on various foods is required in order to confirm this relationship.

5.9.2.3 Particle Size Distributions

The reproducibility of and sensitivity to functional parameters were assessed by comparison of resulting particle size distributions. Experiments with different chewing cycles (9, 18 and 27), demonstrating the reproducibility of the device and that as the number of chewing cycles increased, food particles reduced in size, as expected. The results about the particle size distributions and the comparison with human mastication are available in Section 8.3.4.

5.10 Summary

A linkage chewing robot for use in food evaluation was developed and the mechanism design, construction, force sensing, motion control, simulation and experimental validation were described within this chapter. The design specifications were derived from published in-vivo measurements of human chewing. Whilst a four-bar linkage was capable of following a specified chewing trajectory, a six-bar linkage enabled the molar teeth to maintain movement in a set orientation. A variety of chewing trajectories were achieved by varying the ground link length. The motion of the actuator was designed in order to implement various chewing motions as required for different foods. The 3-D chewing force was measured by a force sensor placed beneath the lower teeth. The operation of the device was made easy via GUIs programmed in LabView. The initial experiments discussed in this chapter show promising results were obtained by the device. More validation experiments have been performed and the robotic device has become a routinely used piece of laboratory apparatus in the pre-processing of foods.

References

1. Lucas, P.W.: Dental functional morphology: how teeth work. Cambridge University Press, United Kingdom (2004)
2. Mongini, F., et al.: Computer-based assessment of habitual mastication. J. Prosthetic Dent. 55, 638–649 (1986)
3. Anderson, K., et al.: The effects of bolus hardness on the masticatory kinematics. J. Oral Reh. 29, 689–696 (2002)
4. Peyron, M.A., et al.: Effects of increased hardness on jaw movement and muscle activity during chewing of visco-elastic model foods. Exp. Brain Res. 142, 41–51 (2002)
5. Foster, K., et al.: Effect of texture of plastic and elastic model foods on the parameters of mastication. J. Neurophysiol. 95, 3469–3479 (2006)
6. Ogawa, T.: Different responses of masticatory movements after alteration of occlusal guidance related to individual movement pattern. J. Oral Reh. 28, 830–841 (2001)
7. Anderson, D.J.: Measurement of stress in mastication. J. Dent. Res. 41, 175–189 (1956)
8. Thompson, T.: How to use and interpret the Coupler Curve and Centrode Atlas (1999), `http://www.cedarville.edu/academics/engineering/kinematics/ccapdf/howto.htm` (accessed in March 2006)
9. Craig, J.J.: Introduction to robotics, mechanics and control, 3rd edn. Prentice-Hall, Englewood Cliffs (2004)
10. Lewis, D.: A robotic chewing device for food evaluation. Master of Engineering Thesis, Massey University, New Zealand (2006)

Chapter 6
Measurement and Reproduction of Mastication Movement*

Abstract. This chapter first briefly introduces the measurement of the mastication movements in human subjects and then describes how a measured mastication movement can be reproduced using a chewing robot. The jaw movement is recorded using an Articulograph (AG500) while a subject chews on a food sample. The definition and conversion between various reference frames established on the subject is discussed. A mapping technique is developed to allow accurate translation of the recorded sensor points to infer the movements of the subjects' molar tooth. A method of aligning the reference frames between the subject and the robot is then described. The inferred molar movement can then be implemented on the robot. A graphical user interface (GUI) for analysis of recorded mastication movements is also presented.

6.1 Mastication Measurement

6.1.1 The Techniques[1]

Jaw trajectories have been studied for over 100 years. Early techniques used photography and light reflection [1]. The most common measures of jaw trajectory involve the production of a magnetic field, the detection of the light movement of infra-red emitting diodes, and videographic techniques. Modern magnetic tracking systems, such as the 'Sirognathograph' and 'Carstens' systems, have become more accurate [1].

Lucas et al. [2] used a Sirognathograph where magnets were cemented to the front of the lower incisors. Sensors were attached to a frame fastened to the subject's head to detect 3D movement of the magnets. A Kinesograph can also be used, where magnetic field variations are created by moving magnets attached to

---

* Reprinted with modification from Torrance JD, Hutchings SC, Bronlund JE, Huang LL and Xu WL (2010) Human jaw motion measurement, analysis and robotic reproduction. Int. J. Intel. Syst. Tech. Appl. **8**: 288 – 302, with permission from Inderscience.

[1] This section was provided by Scott Hutchings, a PhD student of Institute of Food, Nutrition and Human Health (IFNHH), Massey University.

W. Xu and J.E. Bronlund: Mastication Robots, SCI 290, pp. 163–177.
springerlink.com

the teeth [3, 4]. Magnetic devices have also been used to track jaw movement on animals. Yamada & Haraguchi [5] attached a frame to the nasal bone of rabbits, and fixed small magnets to the mandibular bone.

Peyron et al. [6] used a device called the AG100 by Carstens, originally made for studying speech. The device involves the subject placing their head inside a large frame creating a magnetic field. Miniature coils are attached to the middle and lower incisors, and connected to preamplifiers. These coils induce a current inversely proportional to the cube of the distance from the transmitter coil, and therefore their location within the magnetic field can be recorded.

Lassauzay et al. [7], Chew et al. [8], Kohyama and Mioche [9], Kakizaki et al. [10], and Inoue et al. [11] are all examples of studies that have used magnetic field based systems to track jaw movement. Carstens have also developed the AG200 and AG500 (Fig. 6.1) that operate in a similar way but with increasing numbers of degrees of freedom. A range of mandibular parameters can be evaluated using these devices [6].

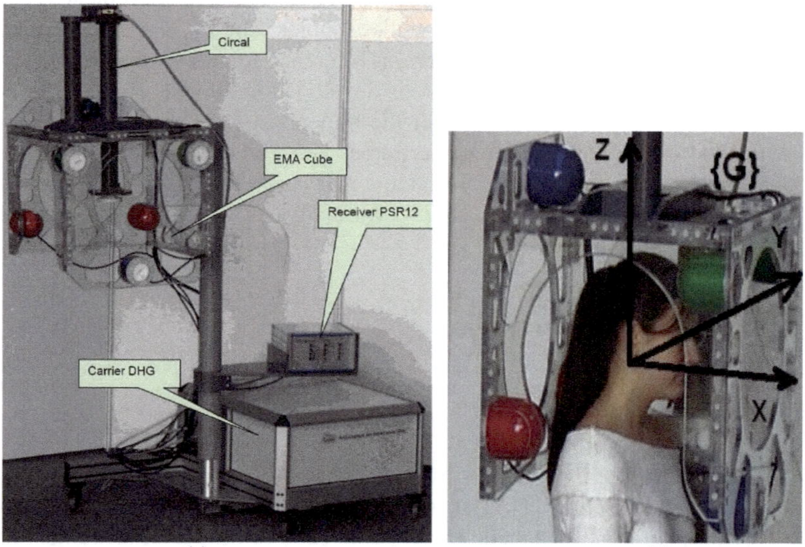

(a) (b)

Fig. 6.1 Articulograph AG500 (a) setup and (b) global coordinate system, $\{G\}$

6.1.2 *The Measurement*

At Massey University an AG500 was set up to measure the jaw movements. The AG500 enables measurement of 5 DOF with each sensor (x, y, z positions and two angles). In order to be able to completely describe the mastication movements in 3D space, a number of magnet sensors are required to be placed on the subject, two behind the ears (Fig. 6.2 a) and three more on the subject's lower teeth

(Fig. 6.2 b). The position of a sensor is recorded in the global co-ordinate system (Fig.6.1 b) central to the device, denoted by ${}^{\text{Global}}\boldsymbol{P}$.

One of the mastication measurements was performed on a healthy 24-year-old male, with the food being a 5 g mixture of oats, nuts and honey. Head movement during the measurements of the jaw movement was calculated and subtracted by a program supplied with the AG500. From the positions of the three tooth sensors we were able to find the three translational and three rotational displacements of the mandible in relation to the head, using the procedure which is described in the following sections.

Electromyograph (EMG) sensors were also placed on the subject's face to measure the power generated by masseter and temporalis muscles during the closure of the mouth. The Articulograph and EMG measurements were synchronized in time. By detecting the time the EMG signal bursts, the closing and opening phases of the mastication movement can be distinguished.

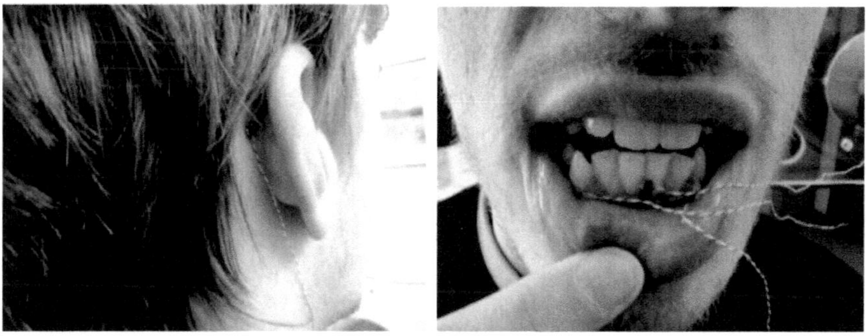

(a) (b)

Fig. 6.2 Sensors placement, (a) two sensors behind the ear, and (b) three sensors on the lower teeth

6.2 Coordinate Systems for the Movement

6.2.1 The Sensor and Mandible Systems

With respect to the afore-mentioned measurement, the three sensors s_1, s_2 and s_3 were affixed on the incisor, left canine and right canine, respectively (Fig. 6.3). At the jaw closed position, which was found by searching the maximum z-axis displacement in the device's global coordinate system, a sensor frame, $\{Sensor\}$ was constructed as follows. It is placed at the incisor, with its z-axis being normal to the plane dictated by three points s_1, s_2 and s_3; y-axis being parallel to s_3s_1 and x-axis by the cross product of y- and z-axis.

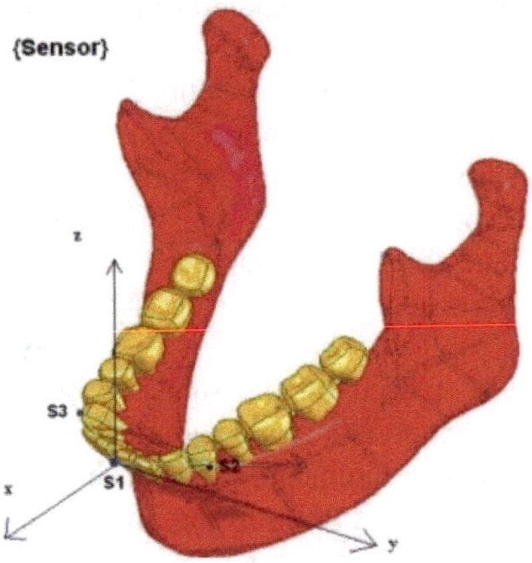

Fig. 6.3 Sensor frame, $\{Sensor\}$ constructed by three lower jaw sensors

The frame, $\{Sensor\}$ can be described in the global coordinate system by:

$$ {}^{Global}_{Sensor}T = \begin{bmatrix} 1.026 & 0.0644 & -0.0899 & 111.2 \\ -0.0792 & 1.000 & -0.1594 & -17.67 \\ 0.0796 & 0.1652 & 1.000 & -67.49 \\ 0 & 0 & 0 & 1.000 \end{bmatrix} \tag{6.1} $$

Its inverse transformation is given as:

$$ {}^{Sensor}_{Global}T = \left({}^{Global}_{Sensor}T\right)^{-1} \tag{6.2} $$

Thus, the position of a sensor can now be described in $\{Sensor\}$:

$$ {}^{Sensor}P = {}^{Sensor}_{Global}T \cdot {}^{Global}P \tag{6.3} $$

The movement of the lower jaw starts from where the mouth is closed. This moving jaw is now denoted by a frame, $\{Mandible\}$, which is coincident with $\{Sensor\}$ at t=0. The frame $\{Sensor\}$ can be viewed invariant and serve as the reference for the jaw movement, while the frame $\{Mandible\}$ is fixed on and moves with the jaw. The movement of the mandible in relation to its initial position is described by:

$$ {}^{Sensor}_{Mandible}T = \left({}^{Global}_{Sensor}T\right)^{-1} \cdot {}^{Global}_{Mandible}T. \tag{6.4} $$

It is noted that the aforementioned frames could be different between the measurement sessions due to the reaffixing of the three sensors, and between the subject and the robot due to the difference in the jaw sizes. In the next section we will present a method to establish the frames by taking into account whole lower teeth and aligning them with those on the robot.

6.2.2 The Skull System

In the robotic model, the jaw movement is defined with respect to a skull frame, $\{Skull\}$. This frame is attached to the robot ground such that when the mouth is shut the molars lie on the x-y plane whose normal is taken as the z-axis with the origin at the incisor (Fig. 6.4).

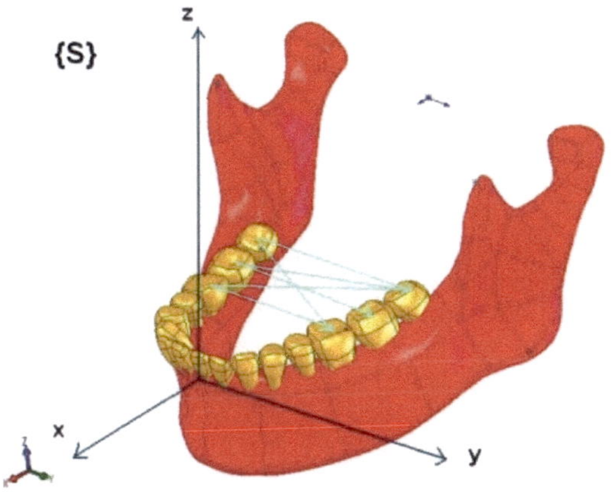

Fig. 6.4 The skull frame, $\{Skull\}$ attached on the robot model

The skull frame can be established by mapping the subject's teeth in the sensor frame. With the three sensors defining $\{Sensor\}$ still in place, a fourth sensor is used to trace around the midpoints of the subject's lower teeth, and its positions are recorded in $\{Global\}$, or further expressed into $\{Sensor\}$ by Eq. (6.3). As shown in Fig. 6.5, there are 5 pairs of two intersecting vectors, and each of them can produce a normal by cross product. The average of the 5 normal vectors is used as the z-axis of $\{Skull\}$. Fig. 6.6 shows how the y-axis of $\{Skull\}$ is established from the measured teeth positions in $\{Sensor\}$. The y-axis is the average of the five vectors formed by connecting the right second molar, first molar, second premolar, first premolar and canine to the left counterpart, respectively. The x-axis completes the right-hand system by taking the cross product of the y- and z-axis. Finally, the origin of $\{Skull\}$ is placed at the origin of $\{Sensor\}$. Therefore, the relation of $\{Skull\}$ and $\{Sensor\}$ is described by a homogeneous transformation as:

$$ {}^{Sensor}_{Skull}T = \begin{bmatrix} 1.000 & -0.0044 & 0.0396 & 0 \\ 0.0044 & 1.000 & 0.0045 & 0 \\ -0.0396 & 0.0012 & 1.000 & 0 \\ 0 & 0 & 0 & 1 \end{bmatrix} \tag{6.5} $$

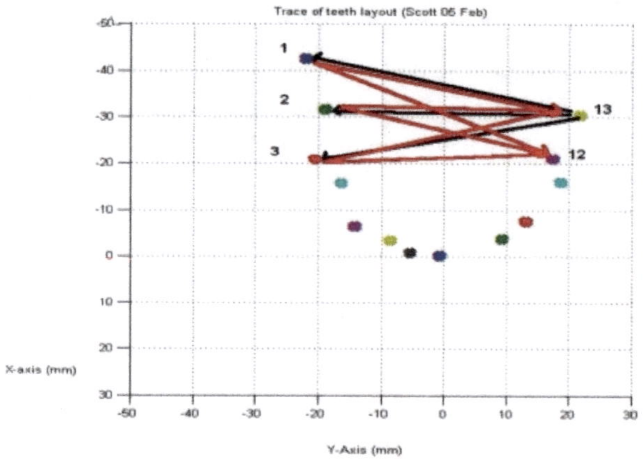

Fig. 6.5 Locations of subject lower teeth in the *x*-*y* plane of {*Sensor*} and the vectors used to establish the *z*-axis of {*Skull*}

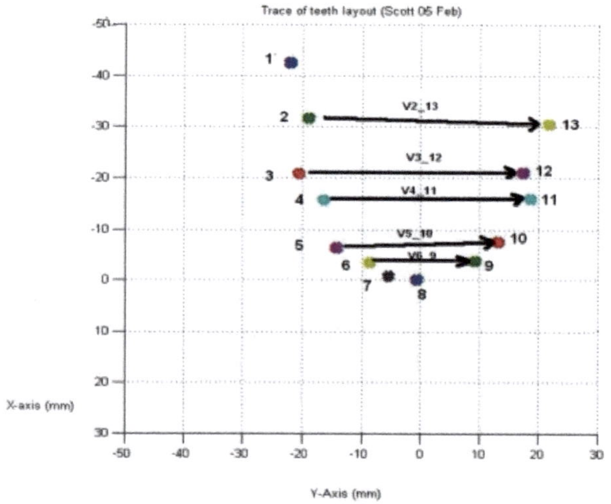

Fig. 6.6 Locations of subject lower teeth, and the vectors defining the *y*-axis of {*Skull*}

The two frames, {*Skull*} and {*Sensor*} have similar orientations with differences in roll, pitch and yaw angles of -0.07, 2.27, and 0.25°. The moving mandible can now be expressed into {*Skull*} by

$$ {}^{Skull}_{Mandible}T = {}^{Skull}_{Sensor}T \cdot {}^{Sensor}_{Mandible}T \tag{6.6} $$

By using more teeth as landmarks, we have a better chance of aligning {Skull} on the subject with the one on the robot, making the robotic reproduction of the recorded mastication movements more accurate.

6.3 Reproduction of the Movements

Because the size of the robotic jaw differs from the subject's, it is only possible to reproduce the movement of one tooth. Because most of the chewing experiments for the recorded mastication movement presented in the example presented here were performed on the left molars, in the following we recreated the movement of this tooth. The location of the left molar is known and fixed in $\{Mandible\}$. We define a new frame, $\{Molar\}$ and express it in $\{Mandible\}$,

$$ {}^{Mandible}_{Molar}T = \begin{bmatrix} 1 & 0 & 0 & {}^{Mandible}LMP_x \\ 0 & 1 & 0 & {}^{Mandible}LMP_y \\ 0 & 0 & 1 & {}^{Mandible}LMP_z \\ 0 & 0 & 0 & 1 \end{bmatrix} \tag{6.7} $$

The moving molar can now be expressed in $\{Sensor\}$,

$$ {}^{Sensor}_{Molar}T = {}^{Sensor}_{Mandible}T \cdot {}^{Mandible}_{Molar}T \,. \tag{6.8} $$

Below we describe the molar's motion with respect to its initial pose at the jaw closed position, defined by a frame, $\{Molar\ Re\}$, which is fixed with respect to $\{Sensor\}$. We have the following relationships:

$$ {}^{Skull}_{Molar}T = {}^{Skull}_{Mandible}T \cdot {}^{Mandible}_{Molar}T \tag{6.9} $$

$$ {}^{Sensor}_{Molar\,\mathrm{Re}}T = {}^{Mandible}_{Molar}T \tag{6.10} $$

Combining Eq. (6.9) and (6.10) yields:

$$ {}^{Molar\,\mathrm{Re}}_{Molar}T = {}^{Sensor}_{Molar\,\mathrm{Re}}T^{-1} \cdot {}^{Sensor}_{Molar}T \tag{6.11} $$

from which we can get the molar movement relative to its initial pose in $\{Skull\}$. From this equation, we can find the results as shown in Fig. 6.7 to 6.9, which were obtained after the raw data for the sensors were processed by a low pass filter. Fig. 6.7 shows the spatial trajectory of the left molar on the robot to be reproduced in $\{Molar\ Ref\}$.

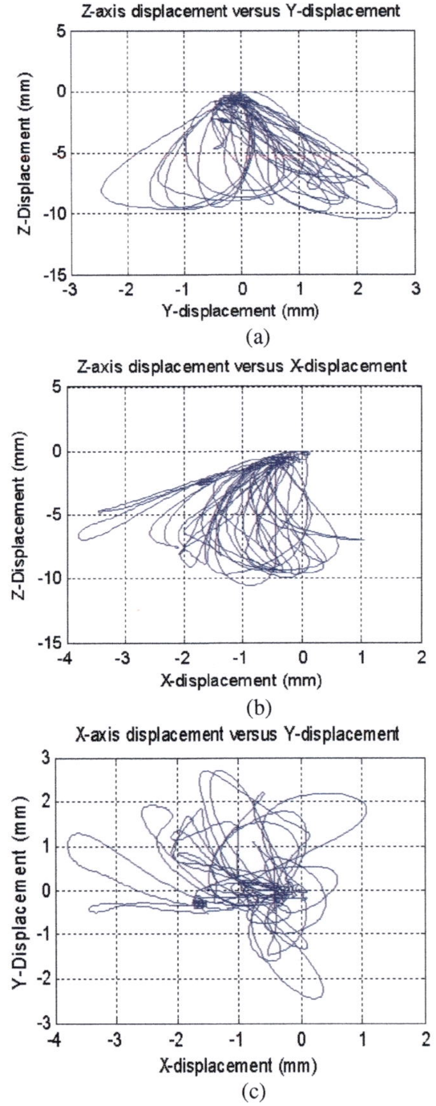

Fig. 6.7 Spatial molar trajectory in {*Molar Ref*} (a) frontal view, (b) side view, (c) top view

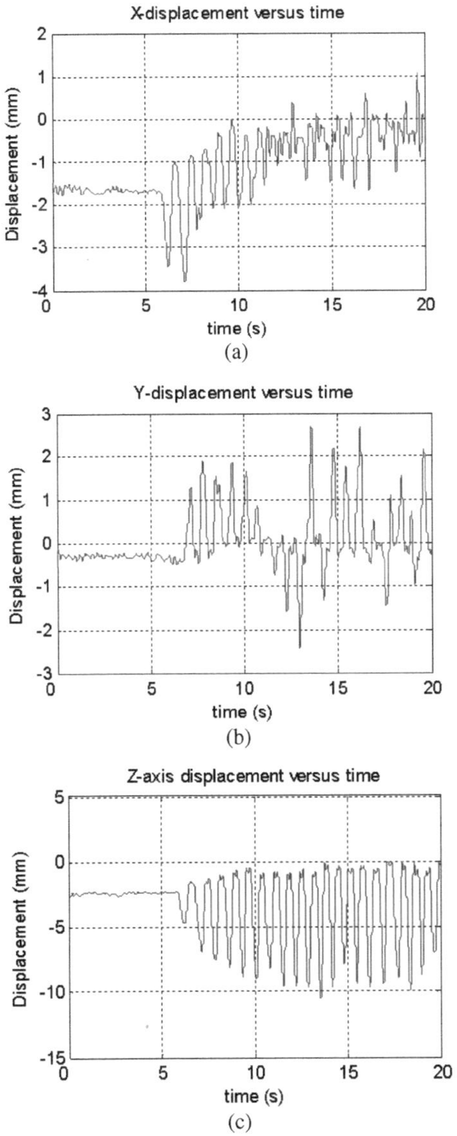

Fig. 6.8 Molar trajectory in time in {Molar Ref} (a) x-time, (b) y-time and (c) z-time

Fig. 6.8 shows the molar movement in terms of time. In the beginning of the chewing measurement, the subject kept the mouth shut and maintained in this state for around first seven seconds. From Fig. 6.8(b) we can see that the subject chewed on the left molars for the cycles of positive y-displacement, and moved to the right to reposition foods during the cycles of negative y-displacement. From Fig. 6.8(c) we can see the occlusal positions where the z-displacement peaks.

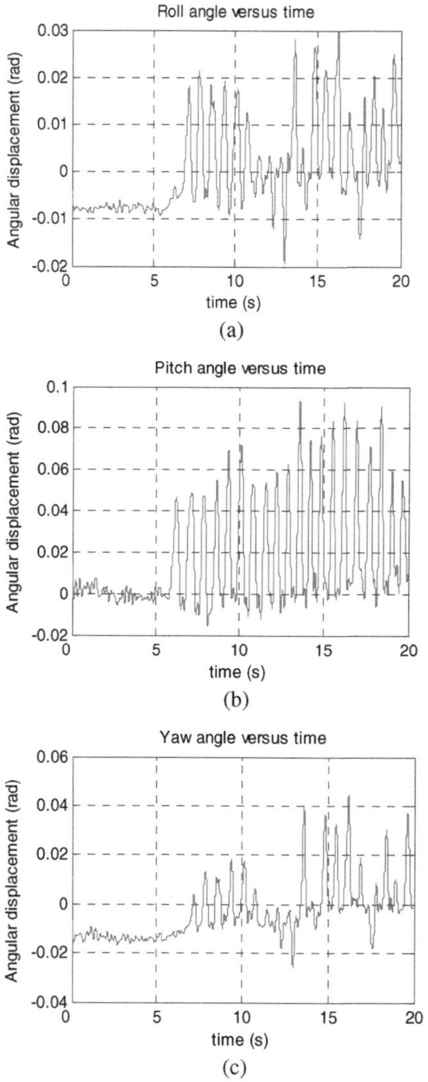

Fig. 6.9 Molar point trajectory in the Molar Ref frame (a) Roll-time, (b) Pitch-tome and (c) Yaw-time

Fig. 6.9 shows the rotational movement of the lower jaw. The pitch about the y-axis is larger than the two other rotations, roll and yaw around the *x*- and *z*-axes. This is because of the lower jaw being pivoted at the TMJs. The presence of angular movements about *x*- and *z*-axes is due to the food repositioning requirements, which is made possible by the flexible structure inside the TMJs.

6.4 GUIs for Analysis of the Movements

To help analyze measurements of mastication movement using the Articulograph, a MATLAB graphical user interface (GUI) was developed. This GUI is based on the measurement procedure described in the preceding sections. The time window for the chewing analysis is 60 s and the sampling interval of the measurement is 0.005 s.

The GUI is divided into three functional modules (Fig. 6.10). The first is "one sensor' for measuring the displacement trajectory of one particular sensor in $\{Sensor\}$, and the second is "one sensor with new frame" for measuring the displacement trajectory of one particular sensor in $\{Skull\}$, and the third is "trace" for analyzing the quality of the measurement by tracing the position of each lower tooth in $\{Skull\}$.

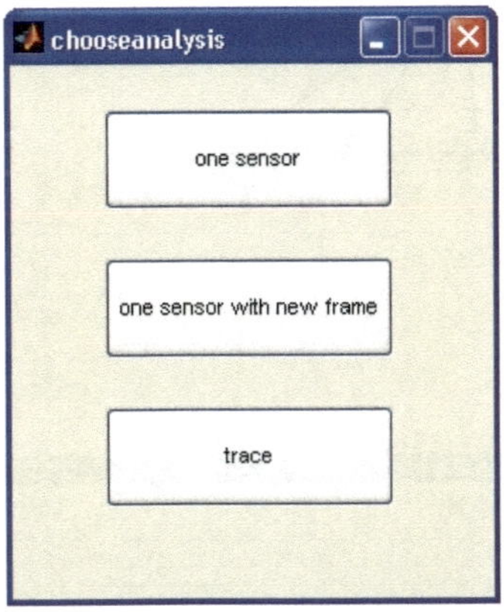

Fig. 6.10 GUI for the analysis of the mastication movement

The GUI was designed following the flowchart as shown in Fig. 6.11. The arrowed lines indicate the procedural progression of each module and the dotted arrows shows where an individual sensor used in the measurement can be selected and its movement analyzed.

For any of the sensors attached on the lower teeth, the GUI can trace its time trajectory in *x*-, *y*- and *z*-axis of $\{Senor\}$ or $\{Skull\}$, and the distance it has traveled, as shown in Fig. 6.12. For each trajectory, its characteristics such as period, peak values, the number of cycles used, jaw opening and closing velocities are also visualized. One can zoom in on any trajectories by pressing the "Zoom" button (Fig. 6.13), view the first three cycles of the movements by pressing the "Start" button (Fig. 6.14) or the last three cycles of the movements by pressing the "Finish" button (Fig. 6.15). One can also overlap the EMG signal onto the trajectories by pressing the "EMG" button, as shown in Fig. 6.16. By pressing the "Sensor" button, one can choose another sensor to analyze.

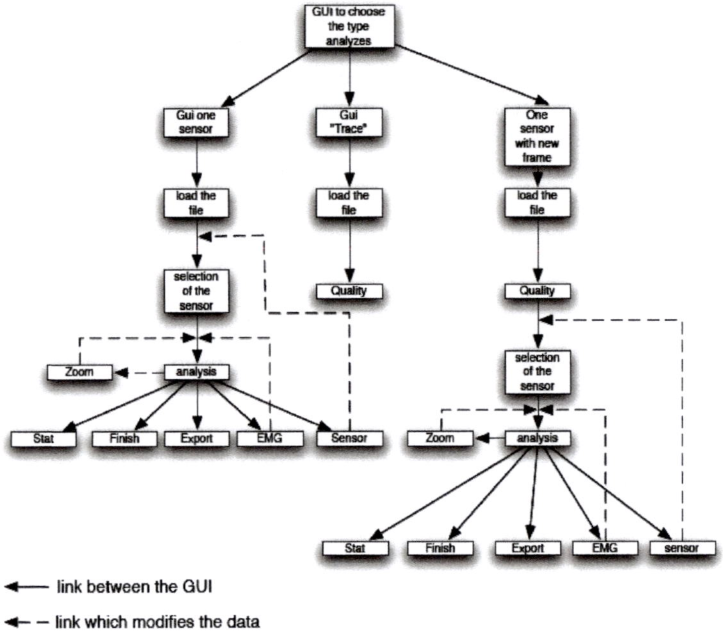

Fig. 6.11 The design flowchart of the GUI

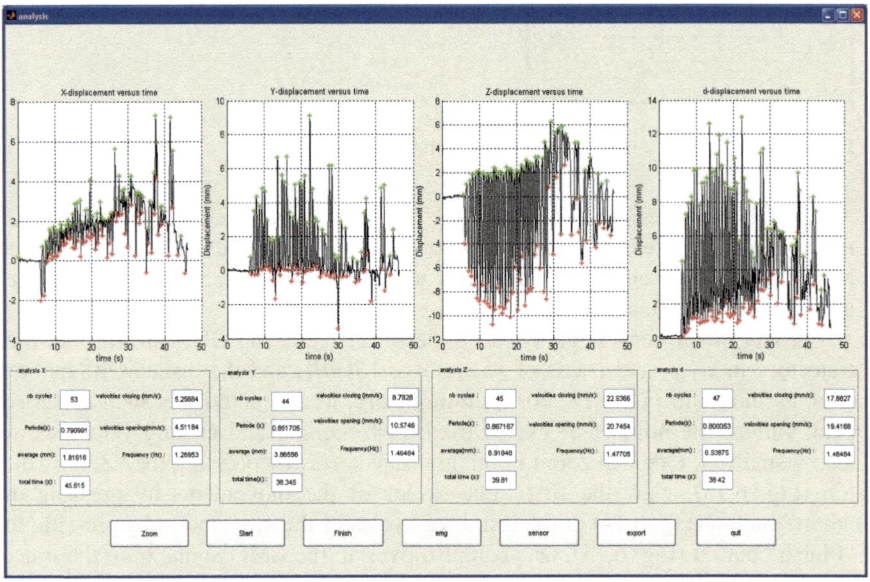

Fig. 6.12 GUI for the analysis of the measured movement of one sensor in the frame $\{Sensor\}$

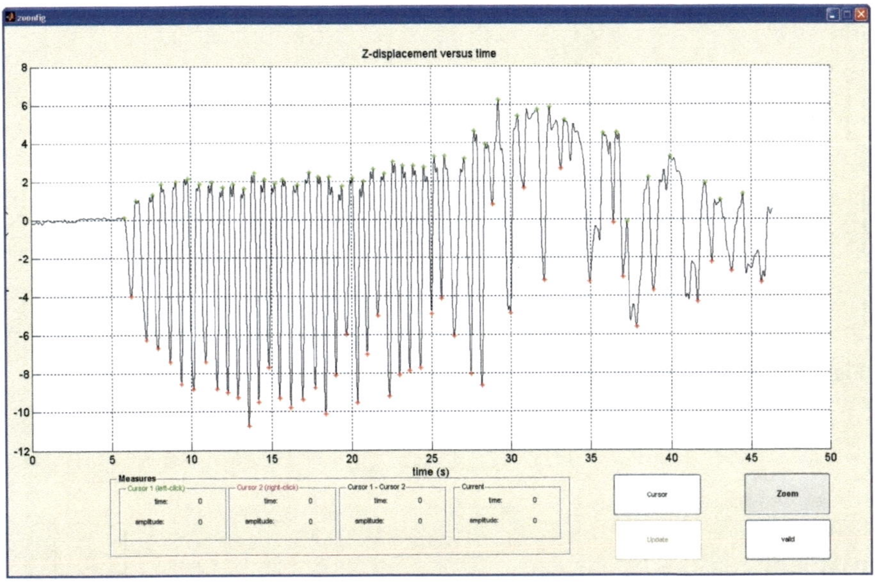

Fig. 6.13 Zoomed z-axis displacement trajectory

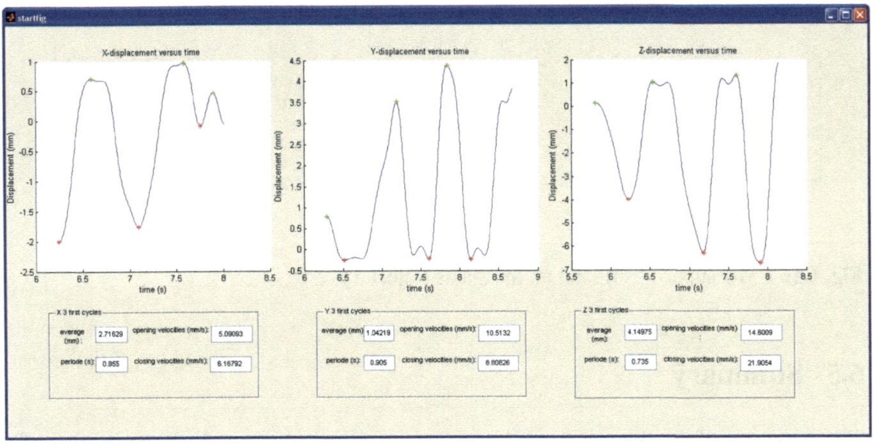

Fig. 6.14 First three cycles of the chewing movement of one sensor

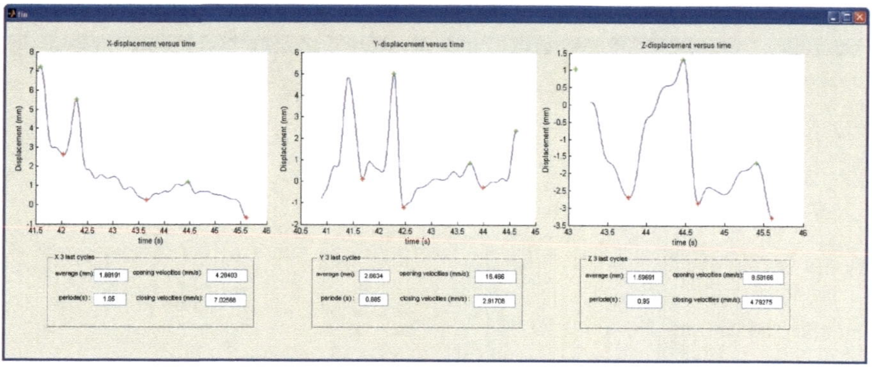

Fig. 6.15 Last three cycles of the chewing movement of one sensor

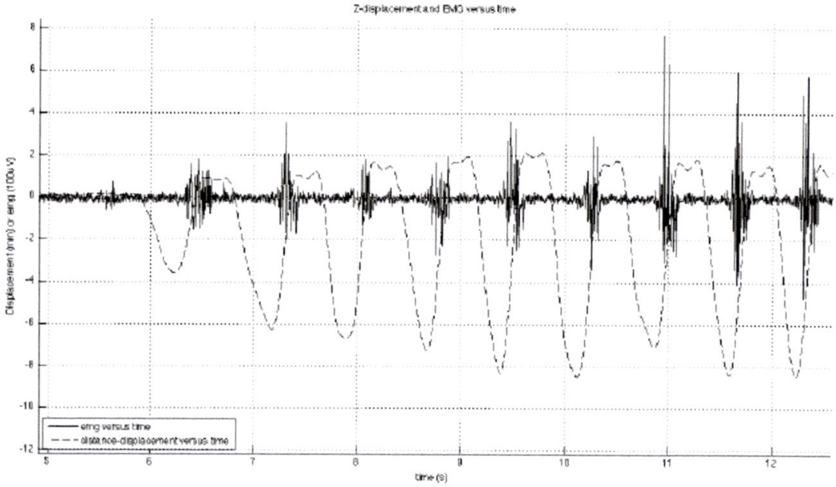

Fig. 6.16 EMG super-positioned on the z-axis trajectory

6.5 Summary

The mapping of a subject's teeth locations has allowed us to align recordings of a human jaw with a robotic jaw using consistent methods based on common landmarks to create the Skull frame on both jaws. Furthermore by locating the left molar in the Skull frame, the motion of this point can be accurately reproduced by the chewing robot. The GUI developed provides a convenient tool to analyze the measured mastication movements to characterise food dynamics during chewing.

References

1. Röhrle, O., et al.: Using a motion-capture system to record dynamic articulation for application in CAD/CAM software. J. Prosthodont. 9, 1–8 (2009)
2. Lucas, P., et al.: Relationship between jaw movement and food breakdown in human mastication. mastication. J. Dent. Res. 65, 400–404 (1986)
3. Horio, T., Kawamura, Y.: Effects of texture of food on chewing patterns in the human subject. J. Oral Rehab. 16, 177–183 (1989)
4. Castro, N.B., et al.: Analysis of the area and length of masticatory cycles in male and female subjects. J. Oral Rehab. 29, 1160–1164 (2002)
5. Yamada, Y., Haraguchi, N.: Reflex changes in the masticatory muscles with load perturbations during chewing hard and soft food. Brain Res. 669, 86–92 (1995)
6. Peyron, M.A., et al.: Masticatory jaw movement recordings: A new method to investigate food texture. Food Qual. Pref. 7, 229–237 (1996)
7. Lassauzay, C., et al.: Variability of the masticatory process during chewing of elastic model foods. Eur. J. Oral Sci. 108, 484–492 (2000)
8. Chew, C.L., et al.: The effect of food texture on the replication of jaw movements in mastication. J. Dent. 16, 210–214 (1988)
9. Kohyama, K., Mioche, L.: Chewing behavior observed at different stages of mastication for six foods, studied by electromyography and jaw kinematics in young and elderly subjects. J. Tex. Stud. 35, 395–414 (2004)
10. Kakizaki, Y., et al.: Coordination between the masticatory and tongue muscles as seen with different foods in consistency and in reflex activities during natural chewing. Brain Res. 929, 210–217 (2002)
11. Inoue, M., et al.: Effects of food consistency on the pattern of extrinsic tongue muscle activities during mastication in freely moving rabbits. Neurosci. Let. 368, 192–196 (2004)

References

Chapter 7
Robotic Chewing Experiments*

Abstract. The Crank Actuation masticatory robot of 6RSS parallel mechanism is applied to the chewing of foods. The forces on food samples during experiments were calculated from the recorded torques measured for each of the six actuators. A series of experiments were carried out on model and real food systems, and the resulting forces were discussed with respect to food texture. To begin with, the robot was compared with traditional uni-axial compression testing by implementing a one dimensional vertical crushing motion on the robot with flat plates used in place of the teeth. Very good agreement was found between the robot and texture analyser force-deformation profiles. The robot was also tested against the measured force profiles recorded for a 2D trajectory used to simulate movement of the 2DOF 6-bar linkage robot. Again good agreement was found, particularly in the vertical dimension. Finally, the robot was used to simulate the trajectory of real time 3D recorded human mandible movements during mastication. Vertical force measurements were consistent with the expected failure mechanism for each food. In each experiment unexpected forces in the non-vertical dimensions were found. These forces could be due to internal friction in robot linkages under loaded conditions. After identification and resolution of the cause of the unexpected x- and y-axis forces, the chewing robot will be suitable for use in analysing the initial textural properties of foods. Research into appropriate force control strategies, food bolus retention and re-orientation of the food on the teeth between cycles is required before the machine can be realistically used to mimic whole chewing sequences and provide insight into the food texture dynamics during bolus formation.

7.1 Introduction

Chapter 4 outlined the design of a Crank actuation masticatory robot of 6RSS parallel mechanism. This device was developed to be able to reproduce human chewing trajectories in 6 degrees of freedom. Chapter 6 describes the techniques and analysis done to record mandible trajectories in humans during chewing of real foods. The geometries of a human subject and the robotic jaw are different. As

---

* This chapter is based on work conducted by Jonathan Torrance, PhD student at the School of Engineering and Advanced Technology, Massey University, New Zealand.

W. Xu and J.E. Bronlund: Mastication Robots, SCI 290, pp. 179–206.
springerlink.com

such the trajectory of the jaw with 6DOF can be faithfully reproduced at only one location on the robot. From the point of view of the magnitude and direction of the forces applied to the food it makes more sense to focus on the first molar on the active side rather than the more commonly reported incisor point. This consideration casts some doubt on some of the analysis of lateral and vertical amplitudes reported in the mastication physiology literature where magnitudes are compared between subjects. Similar jaw geometric differences will exist between the different subjects used in those studies. Comparison of molar trajectories would be a more fundamentally sound basis for analysis.

For these reasons, the trajectories recorded on the human subject were translated from the incisor point to the active molar using the techniques described in Chapter 6. The coordinate system from the recorded trajectory and the robot were then aligned, providing a specified molar trajectory on the robot with the full 6 degrees of freedom (DOF) (x, y, z positions and roll, pitch, yaw angles). In this way the accurate reproduction of the molar and the functionality of the working surfaces of the teeth are achieved. Actuator positions could then be calculated using the inverse kinematics of the robot (Chapter 4.3). By reproducing these human trajectories with the mastication robot, it is then possible to measure information on the resistance forces and deformation of food samples in real time and in three dimensions. Further details of this procedure are given in [1].

The true trajectory of the molars (and hence deformation of the food) will be potentially altered by passive compliance in the robotic design (i.e. backlash) or by the active compliance instigated by the control strategy used. As explained in Chapter 4.9.3, torque limits were set on each actuator, meaning that if the resistance to mandible movement is too high, the position of this actuator will be held (while the others continue to move) until it becomes able to move again within the torque limits. In this way the molar movement and food deformation could vary from the original set trajectory. It should be noted that this partial transfer from motion control to force control allows the fine detail of the molar trajectory during the occlusal portion of the motion to be defined by the gliding movement caused as the active surfaces of the opposing molars passing over each other.

It is interesting to note that the combined motion-compliance control strategy used in the robot is strikingly similar to how the mandible is though to be controlled in humans. Bosman et al. [2] described the broad approach to mastication control in humans. During the portions of movement where there is no contact with the food, motion is controlled through displacement sensor feedback. The vertical and lateral amplitudes of this part of the trajectory are highly variable and are probably related to keeping food in the correct location, evaluating and/or coping with stickiness of the food to the teeth, the size of the food bolus and its textural properties. Despite the widely varying opening and closing trajectories, the trajectory during occlusion is prescribed by the teeth shape. Accurate motion based control during this phase is not readily achievable and so as the teeth meet resistance from the food, force control is applied. It seems that enough force is applied to try and achieve the gliding path prescribed by the molar shape, but not so much as to cause damage to the dental surfaces.

To provide information on the magnitude and direction of the forces applied to the food during chewing, some method of estimating the forces must be employed. This chapter outlines how force measurement was achieved before outlining experiments where the mastication robot was applied to the chewing of real foods. To give some validation of the robot function and the method to estimate the forces, a simplified one dimensional crushing movement with a flat surface (rather than teeth) was used. This trajectory allowed direct comparison of the resulting forces with uni-axial compression tests carried out on a standard texture analyser used by food scientists. Following this, comparison of the forces measured by the 6-bar linkage robot (Chapter 5) was carried out by simulating its two dimensional trajectory in the 6DOF parallel robot. Finally experiments using a real three dimensional trajectory were carried out. Although for this scenario there was no data to compare the chewing robot results with, the simulations demonstrated how the magnitude and directions of the forces on the teeth changed when a real jaw trajectory was used.

7.2 Force Measurement

Rather than using load cells or other sensors to directly monitor force on the food in this work, the measured torques recorded for each of the robots six actuators were used. The Galil DMC-1860 motion control card together with Galil supplied software was used to record the encoder positions and motor torque data. The Galil DMC-1860 can record the control signal sent to each actuator at any time instant. When operating in current source mode, this control signal is directly proportional to the current sent to the actuator, with the factor of proportionality being the amplifier gain set by jumper AG1 on the AMP-19540. With the torque constant of the DC motors also known, the actuator torque can be found dynamically during motion.

The approach used was based on a report by Simpson et al. [3] where a SCARA robot with high gear reduction of 80:1 similar to the our 6RSS chewing robot, or Massey Robotic Jaw, MRJ for short, was used to measure a small force of 3.5N at the end-effector from DC motor currents. The method involves taking consideration for inertia, friction, and position dependent force components. By running the trajectory first without food, all three of these components can be measured and removed from the food runs provided accurate position control throughout the movement is maintained.

One factor that could compromise accurate position control is the occurrence of saturated control signals as the total torque required to both accelerate the jaw and overcome forces due to food exceeds torque limits. For this reason it is an advantage to keep acceleration in joint space to a minimum. However it can be seen in the 6-dof experiments run at full speed that some acceleration can be tolerated, food forces are not extreme and only small positional errors are created allowing accurate force calculations to be obtained.

From the motor torque, forces can be calculated and projected through the 6-RSS linkages. The six force vectors can then be summed giving three force components and three torque components on the active molar expressed in the final

Skull frame. By using transformations the calculated force can be expressed in the Molar frame giving an insight into forces on the robots molar during chewing.

The main problem with this approach is that this calculated force is not the force due to the food alone, but rather the combined forces involved in producing the motion and involved in compressing the food. To estimate the force due to compressing the food alone, several assumptions were made;

1. The actuator motion during the closing part of the occlusion phase could be estimated as linear. This means that actuator acceleration during this phase was assumed to be zero.
2. The jaw link motion during closing occlusion phase could be estimated as linear.
3. All actuators remained in control, where force due to food did not exceed torque limits set on any of the six actuators.
4. Frictional and position dependent force components are unchanged with changing load.

While strictly speaking the first two assumptions are mutually exclusive, due to the kinematic structure of the robot, both are approximately true when the accelerations are of small magnitude. The third requirement needs to be checked after every run as the forces required to achieve the set trajectory are not known till after an experiment is completed. Data logging of position errors together with control signals for all six actuators during the motion ensures that the control signals do not enter the limiting state where position errors will accumulate and commanded motion differs from actual motion. Matlab code was written to search the data for any excessive positional errors or limited control signals and produce a pass or no-pass result. Error limits were selected from a sensitivity analysis of the resultant force calculations on the accuracy of the jaw position data.

By commanding motion at a suitably slow rate, acceleration can be minimised. These assumptions were justified for all three chewing trajectories presented here. With both jaw and individual crank motion approximated as being constant, the only forces to be overcome are those of friction and the force due to compression of the food.

The forces required to overcome friction during this closing occlusion phase were estimated by performing the motion without any food present. The final force due to food was then estimated by removing the forces required to perform free motion running of the trajectory from the forces calculated from the motion performed with the food. Based on these assumptions the force due to food can be calculated from the actuator torques.

The raw data collected by the DMC-1860 consists of 12 fields, six being the control signals for the actuators, and six being the positional errors for the six actuators, all recorded as a function of time. Testing of assumption 3 above was first carried out to confirm no actuators were run in a torque limited state. The forces on the food were then calculated through the linkages shown in Fig. 7.1.

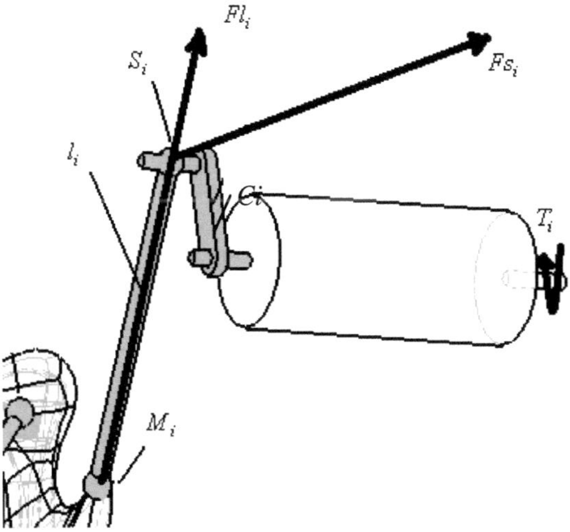

Fig. 7.1 Torques and forces of a representative actuator 'i'

First the control signal is converted into a current value based on amplifier gain of 0.7 A/V.

$$Ci = 0.7Vi \tag{7.1}$$

where Ci is the current through actuator i, and Vi is the control signal in volts for actuator i. The actuator torque T_i is then calculated using the DC torque constants of the actuators.

$$T_i = 25.9C_i \times 66 \tag{7.2}$$

where the torque constants for the motors were 25.9mNm/A, and the gearing factor magnified the torque by the ratio 66:1. From this, the actuator torque through the axis the actuators lie on is known, and could be converted to a force at points S_i where the actuator cranks are joined through spherical joints to links L_i.

$$Fs_i = \frac{T_i}{G_i C_i} \tag{7.3}$$

in which Fs_i is the force at point S_i, and $G_i C_i$ is the length of crank C_i.

This force at points S_i is directed at 90 degrees to the crank link C_i, and the force though links L_i connecting the crank link to the mandible can be found by projecting this force onto the link L_i at a given time instant. It is noted that because of the spherical joints on either end of links L_i, no torques can be transmitted through the link and any force transmitted must be in the direction of the link at that time instant.

$$Fl_i = \frac{Fs_i \bullet {}^{M}M_iS_i}{\left\| {}^{M}M_iS_i \right\|} \tag{7.4}$$

where Fl_i is the force transmitted through link i, ${}^{M}M_iS_i$ is the vector from point S_i to point M_i measured in the mandible reference frame. The six forces could then be summed to give the resultant force acting on the mandible.

$$Fm = \sum_{i=1}^{6} Fl_i \tag{7.5}$$

where Fm is the force at the mandible.

7.3 Experiments for One-DOF Chewing Trajectory

Initial experiments were carried out to verify the force calculation method and its associated assumptions, by comparing results with uni-axial compression tests made on a TAXT2 texture analyser [4]. The uni-axial test is a simple linear crushing motion with a flat probe and platform. The force-distance profile is recorded in one dimension using a load cell. The machine allows the cross head speed to be set, with two force thresholds used to set trigger recording at the start and end points of the motion. Compression was run at 1mm/second, and a force threshold of 0.1N was selected to trigger recording. Although the motion and teeth shape are not representative of the motion occurring during chewing, this method has been widely used to describe food texture [4].

Three samples were tested. A model food consisting of cast silicon rubber was used as this had very well defined properties. The material is completely elastic so the resisting force would increase on compression and it would relax to its original shape after it is released. Two real foods were used including a muesli bar, (a collection of whole and flaked cereals bound together with a sticky sugar or honey based syrup) and uncooked dry noodles. The muesli bar was expected to deform on teeth-food contact after which individual particles can become trapped between the occlusal surfaces of the teeth and fracture or cutting of these components may occur. Dry noodles are an example of a hard brittle food that will fracture under stress.

The vertical 1DOF movement of the Texture Analyser was reproduced on the chewing robot so that the force applied on the texture analyser was aligned with the Skull frames z-axis. Consequently there was no motion in the x- and y-axes of the Skull frame, or rotation about any of the three Skull axes. The velocity was set to be constant at 1mm/s in the positive direction of the Skull frames z-axis. There was no acceleration specified during the movement. The displacement and velocity-time profiles of the 1DOF trajectory are shown in Figs. 7.2 and 7.3 below.

Before the force measurement could be replicated on the mastication robot, some modifications were required. The TATX2 uses two flat aluminium surfaces

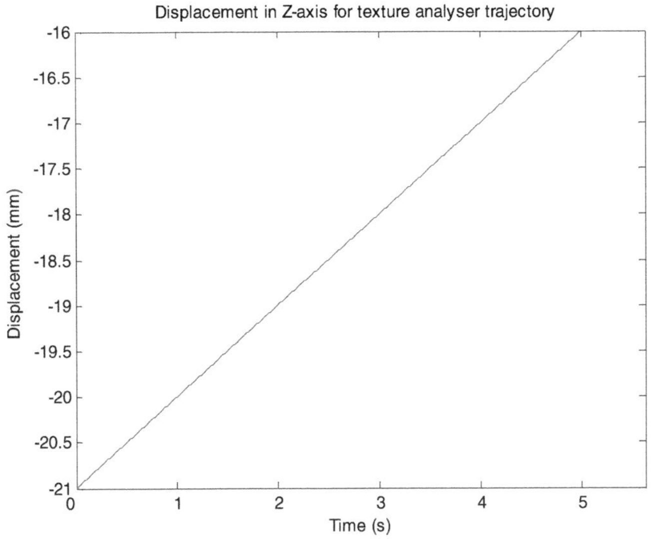

Fig. 7.2 z-axis displacement of the lower jaw in the Skull reference frame for the 1DOF trajectory

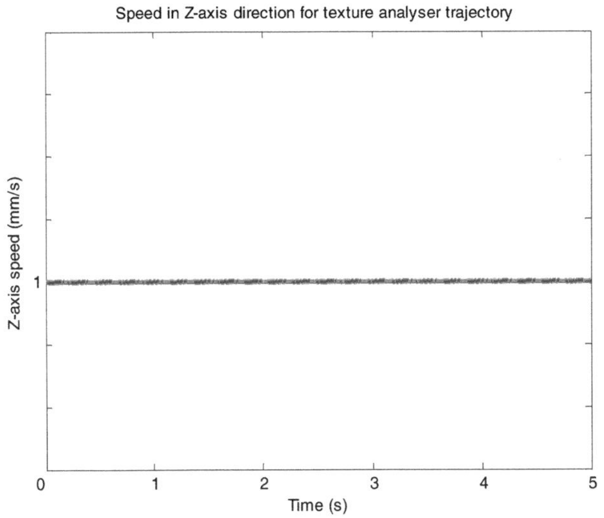

Fig. 7.3 Velocity of the lower jaw plate in the Skull reference frame.

to compress the food sample between. To test comparable geometries, the robots lower set of teeth were removed leaving the flat machined aluminium mounting plate which is comparable to the base used on the texture analyser. The upper set of teeth was replaced with a machined aluminium block (much larger in area than the food sample), with flat surfaces comparable to the upper probe used on the texture analyser.

The texture analyser motion was then programmed to run on the masticatory robot and the forces were recorded as the food was compressed. It should be noted that actuations are required for all six motors to achieve this linear trajectory. Comparisons of the vertical forces are shown for the silicon rubber model food system in Fig. 7.4.

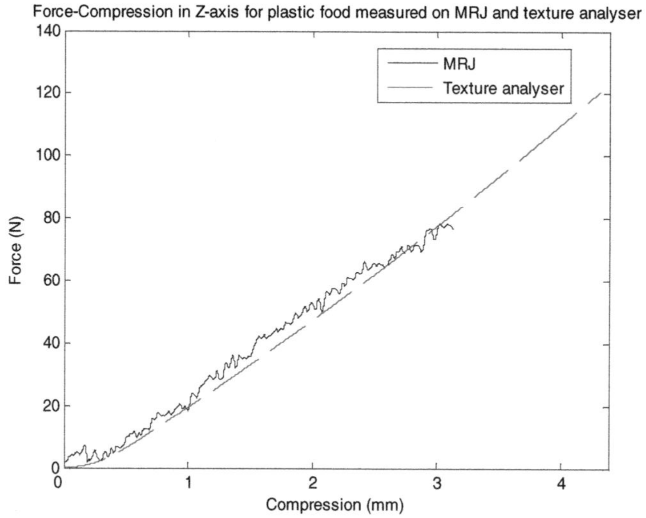

Fig. 7.4 Typical force profile on the 20mm plastic food sample as measured by the chewing robot (MRJ) and the texture analyser for a 1DOF motion

It can be clearly seen in Fig. 7.4 and Table 7.1 that there is very good agreement in the shape and magnitude of the force-deformation curves for the silicon rubber material, suggesting that the force in the chewing robot is being calculated correctly.

Table 7.1 Comparison between the chewing robot and the texture analyser force measurements in the silicon rubber model food system

Plastic food force measured at 3mm compression (N)	
Chewing robot	Texture Analyser
Mean =77.875	Mean=78.26
S.d = 1.4549 (n=4)	S.d = 0.873 (n=3)

'S.d' = standard deviation on 'n' replicate measurements

The force-distance curve measured for muesli bars (Fig. 7.5) shows more complex mechanical behaviour than for silicon rubber. Although the curve is approximately linear, curvature is observed due to the multiple components present in muesli bars which possess different mechanical properties. There is again very good agreement between the uni-axial compression test and the chewing robot. The mean results after 4mm of compression are summarised in Table 7.2.

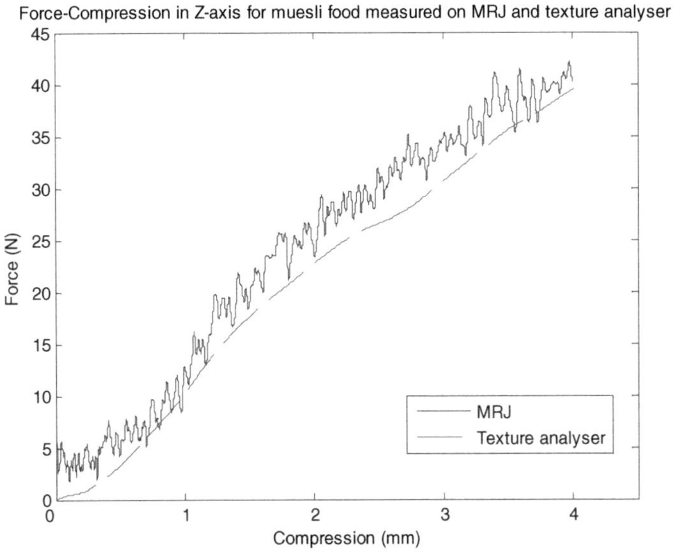

Fig. 7.5 Typical force profile as measured by the chewing robot and texture analyser for 1DOF motion on a muesli bar food sample

Table 7.2 Comparison between the chewing robot and the texture analyser force measurements in the muesli bar food system

Muesli bar food force measured at 4mm compression (N)	
Chewing robot	Texture Analyser
Mean =34.0	Mean=38.5
S.d = 4.98	S.d = 5.31
(n=4)	(n=5)

Uncooked noodles are a hard brittle material which fractures on compression rather than deforming. This can be seen in Fig. 7.6 below where for both the texture analyser and the chewing robot, multiple fracture events are observed over the first 1.5mm. When a fracture event occurs, the resistance of the force is momentarily relaxed, before building again. A peak force is observed of 60-70N after

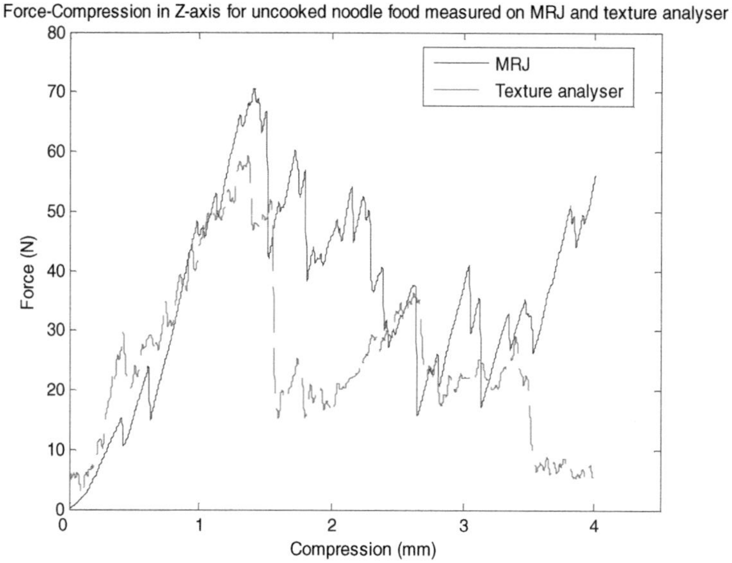

Fig. 7.6 Typical force profile as measured by the chewing robot and texture analyser for 1DOF motion on an uncooked noodle food sample

Table 7.3 Comparison between the chewing robot and the texture analyser force measurements in the uncooked noodle food system.

Uncooked noodle food force measured at max breaking point (N)	
Chewing robot	Texture Analyser
Mean =65.9	Mean=53.8
S.d = 8.00	S.d = 14.7
(n=3)	(n=4)

which a large fracture failure is observed. The force then builds again because by this stage much of the open airspace between the noodles has been filled with broken particles and these fragments begin to be compressed.

Both methods demonstrate this fracture behaviour exhibiting similar peak forces, initial fracturability (force at the first failure event), and strain at failure (product deformation at the point where peak force is observed). The magnitude of the reproducibility within each test method was similar to the differences observed between the chewing robot and the texture analyser. This is shown in Table 7.3.

Overall the results show the chewing robot can function well as a 1DOF texture analyser machine and the force calculation method is valid. These forces applied on the food were contributed to by all of the actuators involved in the robotic movement.

7.4 Experiments with Two-DOF Chewing Trajectory

To validate the chewing robot against more complex trajectories, it was compared against force profiles collected using the 6-bar linkage chewing robot discussed in chapter 5. The 2D linkage robot achieves a proscribed two dimensional molar trajectory with anatomically correct dentition. The linkage chewing machine incorporates a 3D force sensor integrated under the lower jaw which allows the resultant forces on the food to be recorded as foods undergo mastication. The motion was run at approximately 2mm/s in the z-axis direction on both the linkage robot and the 6DOF parallel chewing robot, to ensure any possible influence due to robot dynamics was minimised.

The trajectory of the 2D machine reproduced in the 6DOF chewing robot. The original data recorded using the 2D robot encoder consisted of 370 points for the full cycle. Fig. 7.7 below shows the actual trajectory actuated in the 2D chewing robot. Note that occlusion of the molar teeth occur at z-axis displacements greater than 0. Negative values represent the opening and closing phases of the motion.

Because force measurements are of interest during the occlusion phase, the closing position and angle were matched with the 6DOF chewing robot. Exact reproduction of the non-occlusal phase was not required and so the non-occlusion phase was truncated to allow the trajectory to fit within the 6DOF chewing robots smaller workspace. As this part of the trajectory is not where the food is compressed, it would not affect the forces recorded. The roll pitch and yaw angles of the mandible were all held constant at 0 degrees. Fig. 7.8 shows the full trajectory applied to the 6DOF chewing robot. Only the occlusal phase of the trajectory in Fig. 7.8 involved food compression.

Figure 7.9 shows the implemented trajectory during which forces were measured. Inspection of the velocity and acceleration of the trajectory was important as it is possible that any large acceleration during the movement could lead to errors in the force calculation. Figs. 7.10 and 7.11 show the velocity and acceleration profiles of the mandible movement.

It can be seen that there are a number of sharp inflection points in both the velocity and acceleration plots due to the discrete nature of the data used. Motion was implemented with linear interpolation between points. An attempt was made to smooth the motion by using an Akima spline in Solidworks and taking more points for the trajectory but an improvement was not seen. Despite the spikes, the magnitude of acceleration is small and unlikely to cause problems when calculating forces.

A typical force profile from running the 2D trajectory on both the 6DOF chewing robot and the 2D chewing machine is shown in Fig. 7.12 for the 20mm silicon rubber model food system.

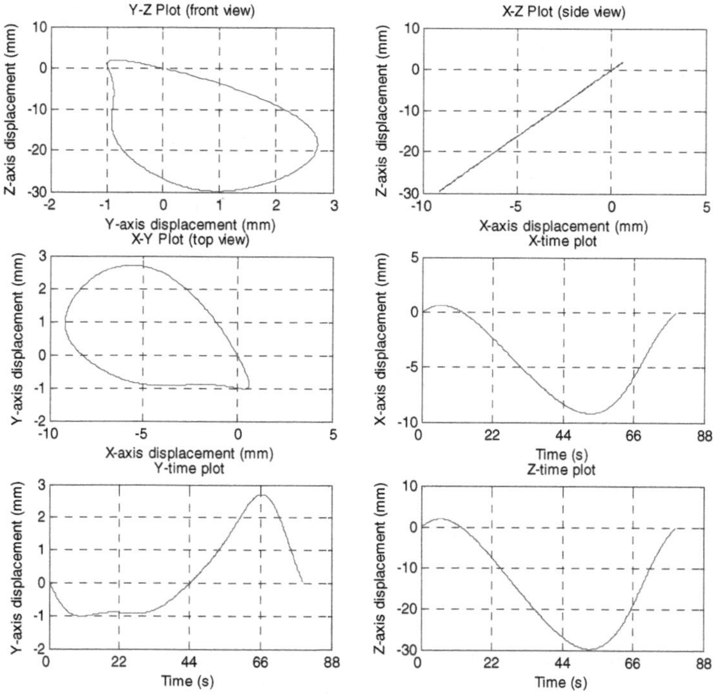

Fig. 7.7 Trajectory implemented on 2D linkage chewing robot during experimental trials

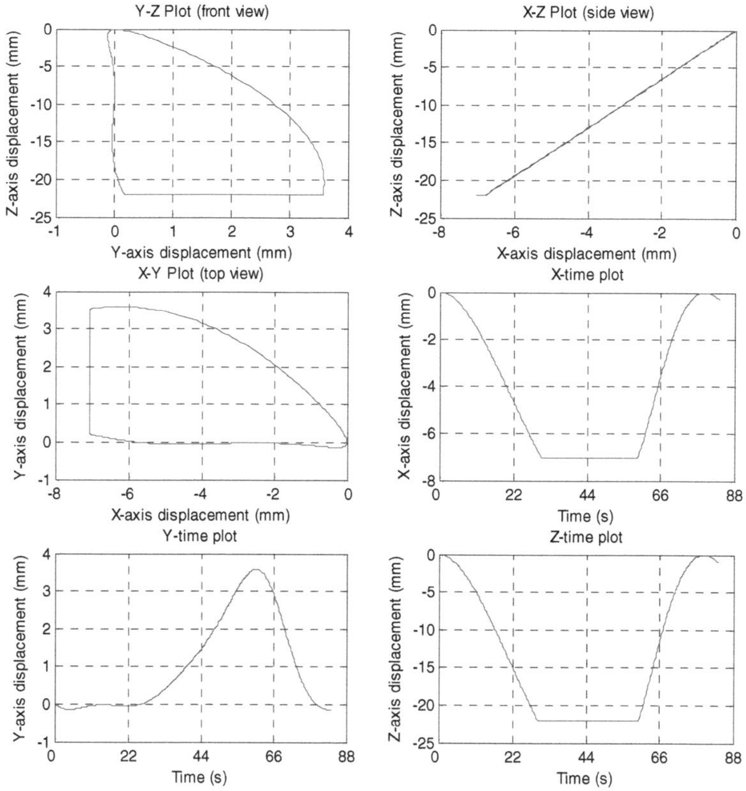

Fig. 7.8 The 6DOF chewing robot trajectory implemented to approximate the 2D chewing robot motion

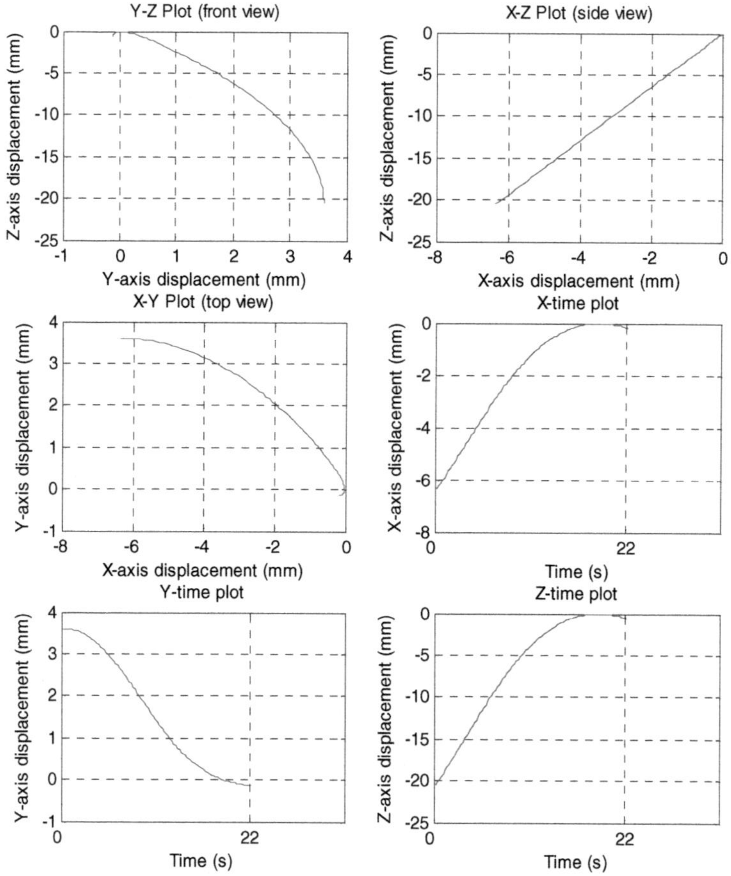

Fig. 7.9 Occluding phase of the 2D trajectory as implemented on the 6DOF chewing robot

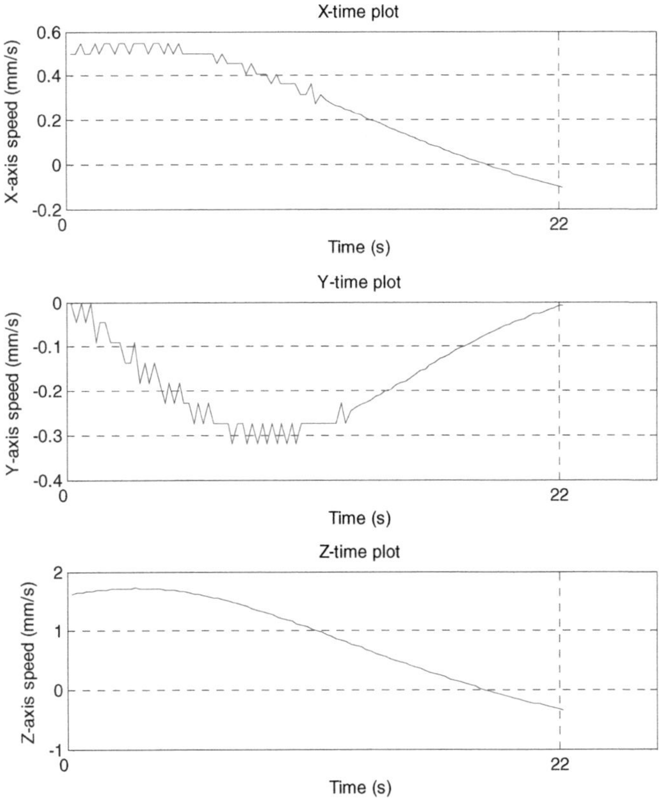

Fig. 7.10 Mandible (molar) velocity implemented in the chewing robot during reproduction of the 2D linkage robot trajectory

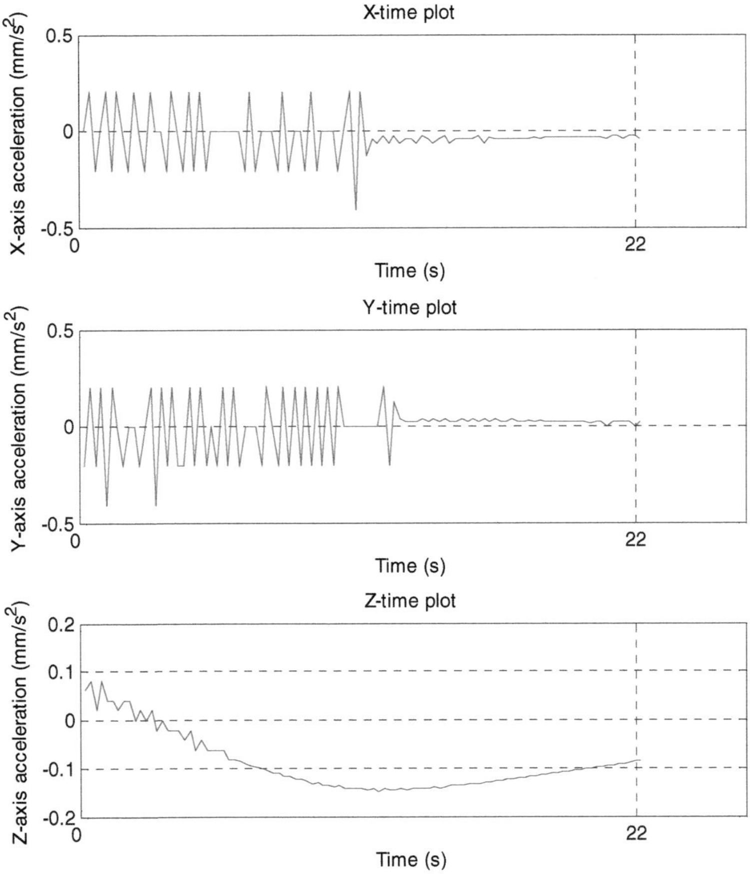

Fig. 7.11 Mandible (molar) acceleration implemented in chewing robot during reproduction of 2D linkage robot trajectory

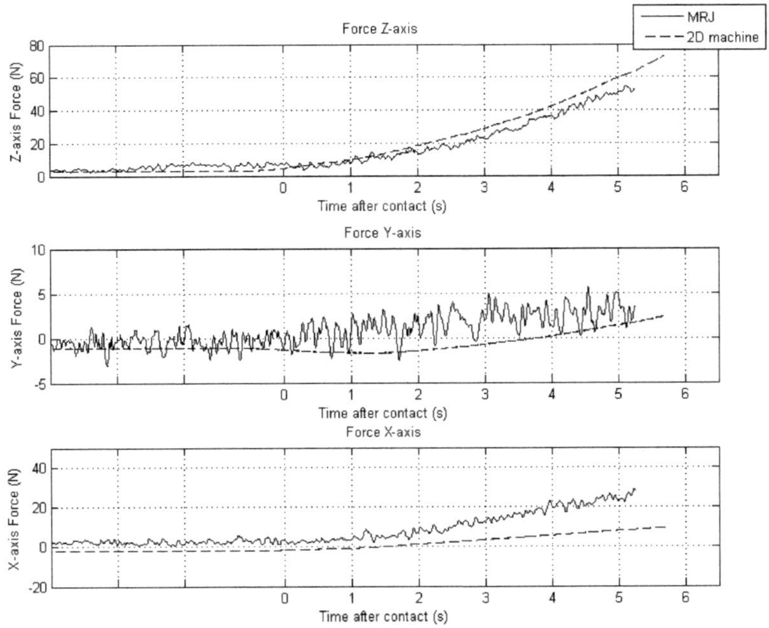

Fig. 7.12 Forces measured during chewing of a 20mm silicon rubber model food using 6DOF chewing robot and the 2D chewing machine

All runs showed similar results, with the 6DOF chewing robot providing slightly less z-axis force, and slightly more x-axis force. Table 7.4 below shows the mean forces measured after 5 seconds for several experiments.

Table 7.4 Comparison of mean forces in each direction from replicate chewing experiments on silicon rubber food models

Silicon rubber model food force measured at 5 seconds.					
x-axis force (N)		y-axis force (N)		z-axis force (N)	
Chewing robot	2D machine	Chewing robot	2D machine	Chewing robot	2D machine
Mean=25.95	Mean=7.56	Mean=4.29	Mean=1.69	Mean=52.48	Mean=58.03
S.d=1.59	S.d=0.78	S.d=0.30	S.d=0.44	S.d=1.53	S.d=1.38
(n=5)	(n=3)	(n=5)	(n=3)	(n=5)	(n=3)

It can be seen that there are significant differences in the forces measured in the x-axis direction, with the 6DOF chewing robot measuring more force. It is unlikely that such large forces would be applied to the food in the x- direction (anterior-posterior) due to the very small mandible movement that occurs in this direction during the food-teeth contact portion of the experiment. It is thought that under loaded conditions (i.e. when the teeth come into contact with the food) the loss of passive compliance combined with the geometric arrangement of the robot causes the propagation of reaction forces in some of the jaw joints. Similar behaviour was observed in the one dimensional experimental trials (data not shown) where significant x-direction forces were observed although the only x-dimensional movement was that caused by mechanical compliance. Further work is required to resolve this issue before the device could be used to characterise the forces applied to food samples. Direct measurement of the forces on the molar teeth in 3D would add some clarification of what is occurring during the loaded operation of the robot. Despite these differences, there is very good agreement in the trend and magnitude of the forces applied on the food.

Fig. 7.13 shows similar experimental results during chewing of biscuits. As expected for brittle food systems such as biscuits, the results show multiple fracture events occurring during deformation. As observed for the silicon rubber food model above, significant forces were applied in the x-direction. The similarity in shape between the z- and x- direction forces reinforces the explanation discussed above. The magnitude of force in the x-direction is highly correlated to the force applied in the z-direction and it may be that as more vertical load is applied, the internal resistance of the robot linkages increase in the x-direction.

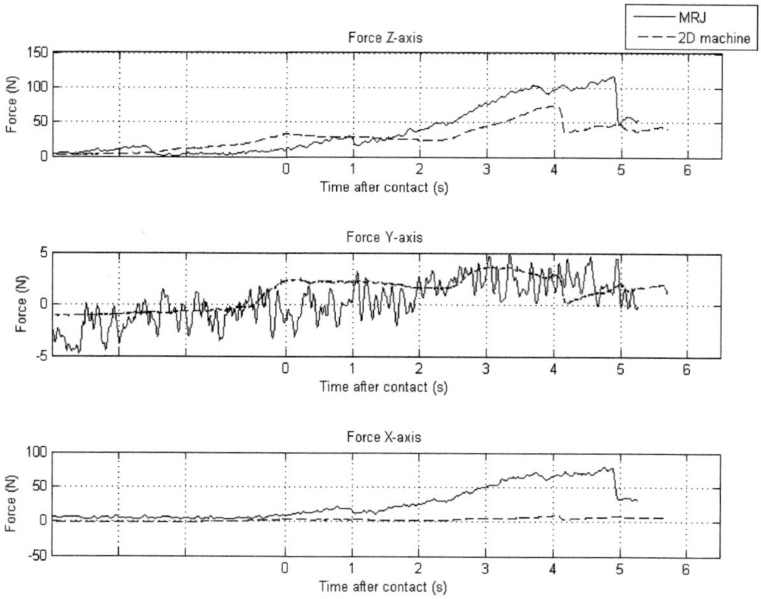

Fig. 7.13 Forces measured during chewing of a biscuit using the 6DOF chewing robot and the 2D chewing machine

Table 7.5 compares the average differences in force in each direction for several replicate experiments. It can be seen that for both machines there is a relatively large spread of forces for the different trials, especially in the z-axis. This is typical of a brittle food where multiple fracture events occur during compressive strain.

Similar experiments were carried out for muesli bar food samples (Fig. 7.14). This experiment showed excellent agreement in the z-dimension between the 6DOF chewing robot and the 2D chewing machine.

Table 7.5 Comparison of mean forces in each direction from replicate chewing experiments on biscuit food samples.

Silicon rubber model food force measured at 5 seconds.					
x-axis force (N)		y-axis force (N)		z-axis force (N)	
Chewing robot	2D machine	Chewing robot	2D machine	Chewing robot	2D machine
Mean=25.95	Mean=7.56	Mean=4.29	Mean=1.69	Mean=52.48	Mean=58.03
S.d=1.59	S.d=0.78	S.d=0.30	S.d=0.44	S.d=1.53	S.d=1.38
(n=5)	(n=3)	(n=5)	(n=3)	(n=5)	(n=3)

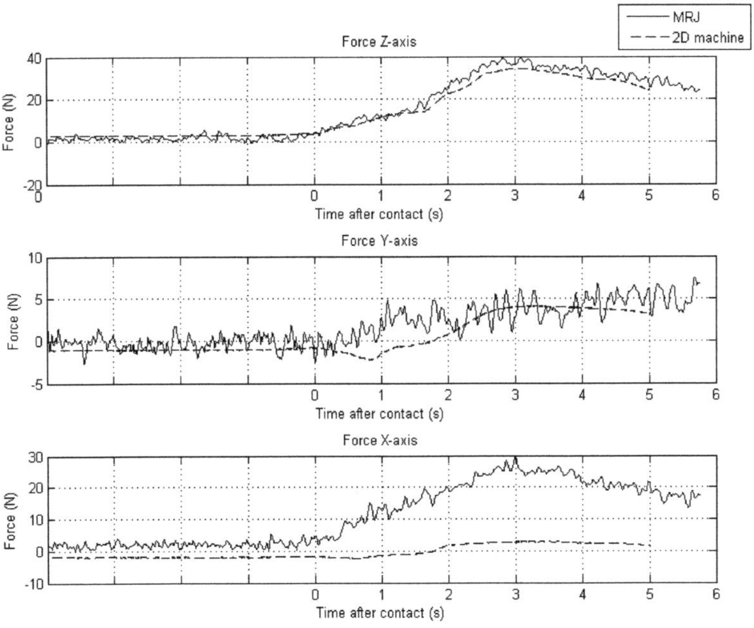

Fig. 7.14 Forces measured during chewing of a muesli bar using the 6DOF chewing robot and the 2D chewing machine

The resistance force due to the shearing lateral movement of the trajectory is evident in the recorded force measurements in the y-axis. There is good agreement in the magnitude of this component of the forces applied to the foods. The problem with measured forces in the x-direction, as discussed above was again present during the experiments with muesli bars. Table 7.6 below shows that there was good repeatability in the measured forces in the y- and z-axes and good agreement between the two chewing machines.

Table 7.6 Comparison of mean forces in each direction from replicate chewing experiments on muesli bar food samples

Muesli Bar force measured at 4 seconds					
x-axis force (N)		y-axis force (N)		z-axis force (N)	
Chewing robot	2D machine	Chewing robot	2D machine	Chewing robot	2D machine
Mean=24.04	Mean=4.31	Mean=3.48	Mean=2.30	Mean=37.31	Mean=29.81
S.d=12.36	S.d=2.11	S.d=1.39	S.d=1.46	S.d=15.10	S.d=8.07
(n=4)	(n=4)	(n=4)	(n=4)	(n=4)	(n=4)

7.5 Experiments with Six-DOF Chewing Trajectory

A jaw trajectory recorded from a human subject using the AG500 articulograph (see Chapter 6) was run on the 6DOF chewing robot with various food samples. The resulting forces were then analysed, both at the chewing surface and for the six actuators. This was carried out as the robot was designed from biological principles and the six actuators mimic the work of the biological muscles of the human subject. The full trajectory implemented on the robot is shown in Fig. 7.15. The plot showing z-axis displacement versus time shows the dynamics of the chewing process. As the number of chewing cycles increase, the z-axis displacement at the top of the chewing motion reduces. This is likely to reflect the breakdown of the food properties so that the molars come closer together due to the reduced resistance imparted by the foods. Close to full occlusion occurs after approximately 10 chewing cycles.

The displacement plot for the y-axis shows that lateral movement during the opening phase of each chewing cycle was most often to the right. After six cycles, a short series of cycles were directed to the left. This could have been done to maintain the position or manipulate the orientation of the food on the occlusal table between cycles. Alternatively it could signal transfer of the bolus to the other side of the dental arch for a few chewing cycles.

The x-axis displacement also changed throughout the chewing process. During the initial few chewing strokes, the amplitude of the anterior-posterior movements was relatively large. The amount of forward jaw movement decreased as the food breaks down and the trajectory of the teeth during the teeth-food contact period moves closer to 0 mm z-axis displacement. During this later stage where much

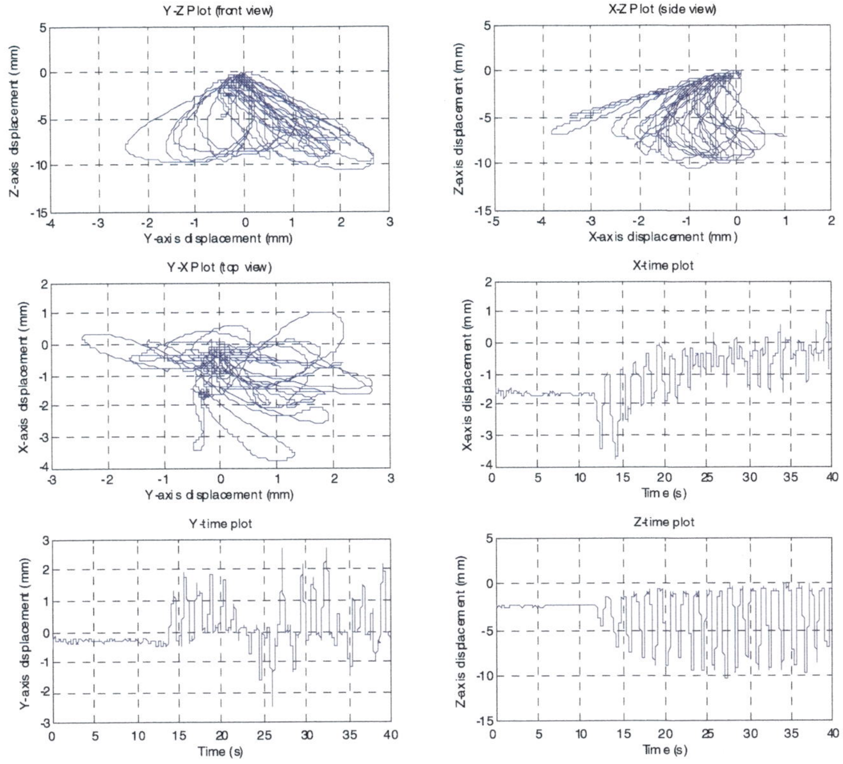

Fig. 7.15 3D human trajectory converted for implementation on the chewing robot. Note that in the first 14 seconds the jaw position was held constant to provide a reference point for the orientation of the jaw.

smaller forward movements occur, it can be seen that the x-axis displacement approaches 0 mm, showing that the molar teeth are correctly aligned for occlusion.

As this is a real chewing trajectory it has 6 degrees of freedom. The roll, pitch and yaw movements are shown in Fig 7.16. The tilting of the jaw forwards as the jaw is opened can be seen in the pitch-time plot. This is the largest rotational movement and the maximum angle (at maximum z-axis displacement) is relatively uniform between cycles. The roll and yaw angles follow a consistent pattern but do exhibit a transfer from positive to negative directions periodically during the mastication duration. These observations correspond to the cycles where measurements of negative x- and y-axis displacements were recorded, indicating that it is most likely that the active chewing side was switched for these few cycles.

For the experiments run on food samples, only a single chewing cycle was implemented. The occlusion movement starting at approximately 20 seconds into the trajectory was selected as it was one of the wider opening movements so larger food samples could be tested. Fig. 7.17 shows the single closing chewing trajectory run on the foods. Looking from the top view (x-y plane) it can be seen that the molars approach the upper teeth from the right and rear of intercuspal occlusion and move slightly beyond the intercuspal position in the y-dimension. Movement in the z-axis does not reach occlusion, suggesting that during the particular bite, for which the measurements were recorded, the food did not fully deform and teeth to teeth contact could not occur. This explains why the maximum x-axis position was about 1mm behind the intercuspal location. If the jaw continued on to reach 0mm in the z-axis, the 1mm or so of forward movement (x-axis) of the jaw to reach full occlusion would occur.

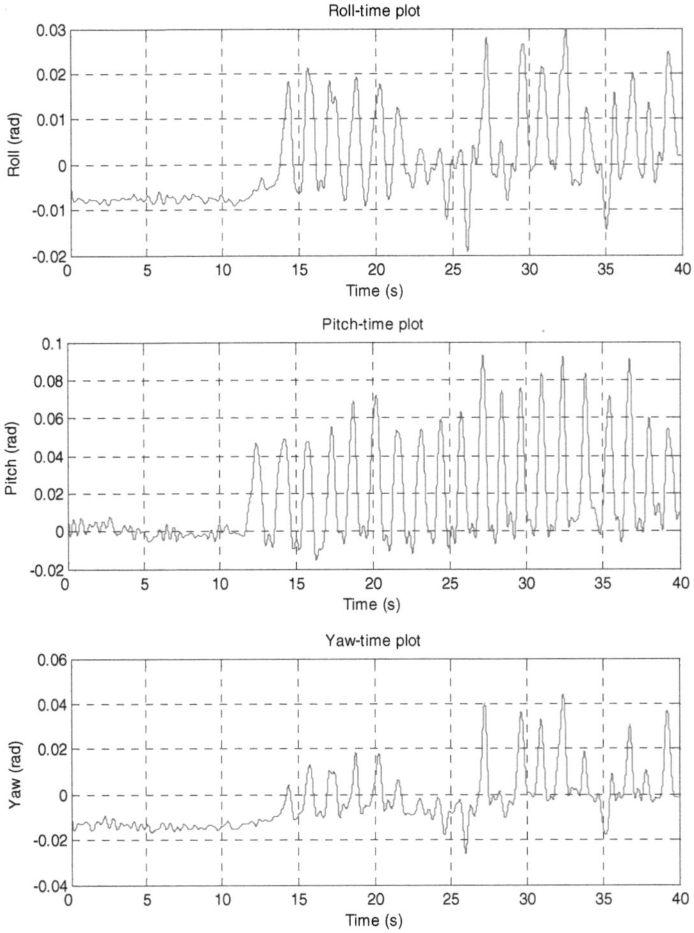

Fig. 7.16 Jaw angle profile for the 3D human jaw trajectory implemented on the chewing robot

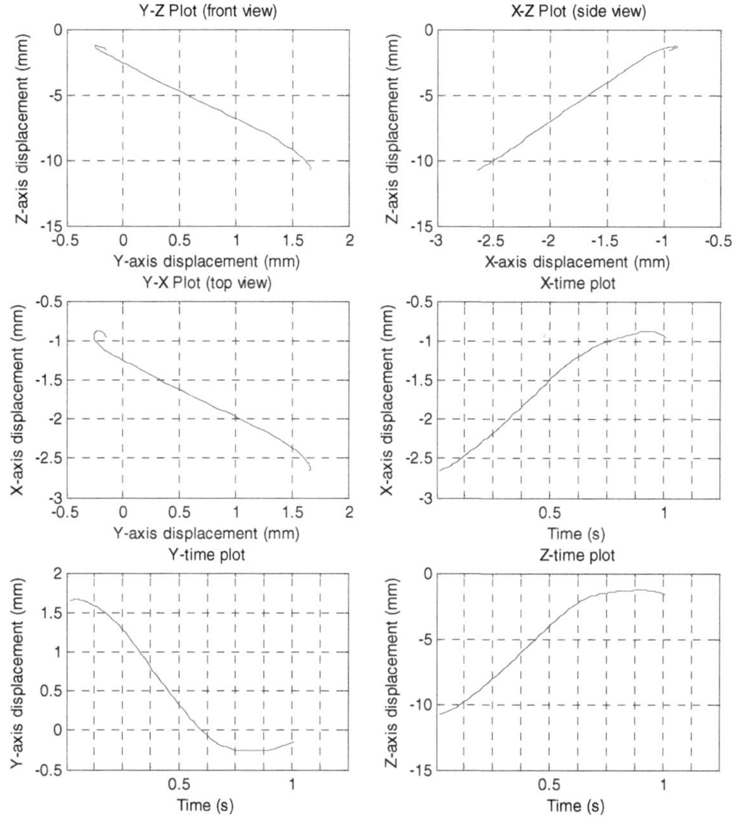

Fig. 7.17 A single chewing cycle implemented on the chewing robot to simulate 3D human chewing movements

The changes in mandible rotation used to simulate the recorded human masticatory cycle during the occlusal phase are summarised in Fig. 7.18. The pitch (rotation about the y-axis) is the motion of the jaw closing, and approached an approximately flat angle at the point where the teeth contact each other. As expected, this angle is the largest of the three rotational dimensions of the jaw. The roll (rotation about the x-axis), represents the angle that the teeth twist in the lateral direction. The teeth become aligned as they approach each other. The yaw (rotation about the z-axis) can be interpreted as the chin pointing left or right and similarly rotates to be back in line with the upper jaw by the time of tooth contact. It can be seen that in the trajectory there is some rotational movement occurring on the teeth before full occlusion occurs, and this will apply more complex stress patterns on the food than that which occurred during the one and two dimensional trajectories implemented in the earlier experiments.

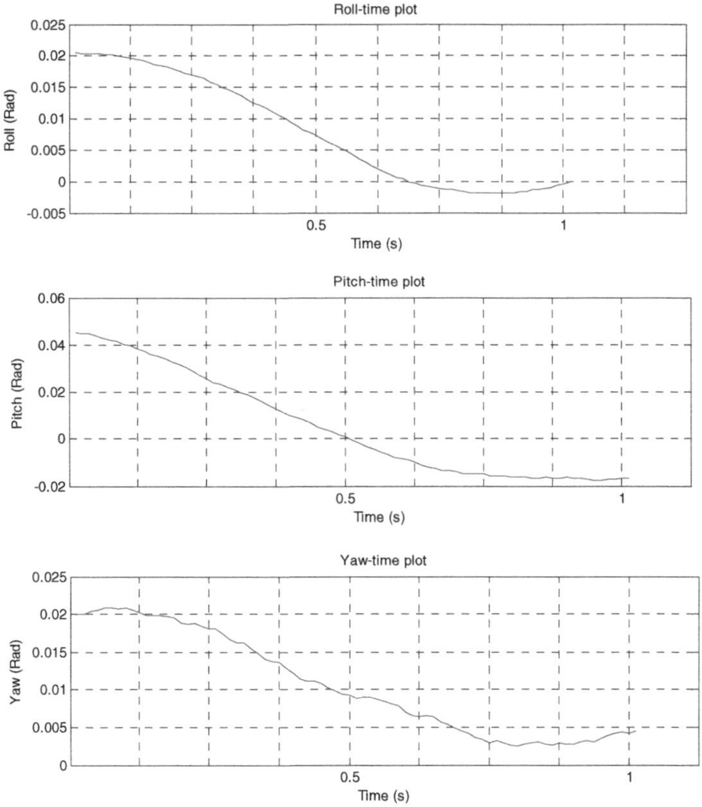

Fig. 7.18 Jaw rotation angle – time profiles during the implemented simulated human chewing cycle

Fig. 7.19 shows the velocity profile implemented during this phase of the trajectory. It can be seen that in each direction, the jaw velocity increased to a maximum before reducing to zero at the point where opening was initiated.

Force results were recorded for three food samples using this section of the trajectory at full human speed. Biscuit, marshmallow and corn thins were used to demonstrate application of the chewing robot to real human mastication simulation. These foods were used as their dimensions were small enough to be able to place the foods between the upper and lower molars at the start of the implemented trajectory. Biscuits were used to represent a moderately hard, brittle food and to allow comparison with the results observed for this product using the 2D trajectory outlined above. The use of marshmallows tested operation of the robot with a soft partially elastic food that exhibited no fracture behaviour. Corn thins are an extruded maize-flour based product with a light porous structure. The material is 'crunchy' and undergoes fracture on compression but is tougher than the biscuit product.

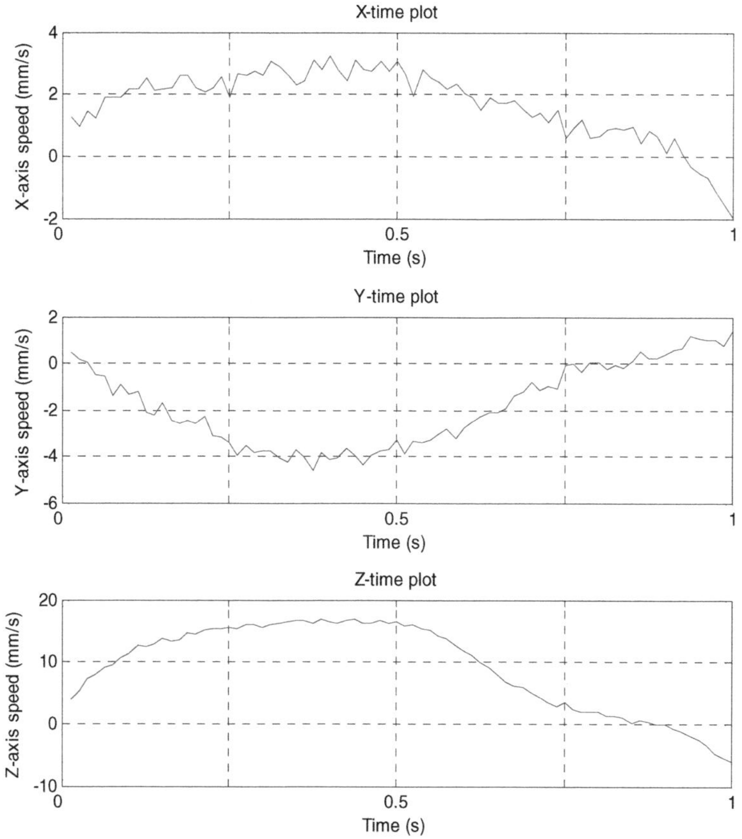

Fig. 7.19 Jaw velocity profile implemented in the single cycle human chewing simulation

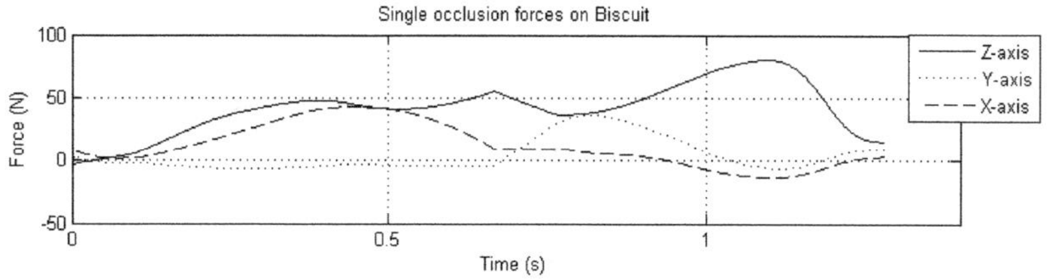

Fig. 7.20 Measured force profiles on the molars during simulated 3D human trajectory applied to biscuit

Fig. 7.20 shows the recorded force profiles applied during chewing of biscuits. The curve is relatively smooth in comparison with the jagged force profile observed for biscuits using the previous 2D experiments (Fig. 7.13). This was due to the much shorter time used for the human trajectory experiments. At faster jaw speed, fracture events occur in very quick succession and as a result, the force relaxation recorded after fracture at low speed will not be as evident. From the results in the z-direction, two larger scale force relaxations were evident during the compression phase of the cycle, indicating failure and that associated large scale structural changes did occur in the sample.

At the start of the cycle the x-axis data was again strongly correlated with the increase in force in the z-direction. The x-axis movement of the jaw is occurring at this time, so force in this direction might be expected. As discussed above however, this could indicate the propagation of some internal forces in the linkages of the robot. It is interesting however that after about 0.5 seconds, this force drops away to almost zero. At the same point there is an increase in y-axis force. A detailed analysis of the forces within the robot is required to resolve whether or not these measured forces are actually being applied onto the food or due to internal friction in the mechanism. This must be carried out before a full analysis of the data in terms of food resistances can be made. The direct measurement of the forces applied onto the teeth may be more appropriate than calculating them from the motor torques.

Fig. 7.21 shows the measured force profiles recorded during a similar experiment using marshmallow as the food system. Due to soft texture of the food, the measured magnitudes of the forces in each direction are low. This could reflect the trajectory used in this experiment where complete occlusion and hence complete deformation of the product was not made. This result highlights an important issue in operation of the robot. If human trajectories are simulated exactly, the forces measured are those that must be applied to achieve this specified degree of deformation on the food. An alternative strategy would be to set the robot to achieve full occlusal trajectory but implement some form of force control that allows incomplete deformation of the food if the resistance of the food is too high. Operation of this second approach using different torque thresholds on the robot could be carried out until the actual human trajectory is matched. This would provide information on the force limits used during human mastication for subjects and how these might vary for different food types or the trajectory used.

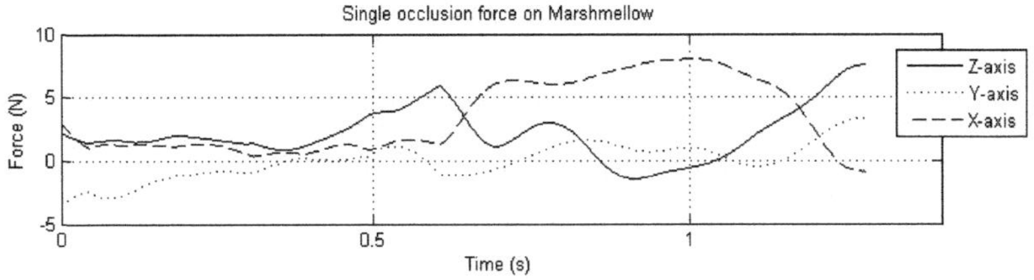

Fig. 7.21 Measured force profiles on the molars during a simulated 3D human trajectory applied to marshmallow

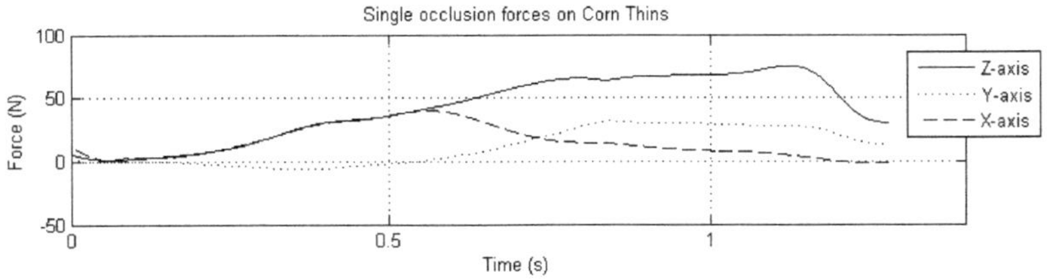

Fig. 7.22 Measured force profiles on the molars during simulated 3D human trajectory applied to corn thins

Fig. 7.22 shows the force profile measured using the chewing robot for the corn thin food product. A similar pattern in which x-axis force is present during the early stages of the motion which is later translated into y-axis direction forces. There is a steady increase in vertical forces until at just after 1 second the forces are reduced as the jaw begins the opening phase of the trajectory. The inflection points on the curve suggest fracture events occurring as observed for the biscuit product. The next stage of development in the robot must resolve whether internal friction forces within the robotic linkages are contributing to the measured x- and y-axis forces when the jaw is under load. Implementation of a miniature 3D force sensor such as used in the 6-bar linkage robot (Chapter 5) could provide direct measurements of the local forces applied on the food.

7.6 Summary

The application of the 6RSS parallel chewing robot to food chewing experiments was described. The force vector applied on the active molar was calculated from the measured torques applied on the six actuators using an analysis of forces through the linkage mechanism. A series of experiments were carried out using model and real food systems. Experiments where the robot jaw was made to follow a one dimensional trajectory showed very good agreement with measured force-deformation curves measured by uni-axial compression testing. As a further stage of validation of the robot, the molar trajectory of the 6-bar linkage chewing robot was reproduced. Good agreement was observed between the z- and y- direction force-profiles measured using both machines, but significant x-axis forces were recorded on the parallel mechanism robot. It is thought that these could be due to internal friction in the linkages under conditions where the jaw is loaded. Experiments were also carried out simulating 3D recorded human trajectories showing expected profiles in the z-direction but unexplained force profiles in the x- and y-dimensions. Further work interpreting these forces or alternatively the direct measurement of forces using a sensor is required. Once this has been completed, testing of various force control strategies can begin.

The work here shows promise for application of the robot to characterise food texture, however a number of future developments are required before simulation of complete chewing sequences can be carried out. One of the key features that must be added to achieve this is retention and manipulation of the food bolus between chewing cycles.

References

1. Torrance, J., et al.: Human Jaw Motion Measurement, Analysis, and Robotic Reproduction. In: Proceedings of the 15th International Conference on Mechatronics and Machine Vision in Practice, Auckland, New Zealand, December 2–4, pp. 286–293 (2008)
2. Bosman, F., et al.: Neuromuscular control mechanisms in human mastication. J. of Texture Stud. 35, 201–221 (2004)
3. Simpson, J.W.L., Cook, C.D., Li, Z.: Sensorless force estimation for robots with friction. In: Proceedings of the 2002 Australasian Conference on Robotics and Automation, Auckland, New Zealand, November 27-29, pp. 94–99 (2002)
4. Bourne, M.C.: Food Texture and Viscosity: Concept and Measurement. Academic Press, San Diego (2002)

Chapter 8
Understanding Food Texture Using Masticatory Robots

Abstract. One of the key applications of masticatory robots is for the characterisation of food texture. Traditional methods to evaluate food texture employ human panels using sensory science techniques, or instrumental measures. Each of these approaches present limitations. Sensory techniques are semi-quantitative and can be used to describe both initial food properties and the textural change during mastication, however subject variability, time and cost can be prohibitive. Instrumental methods are quantitative but are difficult to apply across a wide range of products and are not easily correlated with sensory texture perception. Because masticatory robots are designed to mimic human chewing, they can potentially provide information on the changing texture of foods during mastication and be used as a tool for developing understanding of the interactions between mastication behaviour, food structure and mechanical properties. This chapter summarises the key physical processes that occur to food during mastication in order to define the functionality that mastication robots must incorporate. The current mastication robots are then reviewed with respect to these requirements.

8.1 Introduction

Food scientists are engaged in research aimed at characterizing and manipulating food texture and chemistry in order to retain or enhance quality and preference for both fresh and processed foods. It is natural therefore, that there has been a great deal of effort made into developing tests to characterise texture of foods during mastication. The term 'texture' is used to describe how foods properties respond during the mastication process with respect to all mechanical and tactile (and where appropriate visual and aural) properties of the food [1].

Mastication is a complex process that occurs as the food is crushed and ground by teeth and brought into a condition which is safe to be swallowed. During the chewing process, food particles are positioned on the surface of teeth by the cheek and tongue. As mastication continues, food is ground into fine particles, mixed with saliva, and formed into a bolus. During this process, the structure and properties of the food change dynamically. The extent to which the food is deformed, the forces and the direction of these forces that the food imparts in resisting this deformation, and the rate the food is broken down during mastication are all important aspects of food texture.

W. Xu and J.E. Bronlund: Mastication Robots, SCI 290, pp. 207–236.
springerlink.com

Food texture can be measured using sensory or instrumental methods. The consumer of a food is the ultimate judge of a foods texture and this assessment will utilise all the five senses. As such, one of the key tools in evaluating food texture is by using trained or untrained human panels through sensory science techniques. Carefully defined textural attributes of foods can be assessed using these panels to identify changes or differences in food texture between samples. Because of inherent variability in human subjects, responses will vary from subject to subject, over the course of a day, or due to preference of the subject. As a result multiple subjects with appropriate repetition are required in sensory evaluation. Despite this, the science has developed to a point where reproducible and reliable testing can be performed [2]. Although sensory methods enable direct measurement of perception, there are many disadvantages, such as variability from person to person and variability from time to time, it is generally time consuming, expensive, not subject to absolute standards, only a limited daily sample throughput and the food sample must be acceptable to panel members.

A number of instrumental testing techniques have been developed to overcome these difficulties. These include; empirical tests where a mechanical property is measured in well defined conditions; imitative tests, which simulate aspects of oral processing; and fundamental tests, which measure engineering properties such as Young's modulus [1]. For example, the General Foods Texturometer analyses food texture by compressing a bite-size piece of food two times in a motion that simulates jaw movement, and extracts the textural parameters from the resulting force-time curve [2]. Other universal testing machines, such as the Instron and the TA.XT2 Texture Analyser, use the same principle for texture evaluation but with a modified analysis such as the commonly used food science tool, texture profile analysis, TPA [2]. However, most instrumental measurements of food texture focus on initial food properties and rely on simple one dimensional crushing. They are not able to simulate the complex functions and movements involved during mastication.

Many researchers have investigated the correlation between different instrumental and sensory measures of food texture [3] in an effort to predict consumer response from instrumental measurements. Success has been demonstrated for some food systems but problems do exist. For example high correlations can be found between sensory and instrumental measurements but the regression equations are different. Szczesniak [3] explains that good correlations demonstrate that two different measures are related but not necessarily that two different measures of the same property. Another finding was the need for multiple correlations with a number of instrumental measurements to successfully describe a sensory attribute of a food. The need to characterise the food using a number of tests negates the advantages of instrumental measurements over sensory techniques.

It is often found that different instrumental measurements (or indeed sensory descriptors) are used for different food systems. For example fracturability is a useful term to evaluate properties of brittle materials using TPA whereas it cannot be applied to more plastic materials which deform under compression and exhibit no fracture failure response. Similarly Warner-Bratzler shear measurements have been found to be more useful to characterise meat texture than compression methods [4]. This is due to the differences in size reduction mechanism occurring on the dentition during mastication.

Arguably the most convincing link between instrumental properties and food bolus preparation has been developed through consideration of fracture in foods. Through analysis of crack propagation during fracture of foods, Agrawal et al. [5, 6] suggested that the resistance of foods to be broken down can be described by the fragmentation index $(R/E)^{0.5}$, which is a function of the foods toughness (R) and modulus of elasticity (E). With this index, foods with low fragmentation index can be effectively broken down with 'pestle and mortar' dentition achieved when food is crushed between cusps and fossa of opposing teeth. Foods with high fragmentation index obstruct formation of cracks and are broken down with 'double bladed' dentition where opposing raised edges on the teeth ensure the cracks developed in the food are kept open until the fracture is complete [7].

These difficulties in deriving tests to fully evaluate food properties exist despite the fact we all process and evaluate foods in the same human mastication system. Masticatory robots are one potential technology that could be used in order to quantify textural properties of foods dynamically during chewing. By simulation of oral processing mechanically, it is possible to collect real time data on the breakdown of the food and bolus formation (e.g. the magnitude and direction of forces, work done, cohesiveness etc.). Such data is very difficult to measure in-vivo, resulting in the widespread use of the less quantitative techniques of sensory science.

This chapter sets about reviewing the current state masticatory robots for food texture evaluation and the development required to provide useful information about food texture. The processes occurring in the mouth during mastication are then outlined in order to define what functionality masticatory robots must have. The existing capability of masticatory robots is then evaluated against these requirements and the future needs for robot design are identified.

8.2 Specifying a Mastication Robot for Texture Measurement

The conversion of an ingested food portion into a swallow safe bolus is a complicated process. A number of very good reviews have been made on the physiology and control of mastication [7, 8, 9]. Almost all solid and semi-solid foods are processed in the mouth before swallowing. Mastication involves a number of dynamic processes and the bolus itself is undergoing significant changes in composition and physical properties. As such the perception of texture is an inherently dynamic process. Some of the key changes occurring during mastication include;

- Particle size reduction and its associated increase in surface area.
- Development of inter-particle liquid bridging to provide bolus cohesion and ability to deform during swallowing.

- Lubrication of the outer surfaces of the bolus.
- Addition of saliva including amylase and pH buffering.
- Dissolution of components into saliva.
- Release of aroma and flavour compounds.
- Heating (or cooling) of the food towards oral temperatures.
- Loss of particulate, fat and aqueous components of the food to other parts of the mouth.

Because these processes are dynamic and inter-related, the timeframes associated with mastication are important. As an example, the distribution of moisture in the bolus can vary due to addition of saliva, absorption into particles and loss into the oral cavity. The rates of absorption will increase as total surface area increases due to particle size reduction. The rate of loss is likely to increase as the bolus reaches saturation. Food micro-structure is changing as bridges or networks are dissolved. The quantity and location of the moisture and salivary mucins in the bolus help define its deformability and slipperiness. Absorption of moisture will change the fracture properties of particles. From all these effects it is clear that the time frame of mastication in humans must be matched by chewing robots, or else the relative amounts of these processes will not be consistent with bolus formation and the measured textural parameters are not likely to be representative.

Particle size reduction is well studied and generally conceptualised (and modelled) as a combination of a selection function and a breakage function [8]. The breakage function defines how particles of each size class break down into a distribution of smaller sizes on fracture, and is a property of the food. The selection function describes the probability of whether particles of each size class are to be fractured in the next chewing cycle. It is likely that breakage functions of particles are dependent, not only on the type of food, but also the amount of absorbed moisture and the amount (and possibly rate) of shear and compression occurring during teeth occlusion.

The selection function includes the likelihood of particles that are compressed between the teeth evading fracture (more likely for smaller particles that might survive in fissures between the teeth) as well as the likelihood of a particle being present on the occlusal surface in the first place. Large portion sizes would result in a smaller likelihood of a particle being present on the teeth, as a smaller fraction of the food sample can be acted upon in the limited occlusal surface area. Due to the selectivity of the tongue, larger particles within this mixture will be actively sought out for subsequent size reduction.

A number of factors affect these processes and must be considered in the design of mastication robots. These include; teeth geometry, teeth kinematics, control strategy, sample size and manipulation, bolus retention and losses, addition of saliva and temperature control. Each of these factors are discussed below, and used to specify the functionality required in a robot to simulate food chewing.

8.2.1 Tooth Shape

Post-canine teeth (premolars and molars) are complex, featuring a number of cusps and fossa's. The morphology of the molars has been developed over the course of evolution in adaptation to diet. Humans, being omnivores, partake in a varied diet with food properties ranging over a wide spectrum (soft to hard, brittle to ductile etc.). As such, the shapes of the teeth have evolved to allow efficient mastication of these foods without damage to the teeth. Strait [10] nicely explained how molar structure in animals depends on food properties, emphasising the importance of recognising the compromise in tooth structure required as diet is broadened in different mammals.

Lucas [7] provided an excellent summary of how different features of the teeth can be explained from the fracture mechanics of different foods. Self propagating cracks will form in foods with low fracturability index $(R/E)^{0.5}$ after a point load is applied. This is most efficiently achieved if the point is blunt and tapers at a wide angle, which is reflected in the blunt cusp arrangement of the molars. To hold food during application of the load, fossa or hollows are present in the dentition opposite the cusp. By holding the resulting fragments in this way, multiple food cusp contacts can occur during the course of occlusion.

Cracks initiated in tough foods (high $(R/E)^{0.5}$), will be arrested and so they must be kept open by a blade until fracture is complete. As a result, the two basic features of molars are blunt cusps and blades. Blades must be aligned with corresponding blades on the opposing teeth and contact each other at a point (this point slides along the junction between blades in an action similar to scissors). Cusps do not need to contact each other but are instead operated against opposing fossa's which collect fragments and allow them to be re-fractured during the same mastication cycle.

The importance of these dental features on fracture mechanism illustrates the necessity of using accurate representations of teeth and their orientation in a chewing simulator. It also demonstrates the importance of appropriate kinematics of the molars during mastication.

8.2.2 Teeth Kinematics

A number of different studies have investigated the effect of food properties on jaw trajectories during chewing. Agrawal et al. [11] extended the fragmentation index $(R/E)^{0.5}$ theory to demonstrate that the angle of occlusion during chewing was correlated to the foods fragmentation index. They stated that although the jaw movement during occlusion was guided by the molar cusps, the closing angle varied at the point of tooth-food contact and as such the amount of lateral movement during occlusion varied. A total of 15 different foods including an array of cheese, nuts and carrot were used. The results showed that greater degrees of lateral movements are applied during chewing of foods with low fragmentation index such as Brazil, Hazel, Macadamia and Pea-nuts. Softer less brittle foods such as Mozzarella, Edam and Gouda cheeses were chewed with less lateral movements.

Nakajima et al. [12] also demonstrated that harder foods (such as hard rice crackers and almonds) resulted in more lateral movements during the occlusal phase of chewing than softer ones (apple and chewing gum). Interestingly Nakajima suggested that the closing and opening patterns of the teeth during the occlusal phase of a chewing cycle could be related to common Japanese words, used to describe aspects of food texture. Like Agrawal et al. [11], the recorded jaw movement patterns also showed that during the occlusion phase of a chewing cycle, the jaw movement followed the lateral border movements caused by the molar teeth shape. The position that the jaw trajectory met the lateral border trajectory varied with food properties. Agrawal et al. [11] described this as a variation in the angle of occlusion.

Anderson et al. [13] studied jaw trajectories used for masticating chewing gums of different hardness. Their results also demonstrated that increased hardness resulted in increasing vertical and lateral movements in 26 human subjects. In this case the shape of the trajectory was similar for both hard and soft gum but the amplitudes of the movements varied. It was also clear that the recorded trajectories were the same during the occlusion portion of the cycle, irrespective of the gum properties. The shape of the movement in this phase was due to the molar teeth gliding over each other. It was suggested that the increased lateral amplitude for harder foods was required in order to achieve greater acceleration of the jaw and hence more force when the teeth-food contact occurs.

Peyron et al. [14, 15] showed that increase food hardness increased chewing cycles, muscle activity and the vertical amplitude used during mastication, using model gelatine based food systems. Foster et al. [16] extended this work to include comparison of elastic and plastic model foods of varying hardness. The chewing profiles used for elastic and plastic foods of equal hardness were different. Plastic foods exhibited much less repeatable trajectories and in general had larger vertical and lateral movements than the elastic foods.

From these studies it is clear that a key requirement of any chewing simulation device is the ability to vary the jaw trajectory to reflect those used in humans for different foods. It is also important that the trajectory during occlusion follows the complex path defined by the sliding contacts between the surfaces of the molars. Without faithful reproduction of the occlusal trajectory, the functional area of the teeth will not be correctly implemented and therefore particle breakdown mechanisms will not be representative.

8.2.3 Control Strategy

Lucas [7] provided a good description of the various control strategies used during mastication. A number of sensors present in the mouth provide feedback for control of jaw motion. These were broadly categorized as; sensors that provide feedback on jaw position and others that provide information on the stress that is applied. The primary goal was control of jaw movement but there was modification of this control when there was teeth-teeth or teeth-food contact or anticipated contact.

During the early closing phase of the chewing cycle, displacement sensors are likely to be important. When the mandible reaches positions approaching food contact, force sensors are used to monitor fracture events. It has been reported that these feedback sensors can detect the presence of particles of only 8-15 μm in size. During this phase, the velocity of the jaw movement slows, primarily to avoid teeth damage. Lucas suggested that avoiding damage is more difficult when the jaw is operating under stress control.

Bosman et al. [17] explained that the stress control is present about 25 ms after resistance occurs. The reflex response after this is stronger when resistance is expected. The amount of force applied is under continuous control with a latency of about 25 ms. Brosman et al. stated that during the closing period of the chewing cycle, the jaw movements are directed to reach occlusion (also suggested by [11, 12, 13]). When particles resist this motion, more force is applied to reach occlusion, however during this phase force control is applied. Subsequently complete occlusion may not occur and further breakdown may be required in another chewing cycle.

These reports provide a solid foundation for control of chewing robots. They suggest that a switch from trajectory control to force control and hence potential modification of the occlusal trajectory must be implemented. It is also clear that the velocity of the jaw will vary during the different stages of the trajectory.

8.2.4 Sample Size and Manipulation

As the food sample size ingested by humans increases the chewing time required before swallow also increases [18, 19, 20], the muscle activity used per cycle increases [20] and the degree of particle breakdown reduces [21, 22]. A few studies have quantified the natural bite size of a range of foods including bread, rice, sausages [23], apple [23, 24, 25], banana and biscuits [24, 25]. Recently Hutchings et al. [26] recorded survivorship curves for natural bite sizes on a number of commercially manufactured food bars (Crunchie, Fruit and Nut bar, Muesli bar, Apricot pie bar and Pixie Caramel). It was found that the bite weight varied between subjects from 2 to 16g and bite volume ranged from 2 to 17 cm^3. Although bite size varied with subjects, it was found that for a given subject the bite length taken from the bars was approximately the same.

It is possible to roughly estimate the proportion of a given food sample that might be crushed during a single chewing cycle by considering the molar teeth area. Assuming two molars are used (approx. 2 cm x 0.7cm) and the height of the food before the teeth come into contact with it is 1cm, the maximum volume that can be chewed is of the order of 1.4 cm^3 per cycle. The mean natural bite volume recorded by Hutchings et al. [26] was 8 cm^3, meaning that less than 20% of the food can be acted on at any one time. It follows from this crude analysis, that the sample is actively manipulated across the dental mill over the chewing sequence. It is important to consider this if comparisons are made between mastication robots and human measurements. The total number of chewing cycles and time used to process a food in a human to the point of swallow will be significantly greater than required for masticatory robots where smaller sample sizes are used. To get accurate simulation of human chewing, some mechanism to manipulate the sample across the dentition between cycles is required.

In-vivo this action is achieved by the tongue. The tongue fills up 80-90% of the space inside the dental arches and in this way retains the food particles. The rough surface of the tongue helps grip the food particles during chewing to allow food manipulation [7]. Tongue movements are important during mastication. A good review of the range of motions and functions of the tongue during feeding was carried out by Hiiemae and Palmer [27]. The tongue can twist to align its surface against the rows of post-canine teeth (on either side). After transport to the post-canine teeth during food acquisition, bites of food are tossed onto the occlusal surface ready for mastication. During chewing the tongue continues to ensure inadequately broken down food is selected for further processing by repositioning it when the mandible is at its most vertical portions of the chewing trajectory.

Repositioning of food is achieved through the interaction between the tongue and cheeks. The tongue moves the food laterally to ensure it is moved onto the occlusal surfaces of the mandible and with successive bites pushes the material into the cheek pouch. The cheek muscles (bucinator) contract approximately every three chewing cycles, pushing the food back inward [28]. In addition the tongue and palate receptors are used during the open phase of the chewing cycle to evaluate properties of the food bolus such as particle size. It has been shown that these sensors can discriminate particles size if they are greater than 1-2 mm [7]. It follows that after evaluation of the bolus on the upper palate, particles greater than this size range are actively selected by the tongue for further breakdown.

8.2.5 Bolus Retention and Losses

Masticatory robots will require some sort of mechanism to hold or contain the food on the teeth for chewing. This is achieved by the opposing actions of the tongue and the cheeks which act to ensure the accurate location of the food [27]. This system does not fully retain the food bolus however. For most mastication studies it is normally reported that only 30-60% of the solids in the ingested sample are present in the recovered bolus, e.g. [29]. Losses are low for fibrous products such as meat (12%) and much higher for brittle products (60%) [30].

Particles are lost into the mouth together with moisture, fats and dissolved solids. Particle losses are likely to occur more frequently at the onset of mastication before some cohesiveness of the food bolus is developed to help hold the particles together. Moisture containing solutes, dissolved from the food or juice expressed from the food during compression can also be lost from the bolus. That this occurs, is evident from the ability to taste sweet or salty flavours during chewing. Fats might also be extracted from the food particles in the form of emulsions when mixed with saliva or expressed from very high fat foods. Again, the importance of fat in aiding the transfer of flavour and the residual mouth coating after eating oily foods demonstrates this.

Particle loss has been observed by Flynn [31], who showed that the particle size distribution in the particles rinsed from the mouth following expectoration of the bolus, was significantly different from the bolus itself. This result suggests that once lost from the bolus, the particles do not undergo further size reduction. The loss of particles will also increase the probability of the remaining particles in the bolus being selected for particle size reduction, particle loss being effectively equivalent to a decreasing portion size. It is clear from the magnitude of these solids losses (60-70%) that they should be considered in future development of masticatory robots. For example, a mechanism that effectively washes the bolus during mastication to remove soluble material might be appropriate to simulate the loss of dissolved solids that occur. More physiological studies on the mechanisms and magnitudes of these losses is required before appropriate mechanical systems can be designed to account for this effect in masticatory robots.

8.2.6 Moisture and Temperature Control

Moisture is added to the food bolus during mastication in the form of saliva. Saliva is a pH 6-7 solution containing electrolytes and a range of proteins including enzymes, immunoglobulins, antibacterial proteins, proline rich proteins and mucins. Saliva flow rates vary from 0.3 to 7 ml per minute and are stimulated by the presence of food, irritations in the mouth and even the thought or smell of food [32]. There appears to be no correlation between salivary flow and the number of cycles used to chew a food [33].

It has key functionalities that significantly influence oral processing of foods. It acts as a lubricant between the food and mucosa. It wets the new surface area developed from fracture events to encourage cohesion between particles. It dissolves soluble materials in the food thereby facilitating taste and it buffers food acidity which protects the dentition [7]. Its properties provide anti-sticking behaviour between a food bolus and the teeth or oral mucosa, while still being excellent at wetting food particles. Mucins are the key component in saliva that provides these properties.

Pereira et al. [33] investigated how added moisture affected the chewing behaviour and sensory evaluation of foods. Added fluid significantly reduced muscle activity and the number of chewing cycles used to masticate melba toast, cake and peanuts as well as influencing the perceived texture. These effects were not found for chewing of fatty foods such as cheese, or for wet products such as carrot.

Because saliva plays such a central role in bolus formation and effects particle size reduction, it is important that addition of saliva is incorporated in chewing simulations. It is not clear however, whether the saliva flow is constant over the duration of a masticatory sequence or if an initial pool of saliva already present in the mouth is mixed into the food at the start of chewing. Further experimental work investigating the dynamics of moisture addition in human chewing is required.

8.3 Masticatory Robots – Current Capability

Masticatory robots have already been reviewed in detail in Chapter 1, however not many of these devices have been developed (or can be applied) to provide information on food properties. For this reason, we focus here on devices developed for food evaluation or devices in which their application to foods is reported. Robots such as the Australian Jaw Joint Project (http://www.jaw-joint.com), the speech robot [35] and the WY dental training robots are not discussed here.

The Stewart Platform based dental testing simulator designed by Alemzadh et al. [36-38] is in development for testing failure on dental components and ceramic restorations. Although it is reported to provide accurate simulation of 6DOF chewing trajectories, correct occlusion and bite force control, no data on its application to foods has been reported and it is difficult to gauge how useful it might be for food texture analysis. Similarly the dental testing simulator developed by Conserva et al. [39] was focused on assessing the impact of dental materials on the stresses on dental implants rather than food analysis.

The key robots discussed in more detail are; The Waseda Jaws mastication robot [40], the BITE Master II [41], the Flavour release chewing simulator [42], The Linkage chewing robot [43] and the 6DOF Parallel chewing robot [44].

8.3.1 Waseda Jaws Mastication Robot

Takanobu et al. [40] reported the application of the Waseda Jaws mastication robot to chewing of a Japanese ball shaped biscuit (tamago-bolo). The biscuits were 15mm in diameter and were chewed using the WJ-3RIII mastication robot. This consists of nine artificial muscle actuators, used to move the mandible. Each actuator included a force sensor to measure the tension. In addition, micro pressure sensors were incorporated onto two lower and two upper molars and another three force sensors on the temporomandibular joint. The maxilla and mandible were life sized models of a human skull made of epoxy resin. A more complete description of this robot is given in Chapter 1.

The mandible trajectory consisted of 4 phases, a closing, the first and second halves of occlusion and the opening phase. During opening and closing, the jaw was position controlled, while during occlusion, force control was used. It is not clear from the report as to how the different actuators were recruited to achieve the desired trajectories or application of force. This would not have been a trivial matter considering that the robot is over-defined for the 3DOF movement. Two trajectories were applied; a clenching motion (largely vertical, 0° grinding angle, 15° maximum opening angle) and a grinding motion (more lateral, 7° grinding angle, 15° maximum opening angle). Because the device is capable of movement with only 3DOF, it is likely that some of the rotational movements that occur during human mastication cannot be simulated. Apart from the general descriptions provided of the implemented trajectories, there was no specific information reported as to how they were defined, the velocity profile, whether they were based on measured human recordings or if so, for what foods. The force-time profile is also unspecified for the control during the occlusal phase.

Using this device, a number of measurements could be recorded in real time, including the tensions and torques generated by each of the 9 actuators, the forces applied on the upper and lower teeth and temporomandibular joint. Because the focus of the paper was to demonstrate the application of the device to quantification of mastication efficiency however, no example traces of these variables were presented.

The mastication efficiency of the robot was evaluated using a sieving method on the resulting bolus after varying numbers of chewing cycles. This was quantified as the mass fraction of the whole bolus that was retained on a 1.18 mm sieve. The results showed that the amount of material retained on the sieve decreased as the chewing number progressed, until after about 10 chews the amount retained approached a constant. More lateral grinding motions resulted in more efficient breakdown of the particle size. The masticatory efficiency was also compared with the total work input from the masticatory robot which showed that the more complete breakdown achieved using the grinding motion, also required less energy input. This finding that lateral movement aids particle size reduction in the biscuit is what might be expected according to the hypothesis of Agrawal et al. [11]. A biscuit (assuming it is dry) might be expected to have a low fragmentation index $(R/E)^{0.5}$ and therefore to be chewed with significant lateral motion.

Despite the lack of information on a number of aspects of the WJ chewing robot system (e.g. how the food is retained on the teeth, how the biscuit was realigned on the teeth between cycles, what force was used to control breakdown), the result demonstrate the importance of trajectory and probably teeth shape on the fracture properties of the biscuit. This sort of information would not normally be identified using regular instrumental texture measurement methods.

8.3.2 BITE Master II

Meullenet and Gandhapuneni [41] described the development of a stepper motor controlled x, y, z table based chewing simulator called the BITE Master II (Fig. 8.1). The biting station was a nonacron articulator (Girrbach Dental, Germany) which could mimic condyle movement and subject specific dentures. Four load cells were used to measure the vertical and lateral forces on the teeth. The x, y, z slides were connected to the articulator with spherical joints to allow 6DOF movement, although with this design it is unlikely that true control of movement in 6DOF was possible. Instead the motion could be tuned manually to achieve realistic trajectories for a specific subject through a number of adjustments on the condyle.

Fig. 8.1 BITE Master II (© ELSEVIER 2006 Reprinted from [41] with permission)

The device was used to simulate the chewing of 10 different types of cheeses by 7 subjects who have been previously trained as a sensory evaluation panel. The closing phase trajectory of the first bite was recorded for the 7 subjects while chewing each cheese sample with a BioResearch JT-3 Jaw Tracker Sensory Array. The amplitudes of the vertical, anterior-posterior and lateral movements were recorded using an Electrognathograph. Because the chewing trajectory method used only one sensor on the teeth, true three dimensional location of the mandible must not have been recorded. During chewing of the samples, the panellists were also asked to score the hardness of the cheeses.

Subject specific replicated teeth were used to simulate the chewing of each subject for each cheese using the BITE Master II, controlled so as to reproduce each subject's actual chewing trajectory. The peak force required to observe sample fracture, the energy to fracture (area under the force-deformation curve until fracture) and the total energy (area under the whole force-deformation curve) were recorded from the BITE Master II trials. No information regarding how the samples were held on the moving mandible during the test is provided, suggesting that only one chewing cycle was simulated. To allow comparison with traditional instrumental methods, uniaxial compression tests were carried out on each sample using a TAXT2 texture analyser.

The reaction force measured using the BITE Master II during chewing, were much smaller than those recorded by uniaxial compression tests (e.g. 7-8%). This was attributed to the difference in fracture mechanism occurring between the two tests. In the chewing simulator, fracture occurred at approximately 50% deformation, after which the sample fell apart, offering no more resistance force. In the uniaxial compression, fractured particles were continued to be crushed, hence offering further resistance. A significant negative correlation was found (R^2=0.61) between the first bite duration and the peak forces used to crush the sample as measured by the BITE Master II. This suggests that subjects that choose to chew slowly used higher reaction forces. This suggests that subjects biting at different rates will apply different bite forces and potentially perceive different hardness. These varying rates are also likely to influence the measured and perceived hardness because of the viscoelastic mechanical properties of the food.

Very good correlations were found between both the uniaxial and BITE Master II measured hardness values and the sensory scores (R^2>0.9), with peak force offering better correlations than initial fracture force or total energy. In terms of offering a method to approximate sensory perception of hardness in cheese for food processing industry, the BITE Master II did not offer any better correlations than the more simple and widely used uni-axial compression tests. Despite this, it was concluded that the BITE Master did more closely mimic fracture events occurring in the mouth and is more representative of in-vivo conditions. Further development was suggested to provide a method to hold the sample on the teeth during testing as well as consideration of lubrication (saliva) of the food.

It is interesting that the approach taken to control the robot takes a quite different approach than that carried out on the WJ robot [40]. Rather than to control the amount of force applied on the food during the occlusal phase, the approach taken by Meullenet et al. was to fully prescribe the motion of the molars during occlusion and measure the resulting forces that had to be applied deform the food to achieve this trajectory. It was not clear from the paper whether full contact of the occlusal surfaces of the teeth occurred during the recorded measurements in the human subjects, or if the robot could be configured to achieve accurate occlusion during this part of the trajectory. It might be expected that if significant tooth contact did occur, then the reaction forces would have been higher, even if no food was present.

This study was an interesting approach, which despite some of the shortcomings regarding sample holding and the control of the full 6DOF of mandible movement, did offer promising results and good correlation with sensory perception of food samples. The design also offers the easy adaptation of the robot to allow its use to simulate chewing of recorded subject jaw trajectories with subject specific replicated teeth.

8.3.3 INRA Flavour Release Chewing Simulator

Flavour release is an important factor in taste perception of foods. As the food is deformed and mixed, soluble materials are lost from the bolus and reach taste sensors (sweet, salt etc.) and volatile compounds are evaporated and are transported into the nasal cavity where aroma compounds can be sensed. It seems obvious that the rate of release of volatile compounds from a food bolus is proportional to surface area of the bolus and the rate that the surface of the bolus is replaced due to mixing. Similarly particle size reduction within the bolus will increase the interfacial area between the particles and saliva phases, further increasing transport of volatiles or taste compounds to the bolus surface. Temperature is also an important factor affecting volatile release due to the increase in volatility of organic compounds with increasing temperature.

Several attempts have been made to design mouth simulation systems that allow in-vitro measurements of the rates of volatile release from foods in the mouth. For example van Ruth and Roozen [45] used a model mouth system in which a plunger was screwed up and down in the food by a motor. The sample itself was contained within a vessel that was temperature controlled with a water jacket. Roberts and Acree [46] and Deibler et al [47] used a device termed the retronasal aroma simulator. This device used blades for shearing the food during analysis as well as air flow rates through the device similar to what occurs during human breathing.

Salles et al. [42] extended these relatively simple ideas into a much more realistic mouth simulation device for in-vitro characterisation of volatile release rates from foods (Fig. 8.2). The breakdown of the food matrix strongly influences flavour release. For this reason the INRA chewing simulator focused on more closely mimicking human chewing with respect to the application of shear and mixing of the bolus, the ratio of food to saliva and gas, and the rate that saliva is added to the food.

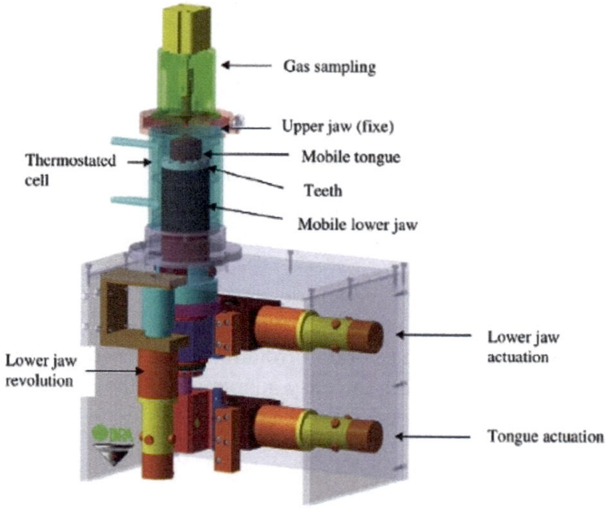

Fig. 8.2 INRA Flavour release chewing simulator (© ELSEVIER 2007 Reprinted from [42] with permission)

The system consists of a cell which contains the food sample in an anulus where it is acted on by an actuated lower ring shaped mandible (Fig. 8.3) and an actuated tongue. The lower mandible can be moved vertically and rotated against a fixed upper ring of engraved teeth. Teeth were engraved into the mandible and upper part to approximate human teeth functionality during occlusion, but in a rotational direction (Fig. 8.3). This rotational motion cleverly solves the problem of food manipulation between chewing cycles, as the food is simply pushed around the ring into an adjacent identical set of molars. Because of the design however, the tooth trajectory can only be actuated in 2 dimensions.

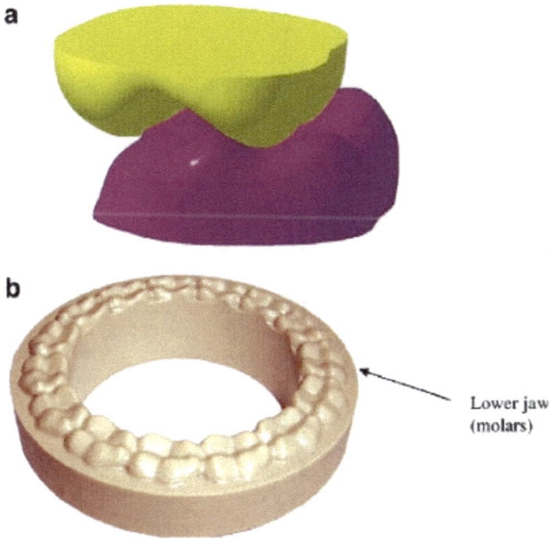

Fig. 8.3 Teeth geometry of INRA Flavour release chewing simulator (© ELSEVIER 2007 Reprinted from [42] with permission)

The tongue component can be controlled to move vertically and has a conical top which, when raised, will push the food into a matching conical roof to the chamber. This part is designed to simulate bolus preparation in humans where the food is pressed between the tongue and upper palate prior to swallowing. The rate and amount of force applied by the artificial tongue can be controlled. Air is pulsed through the chamber at a controlled rate to simulate breathing. The outlet air is then analysed for volatile compounds. The walls of the chamber are temperature controlled using heat tracing and artificial saliva can be dosed into the chamber over the course of the testing sequence. The three motors used to actuate the device were DC brushless motors from which the forces applied could be estimated from the DC current, without the need for load cells.

Salles et al. [42] tested the effect of adjusting the various parameters of the device on the chewing outcome for peanuts. In each case the fraction of bolus that passed through 2 and 4 mm sieves were used to characterise the particle size outcome. The vertical height and applied force of the mandible were also recorded during testing (Fig. 8.4).

During the food-teeth contact phase, the force rapidly increased to a maximum which was held relatively constant until the opening phase. During this time, random small relaxations of the force were observed which are likely to be due to fracture events. The minimum distance between the upper and lower dentition during this period of teeth-food contact reduced, as the number of chewing cycles progressed. This clearly demonstrates that the device's teeth trajectory was allowed to change as a result of implementation of force based control of the device. It is not clearly stated how the mandible movement was controlled but the results tend to suggest that some sort of force threshold was used, where the vertical motion is limited by the specified maximum force applied by the actuator.

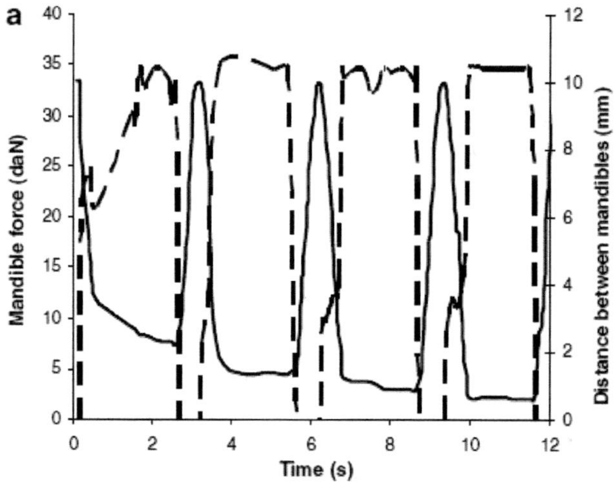

Fig. 8.4 Example force and distance profiles during chewing of peanuts in the INRA chewing simulator. Dotted lines represent force; solid lines represent distance between the mandibles. (© ELSEVIER 2007 Reprinted from [42] with permission).

The results showed that increased shear angle (the rotational motion of the device) resulted in more effective breakdown of the peanuts. This was evident in smaller minimum distances between the dentition during chewing measurements as well as in the particle size results after 4 and 8 chewing cycles. The addition of artificial saliva did not significantly affect the outcomes. These findings are again in agreement with the observations of Agrawal et al. [11] and Takanobu et al. [40] in that more lateral movement provides more effective breakdown of hard, brittle foods. In addition to the shearing angle, the biting force also affected the rate of particle size reduction.

Experiments carried out on four human subjects, as part of the same study, resulted in quite variable particle size outcomes after 4 and 8 chewing cycles. It was found that by manipulating the shearing angle and mandible force, the particle size distributions achieved in the chewing simulator could be made to match the human subjects. However it was suggested that because the outcomes were so subject (and likely food) specific, more physiological mastication studies are required to characterise the range of breakdown patterns used for different food systems.

8.3.4 Linkage Mechanism Mastication Robot[1]

The design of a crank-slider linkage robot has been described in detail in Chapter 5. Further details can be found in Xu et al. [44] and Sun et al. [48]. The

[1] This section is based on work carried out by Richard Sun, PhD student at the School of Engineering and Advanced Technology, Massey University, New Zealand. M.P. Morgenstern from New Zealand Institute for Plant and Food Research also contributed to the work.

chewing robot was developed to simulate human chewing behaviour in terms of kinematics and force application on food. This device consisted of a linkage mechanism, developed to reproduce a full range of chewing trajectories of the first molar, from lateral chewing to vertical chewing in the frontal plane (y-z plane), by varying the length of the ground link (Fig. 8.5). The chewing trajectory in the sagittal plane (x-z plane) was approximated as a straight line with an angle ranging between 0 and 30° to the horizontal plane. This was achieved by manually tilting the base of the linkage mechanism. The chewing robot based on this linkage mechanism was able to match human chewing velocity profiles of any trajectory according to human measurements.

Food Type	Low fragmentation index $(R/E)^{0.5}$	High fragmentation index $(R/E)^{0.5}$
Comminution mechanism	Grinding/crushing	Cutting
Dentition	Blunt cusps	Double blades
Chewing pattern	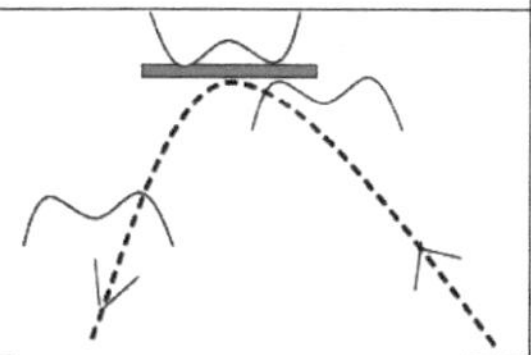	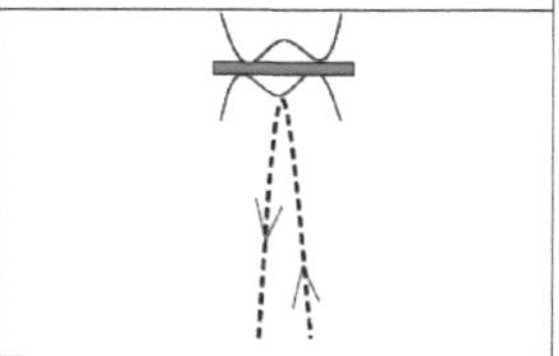
Four-bar linkage	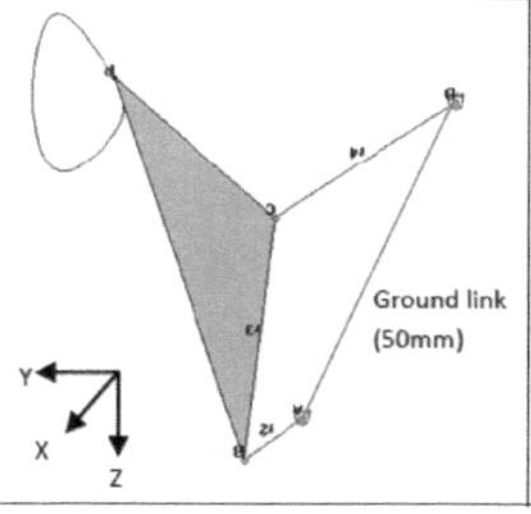	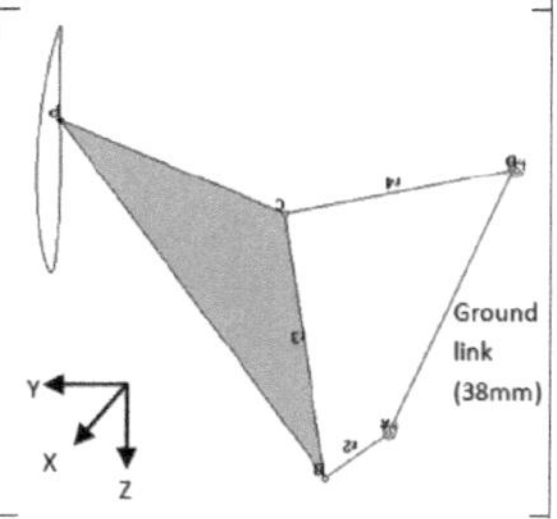

Fig. 8.5 Linkage mechanism with adjustable ground link length to match a full range of chewing motions

The robot could apply forces of up to 150N on the food; encompassing the range of measured chewing forces applied on a single tooth during chewing of foods such as biscuits, carrots and cooked meat [49]. The maximum chewing force applied on the food can be adjusted by varying a spring pre-tightening force. In this way, the robot is very adaptable to provide a range of force profiles during occlusion required for different foods.

The chewing robot was built inverted, with mandible teeth up and the maxilla teeth down for the convenience of collecting chewed food particles. In this way the lower unmoving teeth could hold the food sample by gravity. The upper and lower teeth consisted of two premolars and two molars made of hard durable plastic from a dentistry study model. The teeth were fixed in epoxy resin moulds and attached to a removable plate for easy cleaning. The lower teeth sat on a repositioning table for teeth alignment in case of change of chewing trajectory. The distance between lower teeth and upper teeth could be adjusted.

A cast silicon rubber enclosure around lower teeth (Fig. 8.6) was used to keep the food on the occlusal surface during mastication. The shape of this enclosure was designed to ensure the food was retained on the teeth but without interfering with the mandible trajectory. An ATI Mini 40 force sensor was installed under the lower teeth to dynamically measure both force and torque in three dimensions. The chewing robot is powered by a brushless DC motor, a motor controller and a computer to control and collect the data. A rotary encoder is used to track the position of the teeth. The motor current signal, encoder signal and force sensor signals could be recorded in real time during operation of the device.

Fig. 8.7 shows the chewing force profile for an experiment on peanuts with a pre-tightened spring force of 35N and maximum chewing angle in both sagittal plane and frontal plane. The chewing force in the z-axis (solid line) was applied vertically downwards on the food. The chewing force in the y-axis (long dashed line) was the shearing force in frontal plane. The chewing force in the x-axis (short dashed line) was the shearing force in sagittal plane.

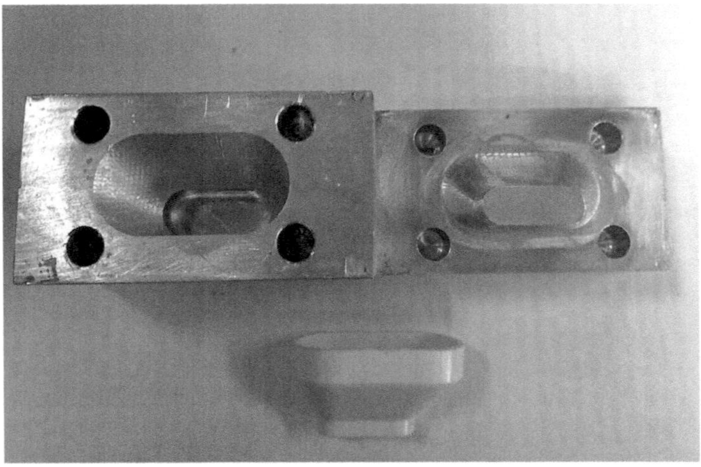

Fig. 8.6 Cast silicon rubber sleeve used to contain the food on the lower teeth (pictured at bottom). The narrow opening is stretched over the lower teeth. The top open section is shaped to allow uninhibited mandible motion. The aluminium moulds (picture at top) are used to cast the sleeve using a two part cured silicon rubber.

While the overall trajectory of the mandible in humans may vary depending on food properties, during the occlusion phase of a chewing cycle the teeth trajectory is guided by the shape of the teeth as the surfaces of the upper and lower molars glide over each other [11-13]. To allow this complex motion during the occlusion phase, the chewing robot incorporates a spring system on the mandible. In this way the vertical component of the robot trajectory can alter to ensure teeth to teeth contact throughout occlusion. The amount of force that must be exerted before the vertical component of the trajectory changes is dependent on the spring constant which can be adjusted by its initial compression (pre-tightening force). If the teeth encounter an object (e.g. a food), the robot must provide enough force to deform the object and achieve the occlusal trajectory. Depending on the pre-tightening force, the full deformation of the food to achieve true occlusion may or may not occur.

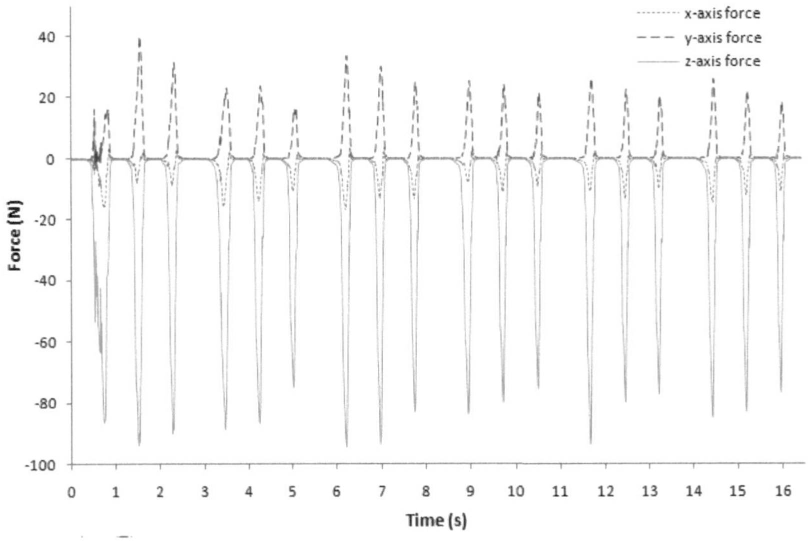

Fig. 8.7 3D chewing force pro files for 18 cycles on peanuts

With continuation of the chewing process, the food thickness was reduced and the compression displacement of the spring was also reduced. As a result, the chewing force generated from the spring dropped. Since the food was repositioned every three chewing cycles, the force profiles had a decreasing trend in a loop of three chewing cycles (Fig. 8.7). The irregular surfaces of the teeth along the forwards movement in the sagittal plane result in component forces along all axes.

The directions of the resultant force (in x-y-z coordinates) during the second cycle of the trial can be seen in Fig. 8.8. As the upper teeth moved towards and compressed the food, the chewing force increased and reached the largest value at the centre of occlusion (intercuspal point). The differences in directions between positional change and force propagation illustrate the importance of teeth shape on mastication mechanisms.

Fig. 8.8 illustrates the changes in trajectory the mandible undergoes as a result of multiple fracture events occurring to the peanut particle. As the trajectory of the mandible progresses, the vertical component of the movement deviates from the free trajectory by compressing the spring. Consequently there is a build-up of force in the z-direction, which results in a fracture event. The compression or shape change of the food due to fracture allows the mandible molars to move closer to the free trajectory. This pattern is then repeated until the trajectory moves to the opening phase.

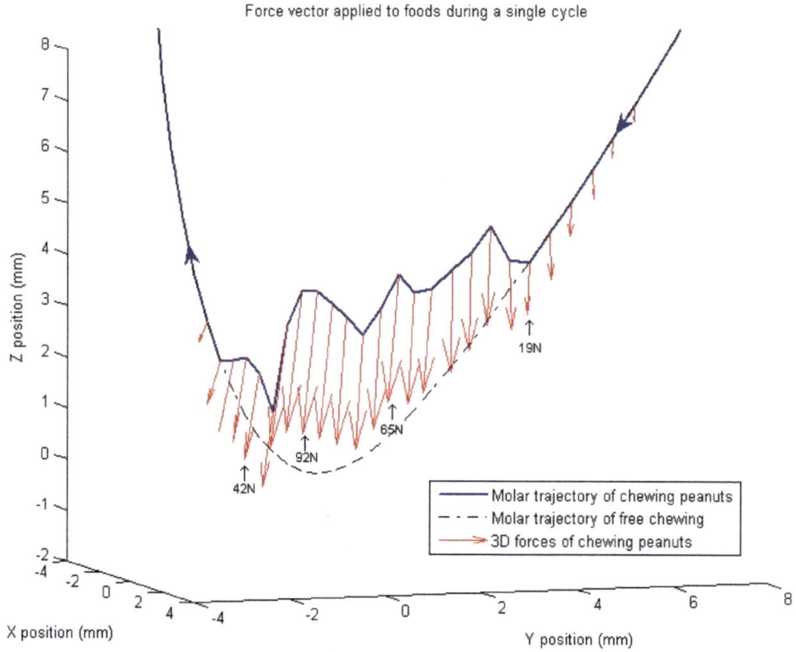

Fig. 8.8 Resultant force vectors projected on the chewing trajectory of one cycle on one peanut

This multiple fracture failure mode is expected for hard brittle particles such as nuts and has been observed in classical texture analysis measurement methodology [2]. Fig. 8.9 shows this fracture mode measured on peanuts using the traditional texture analysis methods. Similar breakage behaviour can be seen in other brittle foods such as hazelnuts [50].

This pattern can also be seen in the resultant forces measured in the chewing robot as shown in Fig. 8.8. While the pattern during fracture is similar between the chewing robot and traditional texture measurement, the magnitude of the forces applied are different. This is most likely to be due to the differences in particle deformation occurring between the two instruments and the directions the forces are applied.

The reproducibility and sensitivity of chewing outcomes to the robots configuration were assessed through comparison of the particle size distributions measured in the resulting boluses collected after each trial. Experimental results with different numbers of chewing cycles (9, 18 and 27) are presented in Fig. 8.10. The results demonstrate the reproducibility of the device along with the changes in particle size distribution occurring over the duration of the mastication sequence. As expected, with increasing numbers of chewing cycles, particles became smaller.

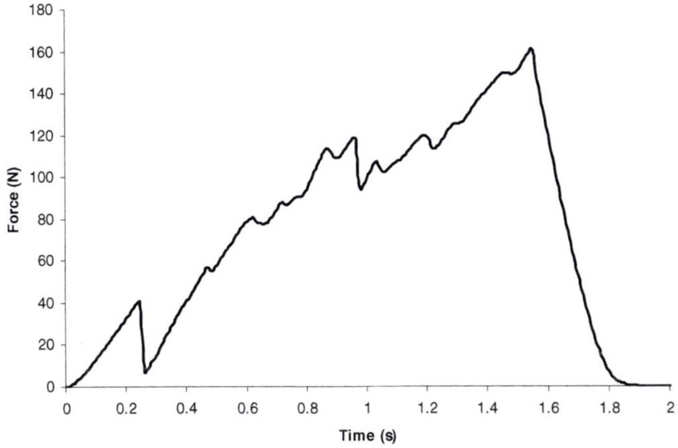

Fig. 8.9 Force-deformation curve measured for roasted unsalted peanuts by compression by TATX2 showing fracture mode failure mechanism

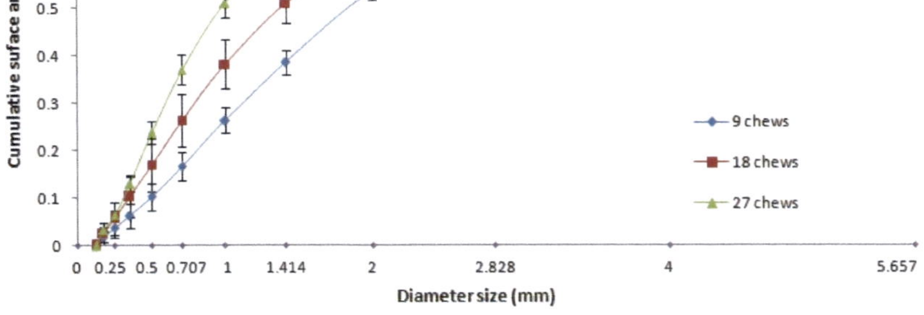

Fig. 8.10 Particle size distributions of peanuts after 9, 18 and 27 chewing cycles

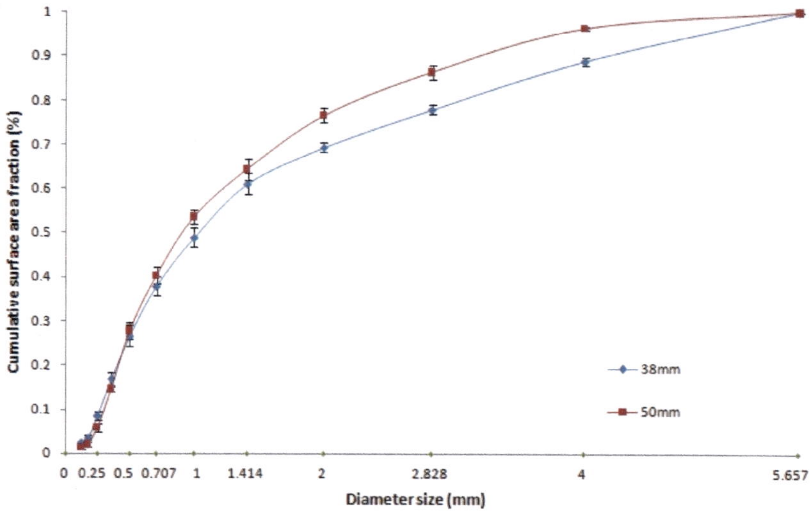

Fig. 8.11 Particle size distributions of peanut fragments after 18 chewing cycles when applying different lengths of ground link: 38mm and 50mm

The influence of adjusting the length of ground link on particle size distribution is presented in Fig. 8.11. When the length of ground link changed from 38 mm to 50mm, the chewing trajectory in frontal plane varied from vertical chewing (compression) to angular chewing (compressive shear) (see Fig. 8.5). Fig. 8.11 shows that particle breakdown was more effective for peanuts with more angular chewing trajectories. This suggests that the application of shear during the compression failure of the peanuts results in a different particle breakage function where more rapid overall particle size reduction occurs. This result demonstrates the importance of teeth shape and trajectory on particle breakage and therefore the limitations in measuring textural information through commonly used one dimensional compression techniques.

The increase in the rate of particle breakdown as the degree of lateral movement increases has been observed for brittle materials using the linkage mechanism robot, the WJ robot [40] and the INRA flavour release robot [42]. The improvement in the rate of particle breakdown with increased lateral movements may be due to a change in the breakage function of the food material as a result of the direction forces are applied during chewing. Fig. 8.12 presents some preliminary data investigating the differences in breakage function observed between the chewing robot and uniaxial compression tests (unpublished data from Morgenstern and Lerebour, Plant and Food Research, New Zealand). Fig. 8.12 shows the area based cumulative size distribution of particles after a single chewing/compression cycle. Although only preliminary, this data clearly shows that much smaller particles form as a result of the combined action of lateral movement and tooth shape. This demonstrates how the robot can be used to develop understanding of the fracture mechanisms of foods during mastication.

The effects of chewing angle in the sagittal plane and chewing force on particle comminution are presented in Fig. 8.13. In general, an increase in both chewing

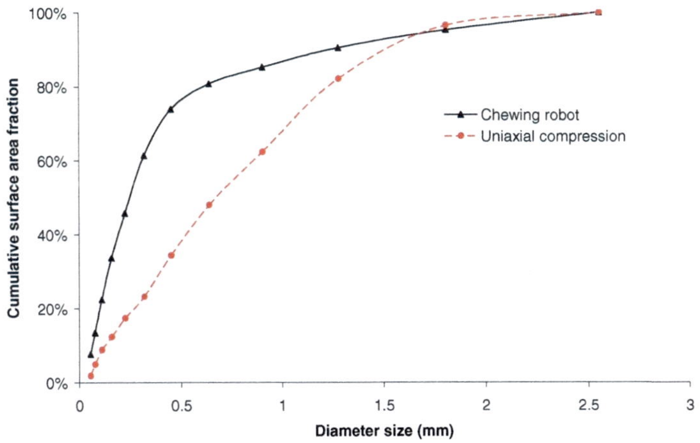

Fig. 8.12 Breakage function of roasted peanuts by Chewing robot and Uniaxial compression testing

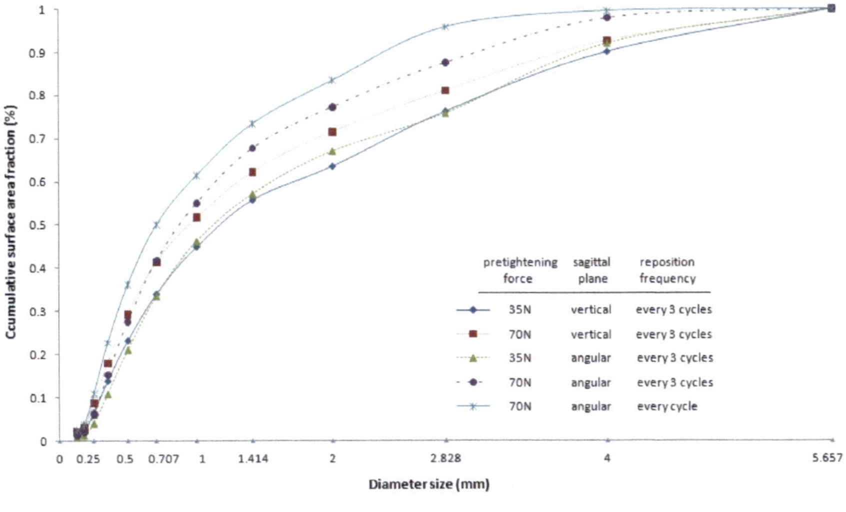

Fig. 8.13 Particle size distributions of peanut fragments after 18 chewing cycles when applying different functional parameters

angle in the sagittal plane and chewing force caused an increase in the extent of particle breakdown. With a spring pre-tightening force of 35 N, the difference in particle size distribution resulting from angular and vertical chewing was not significant. This suggests that under low pre-tightening force, the resulting shear force for angular chewing was insufficient to influence the chewing outcome. With a spring pre-tightening force of 70 N, as chewing angle increased from vertical to angular, higher shear forces were applied and the extent of particle size reduction was greater.

The tongue plays a crucial role throughout the complete food oral processing sequence and one of its key roles is as a mechanical device for food manipulations. For the current chewing robot, food manipulation was achieved manually by repositioning food particles after a given number of cycles. Thus, the particles which were lost from the teeth surface could be positioned back for chewing. In order to study the influence of food manipulation on particle size reduction, an experiment was carried out in which food was repositioned at various frequencies (Fig 8.13).

It was found that when peanut particles were repositioned every cycle compared with every 3 cycles, smaller particles were produced. These results showed that tongue movement during mastication is important for food particle reduction. Therefore, a food manipulation device is required for future improvement to simulate the functions of tongue during mastication.

Tongue movement during mastication involves pushing, twisting and aligning during the opening phase of each chewing cycle. Addition of this functionality to the chewing robot is currently under development. The main aim of this work has been due to the need to process larger samples that are more representative of human mastication. The initial design is shown in Fig 8.14. The mechanism consists of three solenoids to simulate the tongue movement. One solenoid stands vertically to bring the food up to the occlusal surface. The other two solenoids lie horizontally to push towards each other to position the food onto the teeth as well as align the food along the teeth. During an opening phase in a chewing cycle, the solenoids are actuated for pushing. Before occlusion occurs, the solenoids return to their original positions.

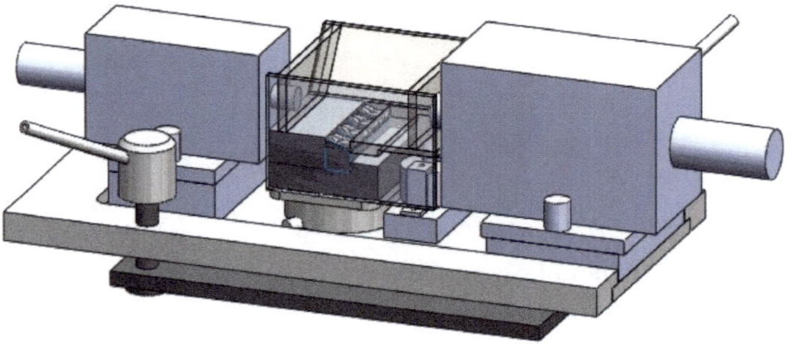

Fig. 8.14 Prototype food repositioning design. This device will replace the current lower teeth and food retention device.

Flynn et al. [31] studied the mastication of peanuts by human subjects. The data (converted into cumulative surface area fraction to allow direct comparison) is presented in Fig. 8.15, compared with the chewing results for peanuts after 18 chewing cycles from the chewing robot. It was observed that the chewing result of peanut was very close to human chewing result under the condition of 70 N spring pre-tightening force, angular chewing and chewing samples repositioned every 3 cycles. There were some minor differences, which could be due to the differences in particle size measurement methods used in the two studies.

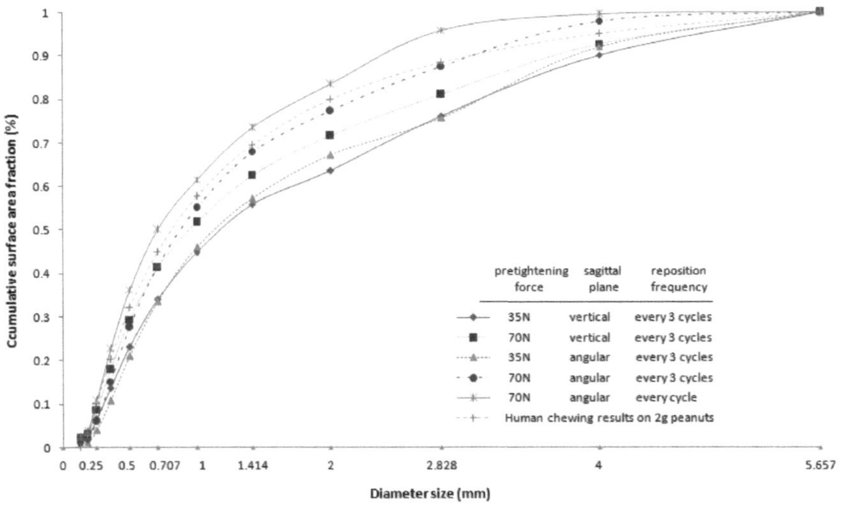

Fig. 8.15 Comparison of chewing robot and human derived peanut boluses

This result again supports the findings of Agrawal et al. [11] who showed that hard and brittle foods (such as peanuts) were chewed with wider lateral movements than softer properties. This analysis shows the potential application of the chewing robot for studying the fracture mechanics of particles as a function of the magnitude and direction of the force in simulated chewing scenarios. The ability to match human chewing outcomes also demonstrates the robots use as a sample preparation device for in-vitro digestion assays.

8.3.5 Six-DOF Parallel Chewing Robot[2]

The development of the 6DOF parallel chewing robot was described in detail in Chapters 3 and 4. The key feature of the robot is that the actuators are positioned

[2] This section is based on work carried out by J Torrance, PhD student at the School of Engineering and Advanced Technology, Massey University, New Zealand.

in order to replicate the main closing phase muscles in the human jaw. The actuators apply approximately linear movement along the lines of muscle action that was derived from human anatomical studies. The resulting parallel design has the advantage that the work done by each actuator could reflect the contribution of work done by the key muscles of mastication in humans.

It was shown that the robot could be controlled to accurately simulate human mastication trajectories. This was carried out by translating the incisor point trajectories measured from human subject trials into molar point trajectories that could then be matched to the robot. Full details of this process were outlined in Chapter 6. Through application of compliance force control, the trajectory of the jaw could be made to achieve occlusion in a way that mimics the hierarchy of control observed in humans. Chapter 7 detailed how the forces acting on the foods could be calculated from the recorded motor torques.

The presentation and interpretation of preliminary experiments on the application of the robot to the chewing of foods was the key purpose of Chapter 7. Because this was discussed in detail, only the broad findings are outlined here. The robot was implemented with a series of 1, 2 and 3D trajectories to allow comparison of the measured force profiles with other alternative measurement methods. Very good agreement between uni-axial compression tests were found using the 1D trajectory. Similarly in the z-direction excellent agreement was found between the chewing robot and the 2D linkage chewing machine outlined in Chapter 5.

3D full speed human trajectories were also applied to a number of foods, each time exhibiting force-curves expected for the food being tested. For each trajectory however there were unexpected forces measured in the x-direction which could be explained by friction forces in the linkage mechanism under loaded conditions. Further work is required to resolve this issue. The implementation of force sensors directly under the active molars might provide more direct measurements of the forces applied to the foods. In this way some ambiguity regarding the calculation of forces from the motor torques can be avoided.

On the whole the experiments using the chewing robot were encouraging and further development of the robot is continuing. In order to accurately implement chewing trajectories it is recommended to use chewing cycles recorded later on in a human sequence, where full occlusion is achieved. With appropriate torque limits applied to the actuator, this would provide a good approximation to the changes in trajectory used during a mastication sequence. As the food breaks down and offers less resistance to the teeth, the trajectory will approach full occlusion.

To allow simulation of whole chewing sequences it will be necessary to add other functionality to the robot. Of primary importance will be a mechanism to retain the food on the occlusal surfaces of the robot. To allow the robot to chew larger portion sizes some way to manipulate the bolus between cycles (like the tongue) will also be required. After these issues have been resolved, secondary features such as temperature control or addition of saliva could be implemented.

The other area for development of the robot is in its control. While the use of PID trajectory control with compliance force control approximates humans, there is an opportunity to implement higher level control systems where the robot trajectory is made to adapt in response to the food properties. This is an exciting area for ongoing research.

8.3.6 Other Oral Simulation Devices

Other devices have also been developed to simulate oral processing of liquid or semisolid foods which are not masticated or for evaluation of food boluses after oral processing. Prinz et al. [51] reported an in-vitro simulator for applying and measuring the changes in viscosity of semisolid foods due to temperature, shear, dilution, and addition of amylase enzymes. It consisted of a vane system which contained the food which is then mixed by rotation of the vane. The extent of mixing was evaluated through imaging a point of laser light projected at the surface of the sample. The rheological properties of the sample were evaluated by measuring the torque required to mix the food. The device was applied to starch based foods where it was shown that the in-vitro simulator could produce similar degrees of mixing as for human oral processing. Although the rates of heat transfer were much slower than in humans, the application of shear could be approximated. An interesting preliminary finding was that even small amounts of amylase could significantly reduce the viscosity of the food sample over the time scale of eating semisolid foods.

Seo et al. [52] developed two different simulation devices to approximate swallowing of food boluses. The successful transport of a food from the oro-pharynx to the stomach is achieved by the application of contact pressure on to the top of the bolus and peristalsis to force the food down the pharynx. The bolus must flow around the closed off trachea and as such deformation of the bolus is important, as was surface lubrication. From these mechanisms Seo et al. identified two important bolus properties for successful swallow; slipperiness (the degree of slide of the bolus against the mucosal surface) and compliance (the degree of ease the shape of a bolus can be transformed).

To allow in-vitro assessment of slipperiness, the food bolus was placed into a 22 mm diameter ring that sat on a sliding platform such that there was direct contact between the bolus and the platform surface. The platform was then tilted at a rate of 1 °/s. The angle of the platform at which the food began to slide and the angle when it had slid 100 mm from the start line were used as measures of slipperiness. While this device is very simple and the actual lubrication properties between the acrylic platform and the food bolus were not representative of the contact between the bolus and oral mucosa, there was excellent correlation between this measure and sensory evaluations for a range of foods including water, pumpkin gruel, yoghurt, tofu, pudding and canned pork.

Bolus consistency was evaluated by pushing a hollow cylindrical probe into a vessel containing the sample. The inner surface of the probe changed in profile and shape along its vertical axis. In this way the food bolus had to change shape from a hexahedral to cylindrical with reducing radius. The force profile was then measured using a texture analyser. High negative correlations were found between sensory and the instrumental measurements of consistency. This suggests that high peak forces characterize low compliance of the food.

While these systems are not chewing robots, they are analogous to them in that they aim to approximate the forces, heat and mass transfer processes occurring in the mouth or pharynx and concurrently measure the changes in sensory properties of the food in real time.

8.4 Summary

In this chapter the application of chewing robots to the understanding food texture was discussed. By reviewing the key processes occurring to foods during mastication in humans, several key requirements for robots were identified. The time frame of human chewing must be replicated, true representations of both tooth morphology and trajectory are required, control of force during the occlusal phase is important to ensure the correct mechanisms of particle size reduction are achieved, temperature control and addition of saliva are important, mechanisms to manage realistic bite volumes are needed, and significant losses occur from a food during chewing but the food must still be retained in some way on the occlusal surfaces.

In view of these requirements, the reports of chewing robots applied to foods were reviewed. Achieving all of these features in a single robot is a tall order and to date most of the focus has been on implementing realistic trajectories with anatomically correct teeth. It is clear from the findings so far, that mastication robots will provide useful information on food texture dynamics which current tools in the food science field can not achieve. The analysis in this chapter helps to clarify the future direction of robotic development in view of their application to understanding food texture.

References

1. Rosenthal, A.J.: Relation between instrumental and sensory measures of food texture. In: Rosenthal, A.J. (ed.) Food Texture: Measurement and Perception, pp. 1–17. Aspen Publishers, Maryland (1999)
2. Bourne, M.C.: Food Texture and Viscosity: Concept and Measurement. Academic Press, San Diego (2002)
3. Szczesniak, A.: Correlating sensory with instrumental texture measurements: An overview of recent developments. J. Texture Stud. 18, 1–15 (1987)
4. Greaser, M.L., Pearson, A.M.: Flesh foods and their analogues. In: Rosenthal, A.J. (ed.) Food Texture: Measurement and Perception, pp. 228–258. Aspen Publishers, Maryland (1999)
5. Agrawal, K.R., et al.: Mechanical properties of foods responsible for resisting food breakdown in the human mouth. Arch. Oral Biol. 42, 1–9 (1997)
6. Agrawal, et al.: Food properties that influence neuromuscular activity during human mastication. J. Dent. Res. 77, 1931–1938 (1998)
7. Lucas, P.W.: Dental Functional Morphology. Cambridge University Press, Cambridge (2004)
8. Lucas, P.W., et al.: Food physics and oral physiology. Food Qual. Pref. 13, 203–213 (2002)
9. Chen, J.: Food oral processing – a review. Food Hydrocolloids 23, 1–25 (2008)
10. Strait, S.G.: Tooth use and the physical properties of food. Evol. Anthropol. 5, 199–211 (1998)
11. Agrawal, K.R., et al.: The effects of food fragmentation index on mandibular closing angle in human mastication. Archives of Oral Biology 45, 577–584 (2000)

12. Nakajima, J., et al.: Masticatory mandibular movements for different foods textures related to onomatopoetic words. J. Med. Dent. Sci. 48, 121–129 (2001)
13. Anderson, K., et al.: The effects of bolus hardness on masticatory kinematics. J. Oral Rehab. 29, 689–696 (2002)
14. Peyron, M.A., et al.: Particle size distribution of food boluses after mastication of six natural foods. J. Dent. Res. 83, 578–582 (2004)
15. Peyron, M.A., et al.: Effects of increasing hardness on jaw movement and muscle activity during chewing of visco-elastic model foods. Exp. Brain Res. 142, 41–51 (2002)
16. Foster, K.D., et al.: Effect of texture of plastic and elastic model foods on the parameters of mastication. J. Neurophysiol. 95, 3469–3479 (2006)
17. Bosman, F., et al.: Neuromuscular control mechanisms in human mastication. J. Text. Stud. 35, 201–221 (2004)
18. Fontijn-Tekamp, F.A., et al.: Swallow threshold and masticatory performance in dentate adults. Physiol. Behav. 83, 431–436 (2004)
19. Gaviao, M.B., et al.: Chewing behaviour and salivary secretion. Eur. J. Oral Sci. 112, 19–24 (2004)
20. Kohyama, et al.: Textural evaluation of rice cake by chewing and swallowing measurements on human subjects. Biosci. Biotech. Biochem. 71, 358–365 (2007)
21. Buschang, P.H., et al.: The effects of bolus size and chewing rate on masticatory performance with artificial test foods. J. Oral Rehab. 24, 522–526 (1997)
22. Lucas, P.W., Luke, D.A.: Optimum mouthful for food comminution in human mastication. Arch. Oral Biol. 29, 205–210 (1984)
23. Yagi, K., et al.: Changes in the mouthful weights of familar foods with age of five years, eight years and adults. Ped. Dent. J. 16, 17–22 (2006)
24. Medicis, S.W., Hiiemae, K.H.: Natural bite sizes for common foods. J. Dent. Res. 77, 295 (1998)
25. Hiiemae, K., et al.: Natural bites, food consistency and feeding in man. Arch. Oral Biol. 41, 429–441 (1996)
26. Hutchings, S.C., et al.: Variation of bite size with different types of food bars and implications for serving methods in mastication studies. Food Qual. Pref. 20, 456–460 (2009)
27. Hiiemae, K.M., Palmer, J.B.: Tongue movements in feeding and speech. Crit. Rev. Oral Biol. Med. 16, 413–429 (2003)
28. Mioche, L., et al.: The intra-oral management of food: a postero-anterior videofluorographic study. Arch. Oral Biol. 4, 267–280 (2002)
29. Jalabert-Malbos, M.L., et al.: Particle size distribution in the food bolus after mastication of natural foods. Food Qual. Pref. 18, 803–812 (2007)
30. Yven, C., et al.: Meat bolus properties in relation with meat texture and chewing context. Meat Sci 70, 365–371 (2005)
31. Flynn, C.S., et al.: Identification of multiple compartments present during the mastication of solid food. Arch. Oral Biol. (2010) (Manuscript submitted)
32. Vingerhoeds, M.H., et al.: Emulsion flocculation induced by saliva and mucin. Food Hydrocolloids 19, 915–922 (2005)
33. Pereira, L.J., et al.: Effects of added fluids on the perception of solid food. Physiol. Behav. 88, 538–544 (2006)
34. Xu, W.L., et al.: Review of the human masticatory system and masticatory robotics. Mech. Machine Theory. 43, 1353–1375 (2008)
35. Flores, E., Fels, S.: Design of a 6DOF anthropomorphic robotic jaw. J. Acoustical Soc. Am. 117, 2547 (2005)

36. Alemzadeh, K., et al.: Prototyping a robotic dental testing simulator. J. Eng. Med. 145, 385–396 (2007)
37. Alemzadeh, K., Raabe, D.: Prototyping artificial jaws for the robotic dental testing simulator. J. Eng. Med. 222, 1209–1220 (2008)
38. Alemzadeh, K., Raabe, D.: Prototyping artificial jaws for the Bristol Dento-Munch robo-simulator – a parallel robot to test dental components and materials. In: Proceedings of Annual International Conference of the IEEE Engineering in Medicine and Biology Society, pp. 1453–1456 (2007)
39. Conserva, E., et al.: Robotic chewing simulator for dental materials testing on a sensor-equipped implant setup. Int. J. Prosthodont. 21, 501–508 (2008)
40. Takanobu, H., et al.: Quantification of masticatory efficiency with a mastication robot. In: Proceedings of the 1998 IEEE International Conference on Robotics and Automation, Leuven, Belgium (1998)
41. Meullenet, J.F., Gandhapuneni, R.K.: Development of the BITE Master II and its application to the study of cheese hardness. Phys. Behaviour. 89, 39–43 (2006)
42. Salles, C., et al.: Development of a chewing simulator for food breakdown and the analysis of in-vitro flavor compound release in a mouth environment. J. Food Eng. 82, 189–198 (2007)
43. Xu, W.L., et al.: A robotic model of human masticatory system for reproducing chewing behaviours. IEEE Robotics Automation Mag. 12, 90–98 (2005)
44. Xu, W.L., et al.: Mechanism, design and motion control of a linkage chewing device for food evaluation. Mech. Machine Theory 43, 376–389 (2008)
45. van Ruth, S.M., Roozen, J.P.: Influence of mastication and saliva on aroma release in a model mouth system. Food Chem. 71, 339–345 (2000)
46. Roberts, D.D., Acree, T.E.: Simulation of retronasal aroma using a modified headspace technique: Investigating the effects of saliva, temperature, shearing and oil on flavor release. J. Agric. Food Chem. 43, 2179–2186 (1995)
47. Dielbler, K.D., et al.: Verification of a mouth simulator by in vivo measurements. J. Agric. Food Chem. 49, 1388–1393 (2001)
48. Sun, C., et al.: A linkage chewing machine for food texture analysis. Int. J. Intell. Syst. Technol. Appl. 8, 303–318 (2010)
49. Anderson, D.J.: Measurement of stress in mastication. J. Dent. Res. 35, 664–670 (1956)
50. Saklar, S., et al.: Instrumental crispness and crunchiness of roasted hazelnuts and correlations with sensory assessment. J. Food Sci. 64, 1015–1019 (1999)
51. Prinz, J.F., et al.: In vitro simulation of the oral processing of semi-solid foods. Food Hydrocolloids 21, 397–401 (2007)
52. Seo, et al.: Sensory and instrumental analysis for slipperiness and compliance of food during swallowing. J. Food Sci. 72, 707–713 (2007)

Chapter 9
Neural Control of a Mastication Robot*

Abstract. The Matsuoka neural oscillator can potentially be employed as the central pattern generator (CPG) for a chewing robot, in order to generate and adapt rhythmic actuations in response to sensory feedback. In this chapter a single Matsuoka oscillator of two neurons is applied to two phase-locked muscles (e.g. masseter and digastric muscles) or for a single robotic joint. Three graphical user interfaces (GUIs) were developed to help design and tune the oscillator. A case study is presented involving a jaw, driven by a couple of opening and closing muscles and commanded by motoneurons. The force of the muscles was described using a nonlinear Hill model while the motoneuron for muscle activities was modelled using the oscillator. Simulations were performed to show the oscillator's ability to generate and adapt its rhythmic outputs with respect to chewing without food (i.e. EMG only for rhythmic muscle activities), with foods (i.e. EMG for rhythmic and additional muscle activities) and with crushable foods (to see how quickly the oscillator to reduce its force commands in order not to damage the teeth). Furthermore, a hardware-in-the-loop simulation system is presented to validate this neural controlling algorithm, where the chewing robot is actuated by two fluidic muscles commanded by the Matsuoka oscillator. The robot had a fixed upper jaw and actuated lower jaw, with position and force sensors used to measure jaw movements and the food resistance. The oscillator, simulated in Matlab, was interfaced to the control valves of the fluidic muscles via a sensory I/O card. The control of the robot is achieved via Simulink with the real-time windows library. Comprehensive experiments were conducted and the results interpreted to show the distinct dynamic behaviours of the neural controlling algorithm.

9.1 Why CPG for Control

Human mastication patterns vary between subjects and with food texture and are continually modified throughout the chewing sequence in response to the food dynamics. This interactive relationship has been utilized to evaluate textural

* The first half of this chapter is reprinted from Xu WL, Fang FC, Bronlund JE and Potgieter J (2009) Generation of rhythmic and voluntary patterns of mastication using Matsuoka oscillator for a humanoid chewing robot. Mechatronics. 19:205-217, with permission from Elservier.

W. Xu and J.E. Bronlund: Mastication Robots, SCI 290, pp. 237–270.
springerlink.com

properties of foods by measurement of masticatory physiology [1, 2]. The challenge in evaluating foods this way is that mastication is difficult to fully characterise and to some extent, the measurements themselves interfere with the natural chewing behaviour. As result, food texture can only be evaluated semi-quantitatively and various hypotheses could not be tested on human subjects easily.

To this end, a chewing robot solution has been proposed [3, 4]. The idea is that while being chewed by a robot, the food properties and textural changes occurring during chewing are evaluated by robotic actuations states, chewing force, and/or jaw movements. Although it could be used to chew foods, the WJ (Waseda Jaw) robot series were developed to work especially with the WY series dental training robots [5, 6]. They did not take into account the biological aspects of the human masticatory system and hence the robotic states can't be used for purpose of food evaluation. The 6RSS mechanism parallel chewing robot reported in Chapters 3 and 4 [7, 8] was based on biomechanical specifications derived from the jaw structure and the muscles of mastication. It was developed especially for food evaluation. These robots may be commanded to chew foods by following recorded masticatory movements and chewing forces, but do not mimic the variations in trajectory and force application in response to the changing food properties in the way a human does.

The chewing of foods by humans is performed by the movement of the jaw due to the muscles of mastication. Alpha-motoneurons (or alpha-MN) innervate a muscle by recruiting a number of motor units and firing them at various frequencies [9, 10]. Electromyography (EMG) measurements have confirmed that a small amount of muscle activity is required for the free rhythmic movements of the jaw, and are produced by the central pattern generator (CPG). Additional voluntary muscle activities are generated in the alpha-MN's if the closing movement is resisted by foods [11, 12]. The harder the food, the larger the muscle activity required [13, 14].

In principal, the muscles of mastication involve anticipatory (or feed-forward) activities for pre-programmed movement depending on individual chewing expectations, rhythmic activities generated by the CPG that are dictated by individual physiology, and voluntary (sensory feedback) activities for overcoming food resistance. To make a chewing robot chew in a human way, these rhythmic, anticipatory and voluntary patterns of muscle activity should be implemented.

A CPG can produce coordinated rhythmic muscle activities without any sensory input. CPG's have been modelled as systems of neural oscillators, such as the Ellias-Grossberg oscillator [15], Matsuoka oscillator [16], RIO [17] and van der Pol oscillator [18]. These CPG models can not only generate robust, the sustained oscillations necessary for many human body movements (such as chewing, locomotion, heart beats and breathing), but can also be modulated to induce gait transition or variations smoothly [19]. Matsuoka oscillator based CPG models have been popular for humanoid control of robots, such as robotic armed and legged devices [20, 21], where the dynamics of the robots are exploited to entrain the oscillators.

In this work we intend to apply Matsuoka neural oscillators to humanoid chewing robots to generate rhythmic actuation patterns and which adapt in response to sensory feedback. The hypothesis is that if each of the joints (as muscles) of the 6RSS parallel mechanism chewing robot [4, 7, 8] are actuated (in future work) using a Matsuoka oscillator, humanoid chewing behaviour could be achieved with the robot.

In the following sections we will present our very first attempt towards the humanoid robotic chewing, with simulations of a Matsuoka oscillator of two neurons serving a two phase-locked muscle system (e.g. masseter and digastric muscles) or a single robotic joint. We also present a hardware-in-the-loop simulation approach to further test the Matsuoka oscillator based controlling algorithm. The simplified chewing robot used was actuated by two fluidic muscles and commanded by a Matsuoka oscillator of two neurons. Position and force sensors were used to determine the mouth opening and food resistances, respectively. The oscillator was simulated in Matlab and interfaced to the proportional control valves of the robotic muscles. The control of the robot was achieved through real-time windows software.

9.2 Matsuoka Oscillator Fundamentals

Here we consider a simple Matsuoka oscillator to generate/adapt the masticatory pattern of two phase-locked muscles. The oscillator, as illustrated in Fig. 9.1, consists of two neurons that are connected in a mutually inhibitory fashion. When one neuron (e.g., for the closing muscle of mastication) is activated the other neuron (e.g., for the opening muscle) is suppressed and versa versus. The firing of the neurons is alternated.

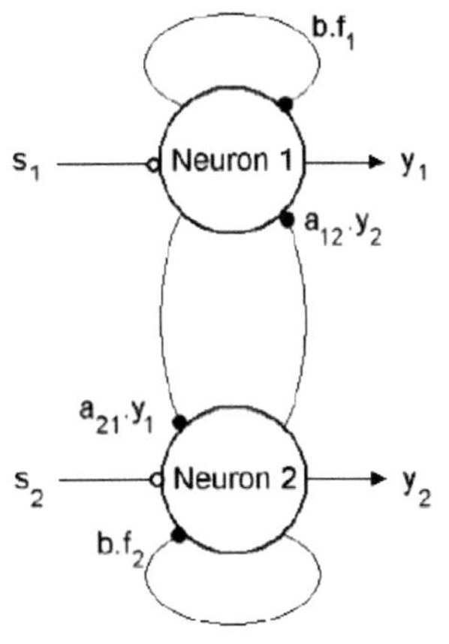

Fig. 9.1 A Mastuoka oscillator of two neurons

Matsuoka [16] showed that the dynamics of the oscillator were governed by:

$$T_{ri}\frac{dx_i}{dt} + x_i = -a_{ij}y_j + s_i - b_i f_i \tag{9.1}$$

$$y_i = g(x_i) \tag{9.2}$$

$$T_{ai}\frac{df_i}{dt} + f_i = y_i^q \tag{9.3}$$

in which i, j =1 or 2 for neuron 1 or 2, x_i is the state of neuron i, s_i the tonic input, y_i the output firing rate, T_{ri} the rise time constant, T_{ai} the adaptation time constant, a_{ij} the weight of the inhibitory connection from neuron j to neuron i, f_i represents the degree of fatigue or adaptation in neuron i, b_i the adaptation firing rate, q_i an exponent for that the adaptation effect is in proportion to a power of the output firing rate, and $g(x)$=max(x,0) is a threshold function.

The above oscillator was implemented in Matlab/Simulink (Fig. 9.2) in which each and every oscillator's parameter could be varied. For the oscillator to produce

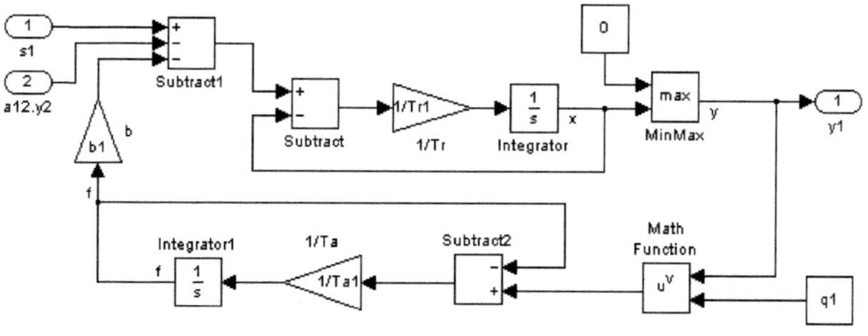

(a) neuron model (neuron 1)

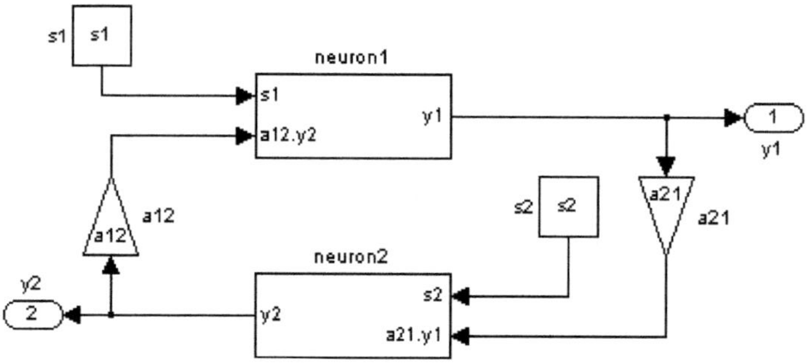

(b) oscillator model

Fig. 9.2 Matsuoka oscillator in Matlab/Simulink, (a) neuron model (neuron 1) and (b) oscillator model

a sustained oscillation, the firing rate constant (b_1 and b_2) must be large and satisfy the criteria provided specified by Matsuoka [16]. After a short transient period, the two outputs of the oscillator become stable but phase-locked due to the mutual inhibition and their alternating firing results from the adaptation in the currently firing neuron and the building-up of the activity in the other neuron. It was found that the frequency of the oscillation is positively correlated to the firing rate constant, and negatively correlated to the rise time constant, adaptation time constant and connection weights, while the amplitude of the oscillation is directly proportional to the increment of the tonic input [22].

9.3 Tuning of a Mastuoka Oscillator

The oscillator (Fig 9.1) may be used to produce up to three rhythmic outputs, i.e., y_1, y_2 and y_1- y_2. For example, for a pair of jaw closing and opening actuators, two phase-locked outputs y_1 and y_2 are required for the motoneuronal control and for a joint of a robotic arm, the output y_1- y_2 is necessary for motion commands [20]. An output of the oscillator consists of both transient and sustained oscillation and consequently, it may be characterized by a number of measurements, such as frequency, amplitude, phase overlapping (of y_1, y_2), transient time and transient peak etc., depending on its application. As the oscillator involves a number of parameters and various measurements, for which no simple analytical closed-form relationships are available, we have developed three user interfaces (GUIs) in Matlab (based on the model shown in Fig. 9.2), to help tune the oscillator for prescribed behaviours of oscillation. The GUIs are more easily comprehended than other design methods, such as describing functions [20] and numerical methods [22]. The first GUI (Fig. 9.3) is useful for simulating the behaviour of the designed oscillator, the second GUI (Fig. 9.5) is helpful in selecting the values of the oscillator's parameters to achieve the specified behaviours, and the third GUI (Fig. 9.6) is to visualise or predict adaptive behaviour of the oscillator with varying parameters.

9.3.1 GUI for Simulation of the Oscillator

The first GUI (Fig. 9.3) is for simulation of the oscillator with known values of the parameters. The values of the oscillator's parameters are supplied directly by typing into spaces (labelled 2) located at the upper right corner of the GUI. Single or multiple outputs of y_1, y_2 and y_1- y_2 can be selected (5), traced (3) and displayed in the window (4). The measurements, including *Period* for the frequency, T_{on} for the duty cycle, *Peak* for the amplitude of the oscillation, *Time of Overlapping* (of y_1 and y_2) and *Transient Time* and *Maximum Peak* can be displayed automatically (6) or measured manually using a combination of cursor and mouse click (8). The radio button *Export>>* (10) is for separate production of the output diagrams (e.g., Figs. 9.4 and 9.6 etc).

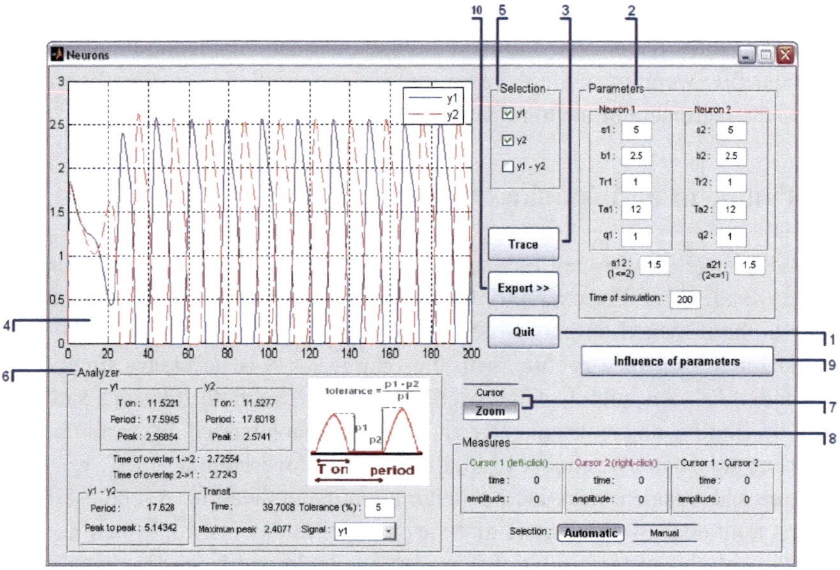

Fig. 9.3 GUI for simulation

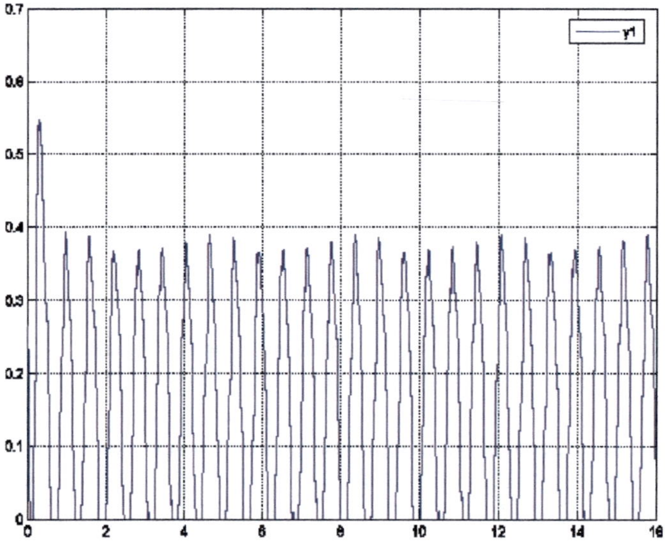

Fig. 9.4 An EMG muscle activity generated by the oscillator

The default example seen in the GUI was taken from Masuoka [16]. The sustained oscillations of y_1 and y_2 match those in the reference, but also involve a transient of around 20 s that is roughly one cycle of the sustained oscillation. This transient dynamics is appreciated in engineering applications of Matsuoka oscillator.

An EMG muscle activity was reproduced for right masseter muscle measurements on a human subject chewing almonds [13], as shown in Fig. 9.4. The oscillator to match the human chewing sequence had the following parameters, s_1= s_2=1, b_1= b_2=2.5, T_{r1}= T_{r1}=0.03, T_{a1}= T_{a2}=0.4, q_1= q_2=0.7, a_{12}= a_{21}=1.5 (found via the GUI shown in Fig. 9.5, to be discussed below). The major properties of the EMG pattern reproduced are T_{on}=0.456 s, *Period*=0.618 s, *Amplitude*=0.380 and Transient Time=0.78 s. The generated EMG agrees with the measured one in most of its characteristics (e.g. frequency, duty cycle and amplitude).

9.3.2 GUI for Influence of the Oscillator Parameters

Matsuoka drew partial and qualitative conclusions on the influence of the oscillator's parameters and its behaviour, as stated earlier [16, 22]. This is very valuable, but not easy for use of selecting the oscillator to generate oscillations for specific applications. A GUI (Fig. 9.5) was developed for this purpose, by means of which the influence of any oscillator's parameter on any particular feature of the oscillator's behaviour can be analysed. This GUI is embedded in the first GUI and accessed via the button *Influence of parameters* (9) as shown in Fig. 9.3. A Matlab program runs behind the scene to simulate the oscillator of interest and predict its characteristics.

In the GUI there are two windows, *Graph 1* and *Graph 2* (labelled 6 and 7) to display up to 8 features of the oscillator, each for 4 features selected via the spaces provided (4 and 5). The features available include duty cycle, period, amplitude, phase overlapping time, transient time and transient amplitude. Any of y_1, y_2 or y_1-y_2 can be chosen as the output to be analysed (4), but the influence analysis can be performed for only one oscillator parameter at a time. The parameter and its range are set in the left middle area *Range* (2) of the GUI. Pressing the button *Trace* (3) displays the variations of the selected feature versus the selected parameter in the *Graph 1* and *Graph 2* windows. Prssing *Export*>> (8) creates higher quality graphs in separate sheets, one graph per sheet.

Using this GUI, a thorough influence analysis was performed. Tables 9.1 to 9.3 give the results with respect to q=q_1=q_2=1, >1 and <1, respectively. The symbols ↑, ↓, ↔, and ≈ used in these tables stand for a measure's increment, decrement, no-change and no-tendency with respect to the increment in value of an oscillator's parameter, respectively.

It is interesting to remark the effect of the exponent constant q=q_1=q_2. When q=1, varying s_1 and s_2 varies the amplitude of the oscillation; varying T_{r1}, T_{r2}, T_{a1}

and T_{a2} changes the frequency (or period) and duty cycle of the oscillation; varying b_1, b_2, a_{12}, and a_{21} changes the shape of the oscillation; and when q <1 or >1, changes in s_1 and s_2 changes the amplitude, period and overlapping time simultaneously. A useful consequence of this is that both s_1 and s_2 may be varied to vary both amplitude and period if q is set to be non-unity.

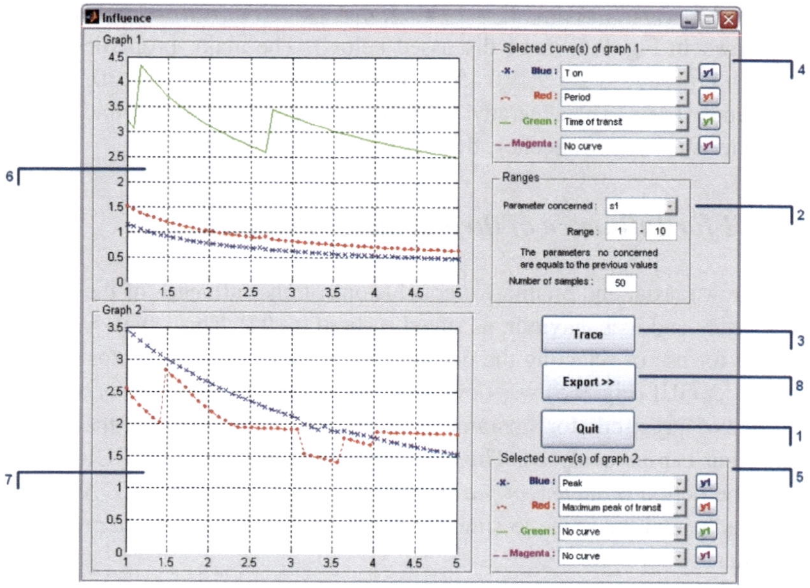

Fig. 9.5 GUI for influence analysis

9.3.3 *GUI for Adaptation of the Oscillator*

In applications such as modelling human mastication, an oscillator can be used as a CPG to generate the rhythmic pattern of muscle activity. This requires fixing a value for each parameter of the oscillator. A considerable amount of experiments have confirmed that the rhythmic pattern is invariant in humans for free jaw movements without foods. While chewing on foods, the rhythmic pattern adapts by the motoneuron quickly in response to the change in food resistance during the chewing process [11-13]. That is to say, if a single oscillator is used to generate the central pattern of muscle activity, one or a few of the oscillator parameters need to be adapted in real time in order to have the varying voluntary muscle activities required. To simulate how the oscillator behaves over time with respect to varying parameter, a third GUI was developed, as shown in Fig. 9.6.

Table 9.1 Influence analysis, q=1

	Ton		Period		Peak		Time of transit		Max peak of transit		Time of overlap	
	y1	y2	y1	y2	y1	y2	y1	y2	y1	y2	y1 => y2	y2 => y1
s1	↑	↓	≈	≈	↑	↔	↓	↑	↑	↔	↓	↑
s2	↓	↑	≈	≈	↔	↑	↑	↓	↔	↑	↑	↓
s1 & s2	↔	↔	↔	↔	↑	↑	↔	↔	↑	↑	↔	↔
b1	↓	↔	↓	↓	↓	↓	↓	↓	↓	↔	↑	↓
b2	↔	↓	↓	↓	↓	↓	↓	↓	↔	↓	↓	↑
b1 & b2	↓	↓	↓	↓	↓	↓	↓	↓	↓	↓	↓	↓
Tr1	↑	↑	↑	↑	↓	↔	↑	↑	↔	↔	↑	↑
Tr2	↑	↑	↑	↑	↔	↓	↑	↑	↔	↔	↑	↑
Tr1 & Tr2	↑	↑	↑	↑	↓	↓	↑	↑	↔	↔	↑	↑
Ta1	↑	↑	↑	↑	↔	↔	↔	↑	≈	≈	↑	↑
Ta2	↑	↑	↑	↑	↔	↔	↑	↔	≈	≈	↑	↑
Ta1 & Ta2	↑	↑	↑	↑	↔	↔	↑	↑	≈	≈	↑	↑
a12	≈	↑	↑	↑	↑	↔	↑	↑	≈	≈	≈	↓
a21	↑	≈	↑	↑	↔	↑	↑	↑	≈	≈	↓	≈
a12 & a21	↑	↑	↑	↑	↑	↑	↑	↑	≈	≈	↓	↓

Table 9.2 Influence analysis, q>1

	Ton		Period		Peak		Time of transit		Max peak of transit		Time of overlap	
	y1	y2	y1	y2	y1	y2	y1	y2	y1	y2	y1 => y2	y2 => y1
s1	↑	↓	↓	↓	↑	↔	=	=	↑	↔	↓	↑
s2	↓	↑	↓	↓	↔	↑	=	=	↔	↑	↑	↓
s1 & s2	↓	↓	↓	↓	↑	↑	↓	↓	↑	↑	↓	↓
b1	↓	↔	↓	↓	↓	↓	=	=	↓	↔	↑	↓
b2	↔	↓	↓	↓	↓	↓	=	=	↔	↓	↓	↑
b1 & b2	↓	↓	↓	↓	↓	↓	↓	↓	↓	↓	↓	↓
Tr1	↑	↑	↑	↑	↓	↔	↑	↑	↔	↔	↑	↑
Tr2	↑	↑	↑	↑	↔	↓	↑	↑	↔	↔	↑	↑
Tr1 & Tr2	↑	↑	↑	↑	↓	↓	↑	↑	↔	↔	↑	↑
Ta1	↑	↑	↑	↑	↔	↔	↔	↑	=	=	↑	↑
Ta2	↑	↑	↑	↑	↔	↔	↑	↔	=	=	↑	↑
Ta1 & Ta2	↑	↑	↑	↑	↔	↔	↑	↑	=	=	↑	↑
a12	=	↑	↑	↑	↑	↔	=	=	=	=	=	↓
a21	↑	=	↑	↑	↔	↑	=	=	=	=	↓	=
a12 & a21	↑	↑	↑	↑	↑	↑	=	=	=	=	↓	↓

Table 9.3 Influence analysis, q<1

	Ton		Period		Peak		Time of transit		Max peak of transit		Time of overlap	
	y1	y2	y1	y2	y1	y2	y1	y2	y1	y2	y1 => y2	y2 => y1
s1	↑	↓	=	=	↑	↑	=	=	↑	↔	↓	↑
s2	↓	↑	=	=	↑	↑	=	=	↔	↑	↑	↓
s1 & s2	↑	↑	↑	↑	↑	↑	↑	↑	↑	↑	↑	↑
b1	↓	↔	↓	↓	↓	↓	=	=	↓	↔	↑	↓
b2	↔	↓	↓	↓	↓	↓	=	=	↔	↓	↓	↑
b1 & b2	↓	↓	↓	↓	↓	↓	↓	↓	↓	↓	↓	↓
Tr1	↑	↑	↑	↑	↓	↔	↑	↑	↔	↔	↑	↑
Tr2	↑	↑	↑	↑	↔	↓	↑	↑	↔	↔	↑	↑
Tr1 & Tr2	↑	↑	↑	↑	↓	↓	↑	↑	↔	↔	↑	↑
Ta1	↑	↑	↑	↑	↔	↔	↔	↑	=	=	↑	↑
Ta2	↑	↑	↑	↑	↔	↔	↑	↔	=	=	↑	↑
Ta1 & Ta2	↑	↑	↑	↑	↔	↔	↑	↑	=	=	↑	↑
a12	=	↑	↑	↑	↑	↔	↑	=	=	=	=	=
a21	↑	=	↑	↑	↔	↑	=	↑	=	=	=	=
a12 & a21	↑	↑	↑	↑	↑	↑	↑	↑	=	=	=	=

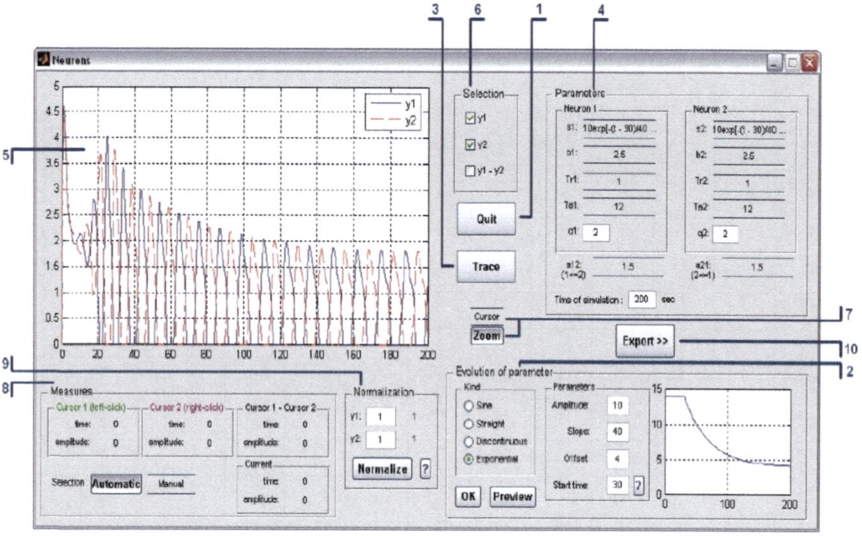

Fig. 9.6 GUI for adaptation analysis

In this GUI any of the oscillator parameters can be selected to vary over time. When this is done, the rest of the parameters remain unchanged. The varying parameter is set in the *Parameters* panel of the GUI (4) and its variation is defined in the *Evolution of parameter* panel (2). There are four options to choose from: Sine (for a sinusoid function), Straight (for a constant or linear function), Discontinuous (for any discrete function specified in data sets) and Exponential (for an exponential function). Once a function is chosen and specified, it can be previewed, and then if satisfactory, confirmed.

The chosen output of the oscillator is displayed on the graph (5). The output can be normalised to any range using the spaces in the *Normalisation* panel (9). The rest of the GUI features are similar to the other GUIs. The output can also be measured at any time point by cursor and mouse click (8).

The default example in the GUI (Fig. 9.6) is for exponential variation of s_1 and s_2 with q_1=q_2=2. Fig. 9.7 shows another output (y_1-y_2) for $s_1 = s_2 = 10e^{-(t-70)/40}$ and q_1=q_2=2. It can be seen from both examples that the oscillator output y_1, y_2 and y_1-y_2 have a decreasing amplitude and increasing period over time. For another example (Fig. 9.8), s_1=s_2 is set at 8 for the period of 80-120 s and 3 for the rest of the time. The result confirmed that the oscillator adapts its behaviour very quickly and changes its stable oscillation from one to another within a cycle. In this example, the oscillation adapted its amplitude at 80 s and 120 s without much transient.

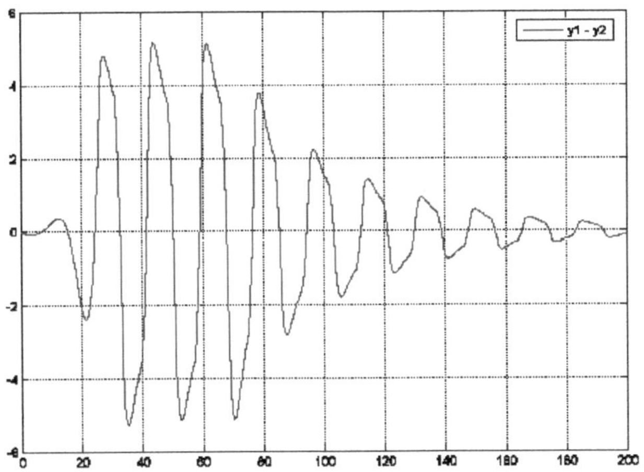

Fig. 9.7 Output y_1-y_2 for s_1=s_2=10exp[-(t-70)/40]

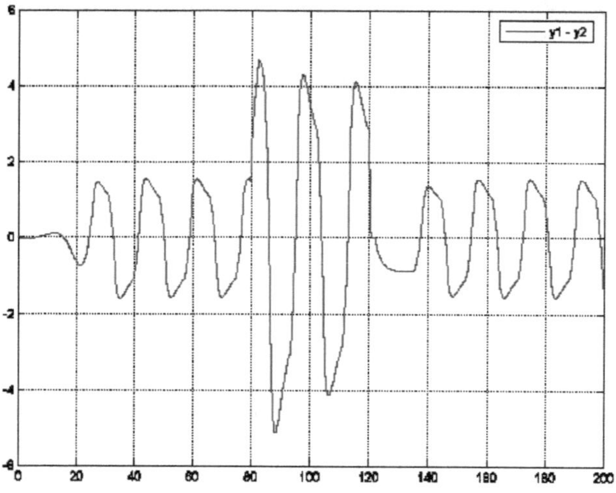

Fig. 9.8 Output y_1-y_2 for s_1=s_2 =3 for 0-80 s, 8 for 80-120 s and 3 for 120-200 s

9.4 Case Study

9.4.1 Purpose and Assumptions

The jaw is actuated alternatively by a group of opening and closing muscles, resulting in rhythmic chewing movements. The masticatory system is a typical neuromotor system whose simplest functional unit is a motor, consisting of an alpha-motoneuron and the muscle fibres it innervates [9]. To increase muscle force, the central nervous system may recruit new motor units and/or increase the frequency of firing of already recruited motor units. For non-clinical applications, muscle activity may be recorded by surface electromyography (EMG). A reproduced EMG envelope of the masseter activity profiled by filtering, rectification and integration has been illustrated in Fig. 9.4.

One of the proposed mechanisms for voluntary control of single muscles is the so called 'servo' hypothesis, which has been applied in this case study using the Matsuoka oscillator [9]. According to this hypothesis, a voluntary movement of muscle is generated using information about the neurophysiological mechanisms of muscle reflexes. The CPG, via gamma-MN's, sets a value of muscle length, and alpha-MN's achieve the set muscle length by varying their activities to balance the variation in external load (e.g. food resistance for masticatory muscles). This hypothesis is criticised because of the large gain in the tonic stretch reflex, but is still attractive in terms of the engineering application of its servo control.

This case study assumes that the jaw is driven by a closing muscle (masseter) and an opening muscle (digastric), gamma-MN's set the rhythmic muscle activities (for rhythmic jaw movement), and alpha-MN's, modelled by the oscillator, generate the voluntary muscle activities. The two outputs of the oscillator (y_1 and y_2) are for the alpha-MN activity of the closing and opening muscles, respectively. The oscillator is adapted by the variation in food resistance (F_f) through the two identical tonic inputs s_1 and s_2. Fig. 9.9 illustrates the case study problem.

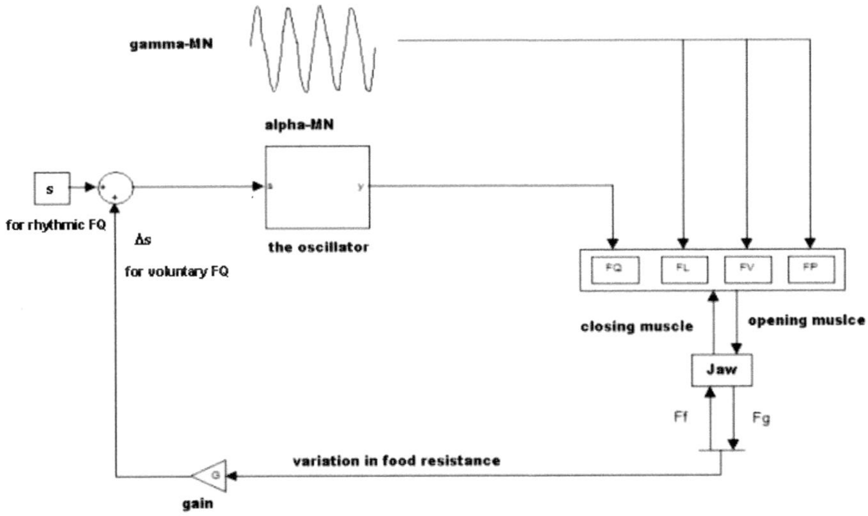

Fig. 9.9 Rhythmic and voluntary muscle activities in a simplified jaw system

9.4.2 Modeling of the Masticatory Muscles

Both masseter and digastric muscles are represented by the nonlinear Hill model and the force of the muscle is calculated by [23]

$$F = F_{\max}(FL \times FV \times FQ + FP) \tag{9.4}$$

in which $F_{\max}$ is the maximum force of a muscle; *FL*, *FV* and *FQ* are force/length factor, force/velocity factor and activation factor, respectively; and *FP* is force scaling factor for the parallel elastic element. It is noted that *FL*, *FV* and *FQ* contribute to the active and *FP* to the passive part of the muscle force.

While *FQ* depends on the activity in the alpha-MN, which is generated by the oscillator or supplied by measured EMG recordings; *FL*, *FV* and *FP* are calculated assuming instantaneous sarcomere length (or given by gamma-MN activity) of the muscle fibres [23],

$$FL = 0.4128Ls(t)^3 - 4.3957Ls(t)^2 + 14.8003Ls(t) - 15.0515 \tag{9.5}$$

$$FV = \begin{cases} \dfrac{12.5 - [Vs(t)/2.73]}{12.5 + [Vs(t)/0.49]} & for\ Vs(t) \geq 0 \\ 1.5 - 0.5\dfrac{12.5 + [Vs(t)/2.73]}{12.5 - 2[Vs(t)/0.49]} & for\ Vs(t) < 0 \end{cases} \tag{9.6}$$

$$FP = 0.0014\exp[6(Ls(t) - 2.73)/2.73] \tag{9.7}$$

in which $Ls(t)$ and $Vs(t)$ are the instantaneous length and shorting velocity of the sarcomere, respectively. They can be found using equation 9.8 below [23]:

$$Ls(t) = [Lm(t) - (Lm_i - Lf_i)]\frac{Ls_i}{Lf_1} \tag{9.8}$$

where Lm_i, Lf_i and Ls_i are the initial muscle length, muscle fibre length and sarcomere length, respectively, whose values are given in Table 9.4, and the instantaneous muscle length Lm(t) can be scaled from the jaw movement as in equations 9.9 and 9.10 below.

Table 9.4 Muscle parameters for nonlinear Hill model (extracted from [24, 25])

Muscle		Initial muscle length Lm_i (mm)	Initial fibre length Lf_i (mm)	Initial sarcomere length Ls_i (µm)	Force max F_{max} (N)
Masseter	Superficial part	45.6 ± 4.2	24.6 ± 4.1	2.47 ± 0.27	272.8
	Deep part	25.3 ± 3.9	18.0 ± 2.8	2.44 ± 0.22	139.6
Temporalis	Anterior part	49.7 ± 3.9	27.1 ± 3.8	2.35 ± 0.14	308
	Posterior part	52.2 ± 4.7	25.7 ± 3.3	2.31 ± 0.12	222
Medial Pterygoid	Anterior part	37.9 ± 6.0	13.5 ± 1.9	2.48 ± 0.36	240
	Posterior part	44.4 ± 5.5	12.4 ± 1.5	2.54 ± 0.38	
Lateral Pterygoid	Inferior head	32.6 ± 3.1	23.0 ± 2.7	2.83 ± 0.10	112.8
	Superior head	31.3 ± 2.9	21.4 ± 2.2	2.72 ± 0.11	38
Digastric	Posterior belly	54.9 ± 5.5	20.5 ± 2.6	2.72 ± 0.14	Unknown
	Anterior belly	43.2 ± 4.9	21.4 ± 4.5	2.75 ± 0.21	46.4
Geniohyoid		40.0 ± 5.5	34.3 ± 5.1	2.65 ± 0.33	38.8
Mylohyoid	Posterior part	42.6 ± 5.3	28.2 ± 3.3	2.89 ± 0.19	21.2
	Anterior part	23.0 ± 2.8		2.80 ± 0.15	63.6

For the masseter:

$$Lm(t) = \frac{0.21Lm_{max}D}{D_{max}} + 0.79Lm_{max} \tag{9.9}$$

in which Lm_{max}, D and D_{max} are the masseter length when the jaw is fully opened, instantaneous opening of the jaw, and maximum opening of the jaw, respectively.

At the full open position the sarcomeres of the masseter are assumed to stretch to 4.0 μm, and then using Eq. (9.8) Lm_{max} can be found.

For the digastric:

$$Lm(t) = \frac{-0.26 Lm_{max} D}{D_{max}} + Lm_{max} \tag{9.10}$$

in which Lm_{max} is the digastric length when the jaw is fully closed. The digastric sarcomere was assumed to be decreased to 2 μm at the full close position of the jaw. Lm_{max} for the digastric can be found by Eq. (9.8).

9.4.3 Simulations

The jaw displacement used in the simulation had an amplitude of 20 mm and a frequency of 2 Hz; the jaw mass was set at 230 g; F_{max} for the masseter was 272.8 N; F_{max} for the digastric was 46.4 N; the masseter (closing muscle) contracts about 20% of its length when jaw is fully closed; and the digastric (opening muscle) contracts about 26% of its length when jaw is fully opened.

The first experiment did not involve any food (or nil food resistance). The alpha-MN oscillator needs to generate a pattern for *FQ* so that the specified jaw movement is satisfied. The oscillator tuned using the GUIs has the following parameters: s_1=s_2 =5, b_1=b_2=2, q_1=q_2 =1, a_{12}=a_{21}=1.2, T_{r1}=0.06, T_{a1}=0.7, T_{r2}=0.06, and T_{a2}=0.7. The feedback gain for adjustment in the tonic input was *G*=2.17. Figs. 9.10 - 9.12 show the jaw movement, masseter and digastric *FQ*s and masseter and digastric forces acting on the jaw, respectively.

It can be found from the results that a transient phase exists in the beginning as the jaw initially stands still, the digastric *FQ* is relatively larger (meaning more activities are required) than masseter's because the digastric is weaker and the maximum muscle force of 2.5 N was due to a combination of the gravitational, inertial, and masseter and digastric forces acting on the jaw. It is worth noting that instantaneous masseter and digastric forces are coupled and while closing, the digastric drags the masseter and while opening the masseter pulls the digastric.

When a constant food force is suddenly added at 10 s and removed at 25 s (Fig. 9.13), the alpha-MN (or oscillator) reacts immediately and adapts the closing muscle force within a cycle of chewing. The muscle takes one or two cycles to get to the stable force required to chew the food or get back to the lower force for free rhythmic chewing, as shown in Fig. 9.14. Due to the inertia and nonlinear behaviours of the muscles, the jaw movement is disturbed at the times the force is applied and removed (Fig. 9.15). These simulation results are in good agreement with the experimental results in that the closing muscles adapt their force within a chewing cycle when the food is suddenly put into or removed from the mouth [13, 14]. It is noted that the digastric force does not vary very much during the change in the food resistance.

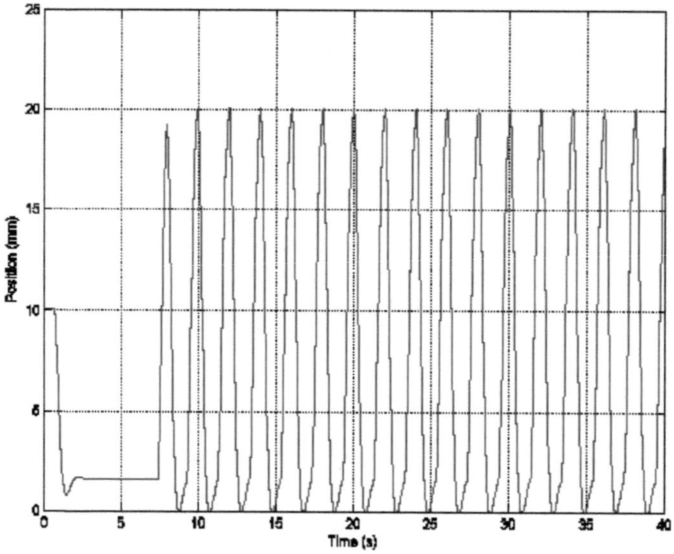

Fig. 9.10 Jaw movement without food

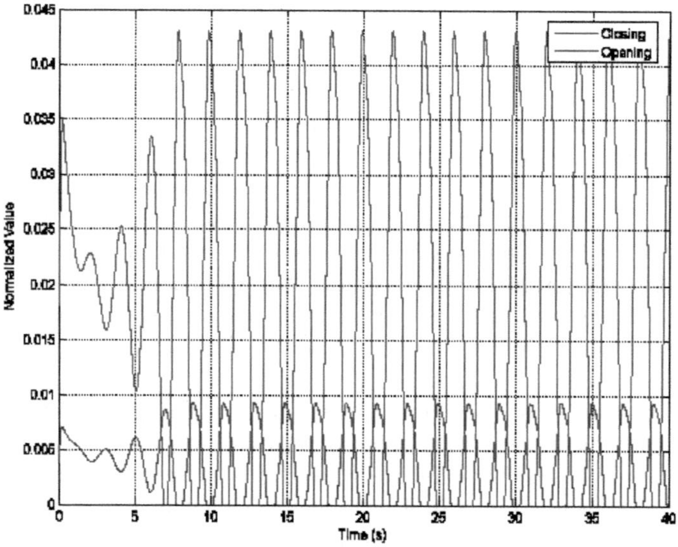

Fig. 9.11 Normalised *FQ*s without food resistance

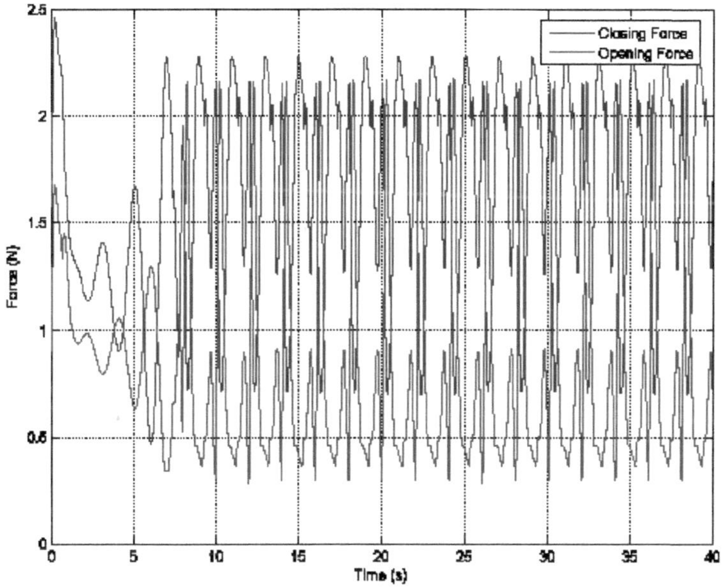

Fig. 9.12 Muscle forces applied on the jaw without food resistance

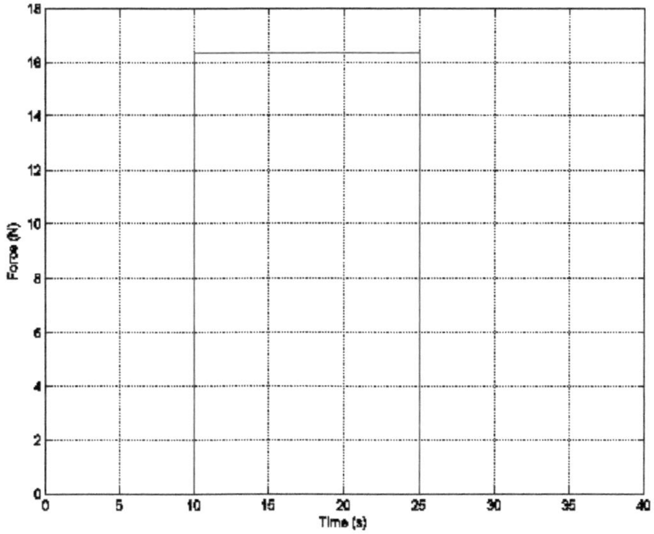

Fig. 9.13 A step change in food resistance

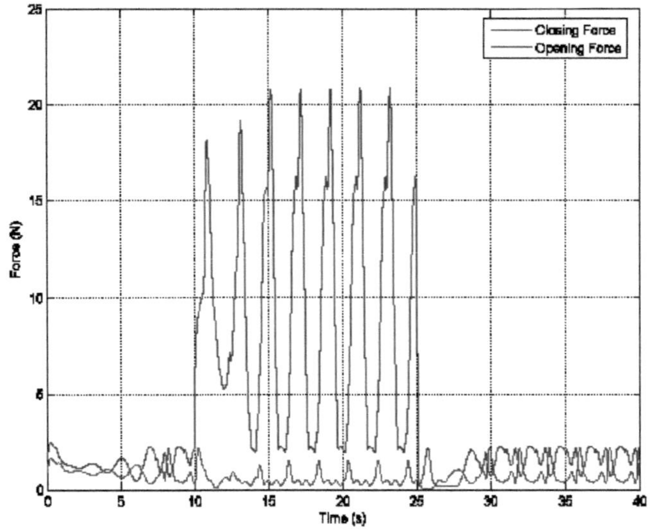

Fig. 9.14 Muscle forces at a step food resistance

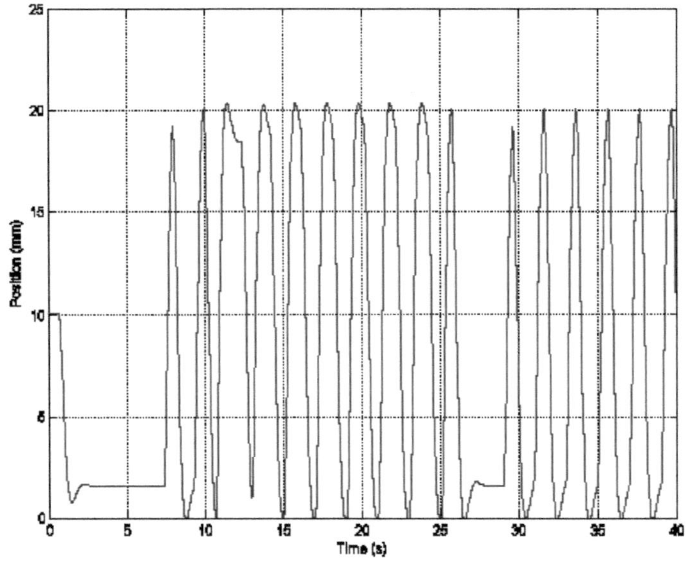

Fig. 9.15 Jaw movement at a step food resistance

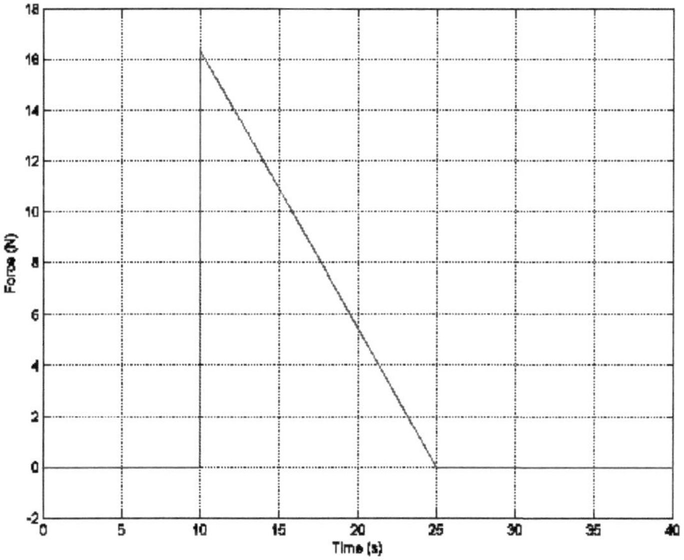

Fig. 9.16 A force profile simulating the decreasing food resistance

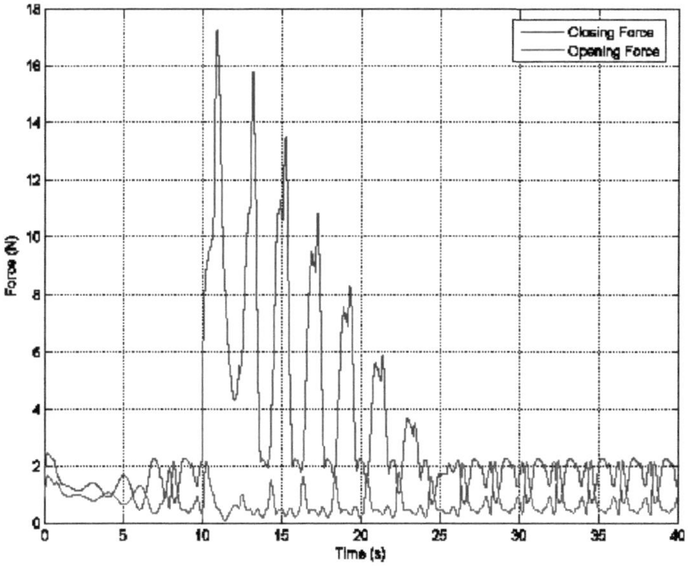

Fig. 9.17 Muscle force in response to the decreasing food resistance

While being chewed, certain foods behave with a gradually decreasing resistance. This scenario is simulated using a force profile (Fig. 9.16) where at time 10 s, a food is inserted suddenly and then gradually reduced to nil at 25 s. Fig. 9.17

shows clearly that the alpha-MN (or oscillator) reacts immediately and adapts the closing muscle force in direct proportion to the food resistance.

It should be noted that the chewing force (or to be exact, occlusal force) changes from cycle to cycle. The magnitude of the force peak of a cycle decreases as the chewing goes on. The force profile within one occlusal phase of a cycle is definitely nonlinear, subject to lots of factors involving foods, jaw/teeth movement and jaw muscle activities. The force profile (Fig. 9.16) that illustrates the decrement of the peak forces in magnitude is used to test the quick response and adaptation of the Matsuoka oscillator. Note that Figs. 9.13 and 9.16 present the force on the masseter and digastric actuators and not the actual occlusal force applied to the food.

9.5 Hardware-in-the-Loop Simulations

Further testing was carried out on the Matsuoka oscillator using physical hardware in place of the Hill muscle models outlined above.

9.5.1 Simulation Setup

The setup was organised in a closed-loop for the hardware-in-loop simulation [26] as shown in Fig. 9.18. The chewing robot consists of an fixed upper jaw and a lower jaw that is actuated by two phase-locked muscles. The Matsuoka oscillator served as the CPG to generate actuation commands (y_1 and y_2) for the two muscles. The oscillator was simulated in Matlab via a real time windows library and interfaced to the proportional valves of the muscles via a sensory I/O card. The chewing robot was instrumented with displacement and force sensors to provide feedback to the CPG via the tonic inputs (s_1 and s_2) for the adaptation of the actuation commands.

Two MAS-20 fluidic muscles from Festo were used for the robot, one for opening the jaw and the other for closing it. The muscle has a nominal length of 150mm and contracts by 25% under a full pressure of 6 bar with no load applied. In theory, one such muscle can produce 1.5 kN of force under full pressure, which is more than enough for chewing in which the jaw exerts approximately 500 N at maximum. The fluidic muscles were operated by two proportional valves (MPPE-3-1/8 from Festo). The valves control the output pressure linearly following a setpoint voltage applied to them, so that the movement of the jaw can be controlled accurately. The valves used a 24V supply and the set point voltage ranged from 0 to 10 V. The pressure output was monitored via an electrical signal.

For generating the feedback signal for the oscillator, two sources of information are used, one being the position of the lower jaw and the other the resulting forces on the upper jaw. To gather the positioning signal, an optical encoder was attached inside the base. The corresponding linear strip was attached to the moving bar that holds the mandible. The optical encoder was an EM1 from US Digital with a LIN-250-10-1 strip with a resolution of 250 increments per inch (10 increments per mm). The signal can be directly processed and sent to Matlab via an interfacing card.

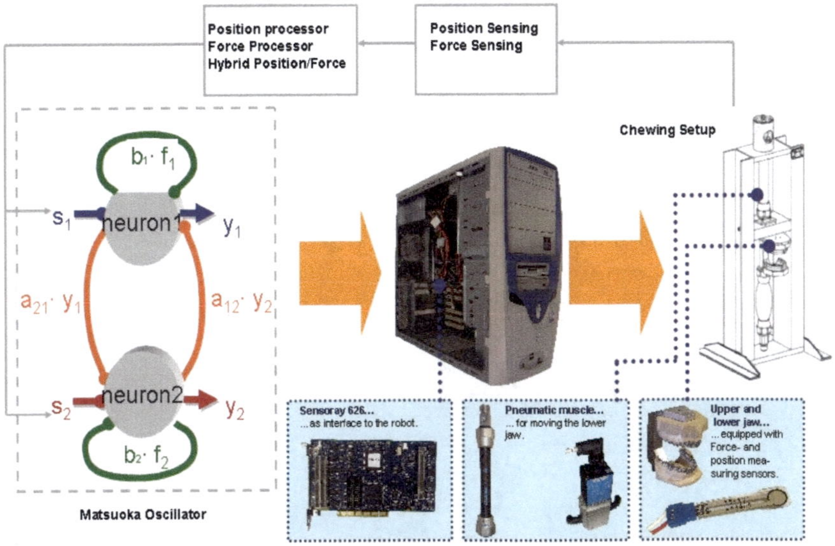

Fig. 9.18 Concept of the hardware-in-loop simulation approach

For gathering information about the force, three Tekscan Flexiforce sensors were attached between the upper jaw and the upper jaw base. The sensor can measure a force of up to 45kg (100 lbs) with an error of about 5% and has the advantage of being only 0.2 mm thick. Depending on the physical load on the sensor, it changes its electrical resistance within a range from 20 kΩ to 20 mΩ. To record this output a voltage of 5 V was applied and the signal amplified.

To get the output signals from the force sensors, a simple electronic circuit was needed. It consists of one operational amplifier, used as a linear, inverted amplifier (Fig. 9.19a). Voltage regulators were used to generate the voltages needed by different electrical parts of the setup. The proportional valves require +24 V, the force sensors -5 V and the operational amplifiers +9 V and -9 V. The optical encoder needs an additional +2.5 V supply as a reference. A board accommodating the above circuits and regulators was designed with Eagle and produced by milling the circuit path out of a copper coated plate (Fig. 9.19b). The board was powered by a single power supply with +24 V, GND and -24 V.

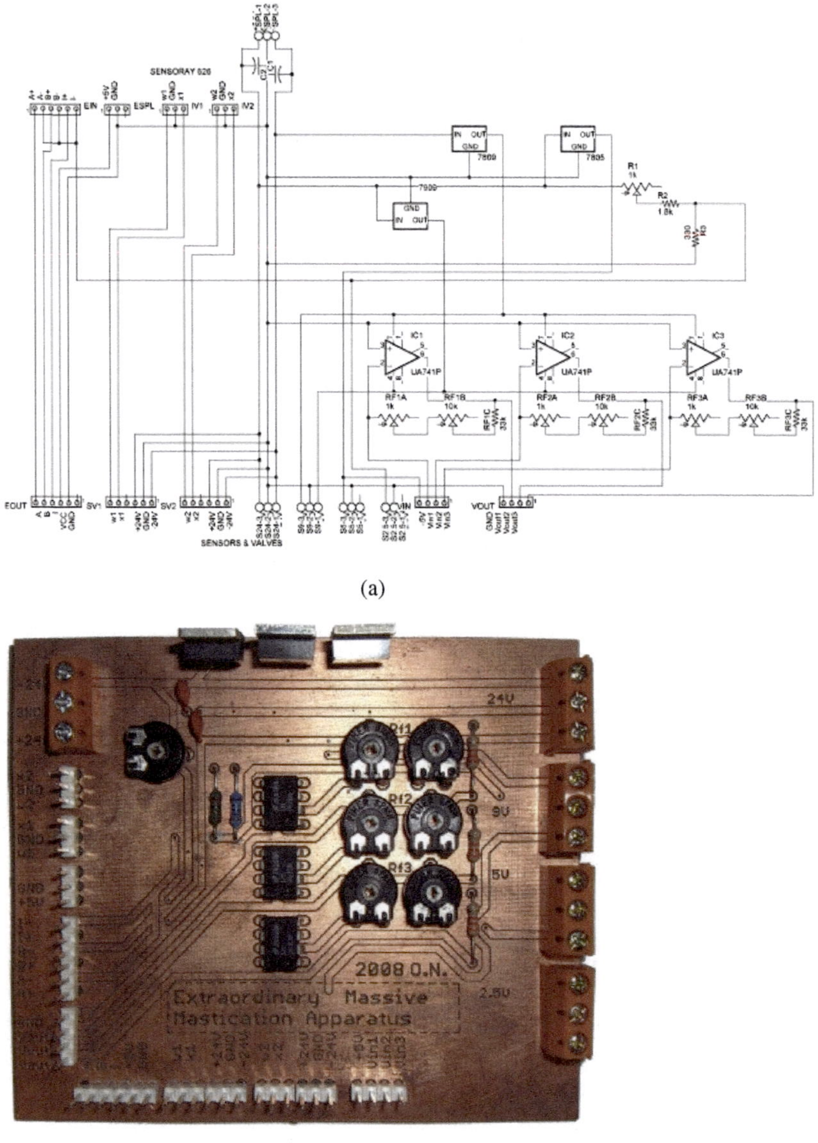

(a)

(b)

Fig. 9.19 Circuit board for force sensors and power regulator, (a) circuit schematic and (b)circuit board

A PC was used to control the chewing setup and run the CPG model. It used Windows XP with Matlab 2006b and Simulink 6.5. To interact with the physical device, one Sensory 626 I/O board was added to the system. This board supports 48 digital I/O channels, 6 encoder channels, 16 A/D converters and 4 D/A converters. Tables 9.5 and 9.6 show how the board was connected to the circuit board. Using the board with Matlab is straightforward as it is directly supported by Simulink with its Real-Time Windows Target library.

9.5.2 Implementation of the Algorithm

To be able to control the chewing robot directly, it is essential to run the Matsuoka model in real time with direct access to the Sensory 626 I/O board. In this study Real-time Windows Target was used as the development platform where RTWIN.TLC was chosen as the 'System target file' under the tab 'Real-time Workshop'.

Table 9.5 Connection of J1 with circuit board

J1 - Analog inputs & outputs 50 pin IDC ribbon cable connector					
Pin	Function	Designation	Pin	Function	Designation
1	Shield (Chassis)	*not connected*	2	GND	*VOUT GND*
3	-AD0	*not connected*	4	+AD0	*VOUT Vout1*
5	-AD1	*not connected*	6	+AD1	*VOUT Vout2*
7	-AD2	*not connected*	8	+AD2	*VOUT Vout3*
9	-AD3	*not connected*	10	+AD3	*IV1 x1*
11	-AD4	*not connected*	12	+AD4	*IV2 x2*
...			...		
17	-AD7	*not connected*	18	+AD7	*not connected*
19	GND	*not connected*	20	GND	*not connected*
21	-AD8	*not connected*	22	+AD8	*not connected*
...			...		
35	-AD15	*not connected*	36	+AD15	*not connected*
37	GND	*not connected*	38	GND	*IV1 GND*
39	GND	*not connected*	40	GND	*IV2 GND*
41	Sense1	*not connected*	42	DAC0	*IV1 w1*
43	Sense2	*not connected*	44	DAC1	*IV2 w2*
45	Sense3	*not connected*	46	DAC2	*not connected*
47	Sense4	*not connected*	48	DAC3	*not connected*
49	GND	*not connected*	50	Shield (Chassis)	*not connected*

Table 9.6 Connection of J5 with the circuit board

J5 - Encoder 26 pin ribbon connector					
Pin	Function	Designation	Pin	Function	Designation
1	Encoder (0A) A-	*not connected*	2	Encoder (0A) A+	*not connected*
...			...		
9	GND	*not connected*	10	Encoder (1A) A-	*EIN A-*
11	Encoder (1A) A+	*EIN A+*	12	+5V	*ESPL +5V*
13	Encoder (1A) B-	*EIN B-*	14	Encoder (1A) B+	*EIN B+*
15	GND	*ESPL GND*	16	Encoder (1A) I-	*EIN I-*
17	Encoder (1A) I+	*EIN I+*	18	+5V	*not connected*
...			...		
25	Encoder (2A) A-	*not connected*	26	Encoder (2A) A+	*not connected*

The core of the program for the controlling algorithm consists of three subsystems, **Feedback**, **Neuron Network** and **Interface** (Fig. 9.20). For debugging purposes two **ToWorkspace** blocks save the actual pressure, measured by the proportional valves.

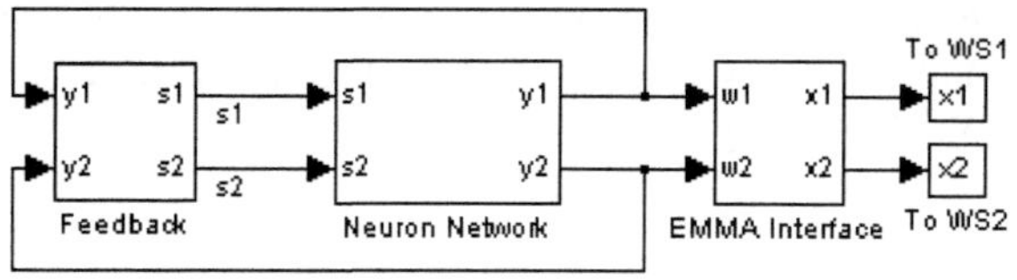

Fig. 9.20 Three subsystems for the controlling algorithm

Feedback (Fig. 9.21) generates the input signals $s1$ and $s2$ for the neuron network or oscillator (Fig. 9.18). The subsystems **getPosition** and **getForce** gather the sensory input and pass them to the corresponding analyser subsystem **position processor** and **force processor** that create two input signals. **Merger** combines the two signals *s_position* and *s_force* to one by:

$$s = (1 - closeLoopMode) \bullet s\_position + closeLoopMode \bullet s\_force \quad (9.11)$$

in which *closeLoopMode* has a value in between 0 and 1, where '0' means fully position controlled and '1' fully force controlled.

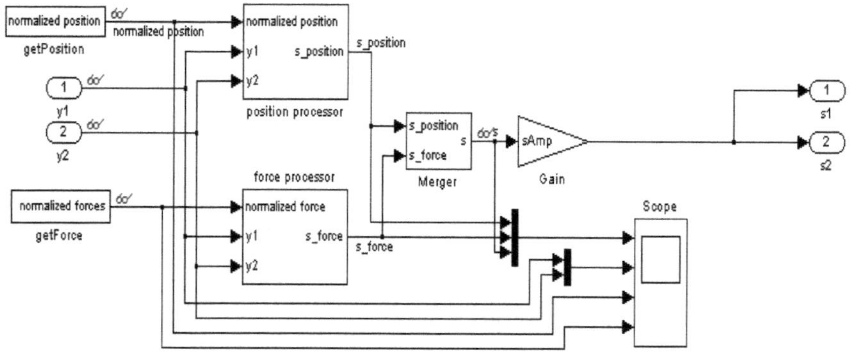

Fig. 9.21 The Feedback subsystem

The system **getPosition** (Fig. 9.22a) gathers the position sensor input via an **Encoder Input** block, and **getForce** (Fig. 9.22b) captures the force sensor data via three **Analog Input** blocks. The types of both blocks can be found in the Real-time Windows Target library of Simulink. The position and force signals get normalised with the use of the min/max data collected during the calibrating process.

The function of the **position processor** (Fig. 9.23) is to prepare a useable input signal for the oscillator out of the position information gathered from the **getPosition** subsystem. This processor analyses the movement of the jaw during each cycle of the oscillator output signals y_1 and y_2. It measures the maximum position of each y_1 cycle, which lasts from one rising edge of y_1 to the next. Using the same principle, the minimum position is measured during each cycle of y_2. These signals are filtered by a lookup table which prevents the output value from getting either negative or too high. The larger signal is determined by the **Max** block and passed to the subsystems output.

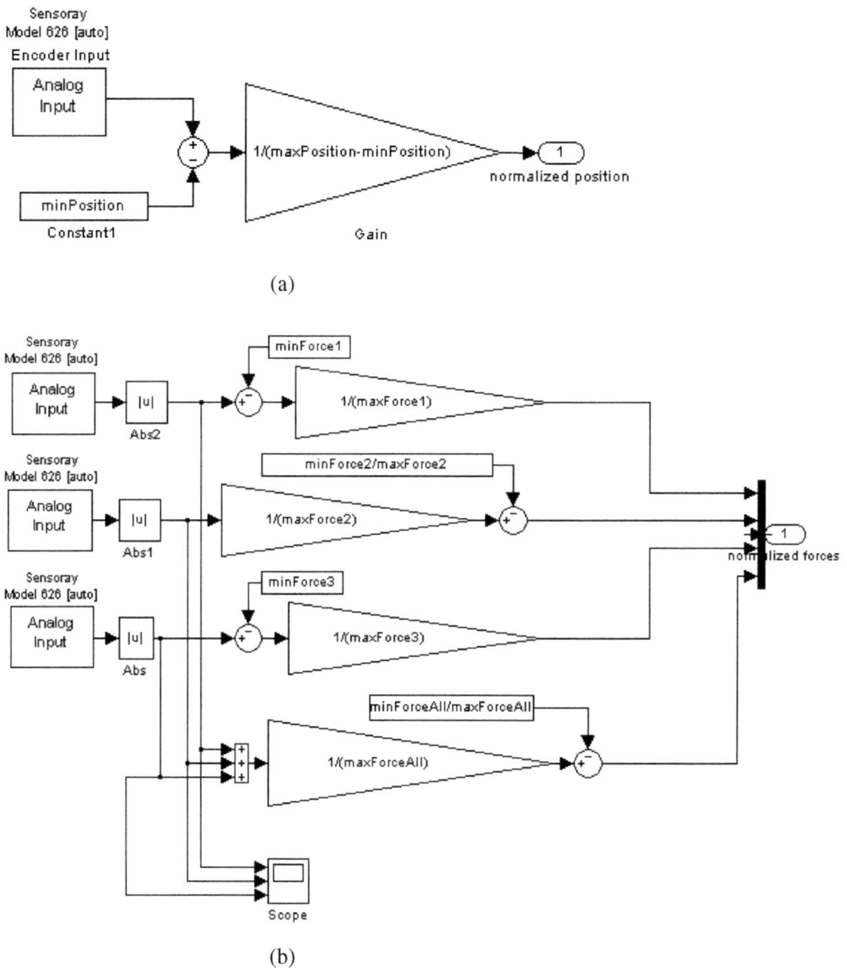

Fig. 9.22 Two sensor subsystems, (a) get**Position** and (b) **getForce**

To get the rising edge of a signal, it is passed through a **Detect first rise** block together with the value of the previous step, generated with a **Memory** block. The position signal is passed to a **MinMax Running Resettable** block with the trigger signal wired to the reset input. With this setup the maximum position signal for each cycle of y_1 and the minimum for the cycle of y_2 were evaluated.

The **force processor** in general works like the **position processor**, as shown in Fig. 9.24. The interface is the last subsystem of the model (refer to Fig. 9.20). The two signals y_1 and y_2 are amplified to get a range of 0 to 1, where '0' means 'no pressure' and '1' 'full pressure'. Both signals get an offset added so that the muscles always have a minimum pressure that prevents them from bending. As it is the interface to the proportional valves, it needs two **Analog Output** blocks of the Real-Time Windows Target library, as shown in Fig. 9.25. For the purpose of analysis, two **Analog Input** blocks were used to gather the output pressure, measured by the proportional valves.

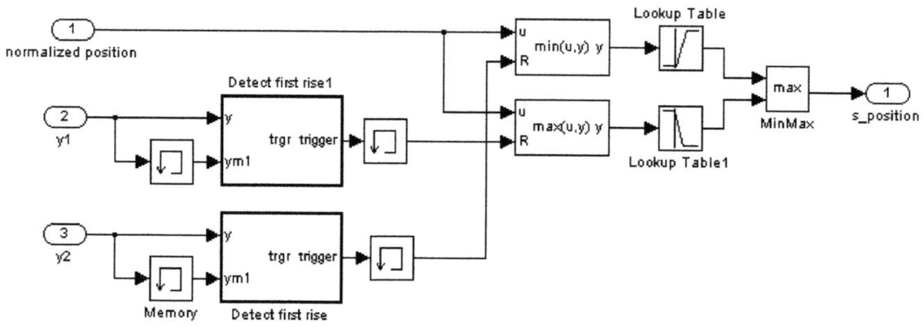

Fig. 9.23 Subsystem position processor

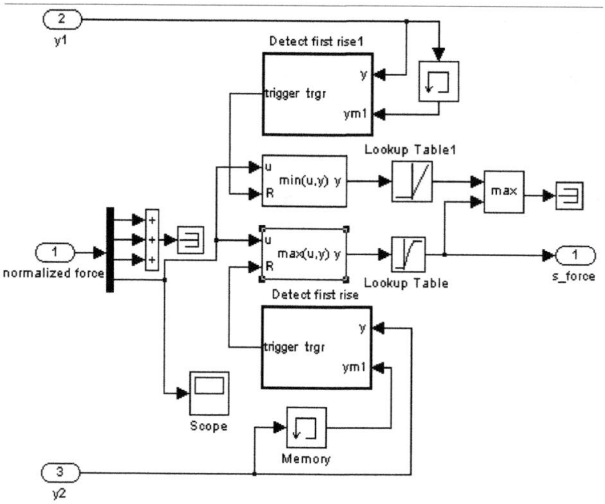

Fig. 9.24 Subsystem force processor

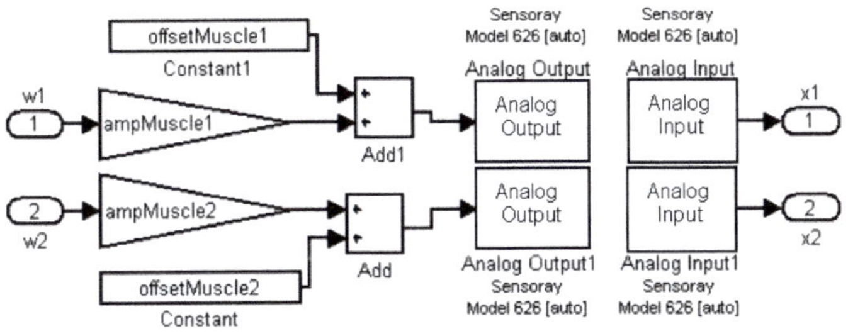

Fig. 9.25 The **Analog Input** and **Analog Output** subsystems

9.5.3 Experimental Results

To compare the chewing behaviour of the simple chewing robot to that of the humanoid control with various sensory feedback signals, different artificial food samples were used. The robot was mostly fed with different types of foams but brownies and cookies were also carried out. One of these foam samples can be seen in Fig. 9.26, where the force generated by the muscles is sufficient to compress the sample. For the test series describe below however, a sample made from elastomer plastics Elastosil M4503 was used. The sample has properties that are known and deforms almost linearly to the force applied to it.

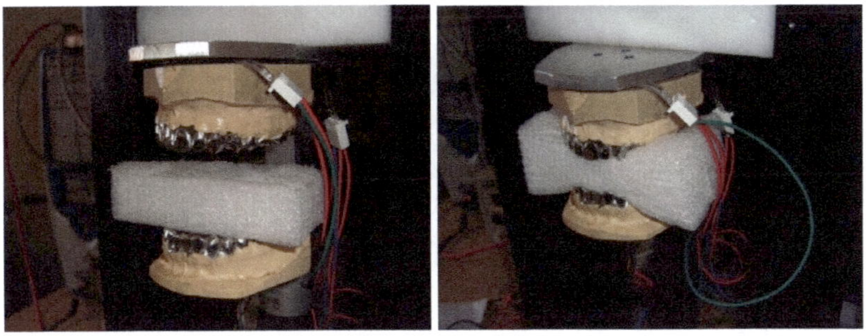

Fig. 9.26 The robot chewing on a foam sample

9.5.3.1 Position Controlled Operation

For the robot under positional feedback control, the oscillator's input signal, (*s_position*), was created purely by the **position processor**. As shown in Fig. 9.27, the dashed line is the normalised position of the jaw where 0 means fully closed and 1 fully open. The solid curve is the oscillator's input signal. The graph covers the time from 10 to 20 s.

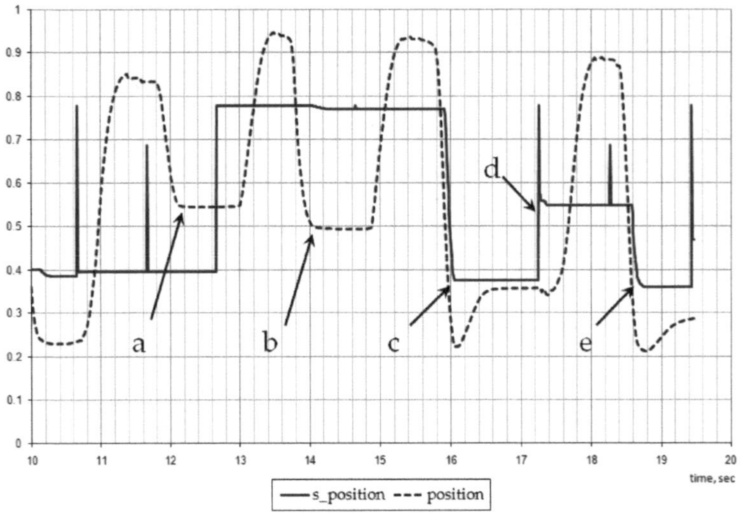

Fig. 9.27 Normalised position and generated input to the oscillator over time

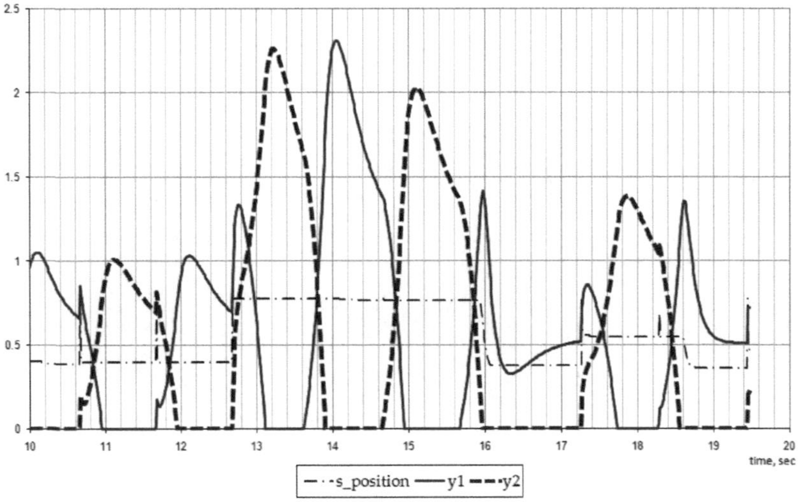

Fig. 9.28 Position controlled chewing: outputs (y1 and y2) and input (dash-dotted) of the oscillator

By the second cycle, the food sample was inserted into the mouth which, as a result, could not close completely (marked a in the figure). The output signal then rose to make the robot chew harder. In the next chewing cycle (b), the displacement of the jaw is larger, as the robot chews with more power. After that cycle, the artificial food was removed from the mouth so that the jaw could now close fully (c). The system took some time to return to its normal rhythm (d and e).

Fig. 9.28 shows the response of the neural oscillator over the same duration. The dashed-dotted curve represents the oscillator's input signal s as described above. The solid and dashed curves represent the oscillator's output signals y_1 and y_2. It is remarkable that the oscillator adapted its output within milliseconds after the change in s. On the other hand it took almost one second to return to the idle oscillation again after the drop in s.

Fig. 9.29 shows a second test series with the same food sample. This time, the oscillating period of the oscillator was shortened by doubling the values of the oscillator's parameters T_{a1}, T_{a2}, T_{r1} and T_{r2}. Also, the startup of the robot can be seen as the graph starts at 0s. The graph covers the input signals and the two output signals y_1 and y_2. It should be noted that the system established a steady oscillating rhythm after about 1 s (marked in the graph with a). The artificial food was applied after the second stable cycle and the Matsuoka oscillator reacts immediately after recognising the change (b). After one cycle of chewing, it was removed again and the oscillator took about one second to come back to its idle chewing state (c).

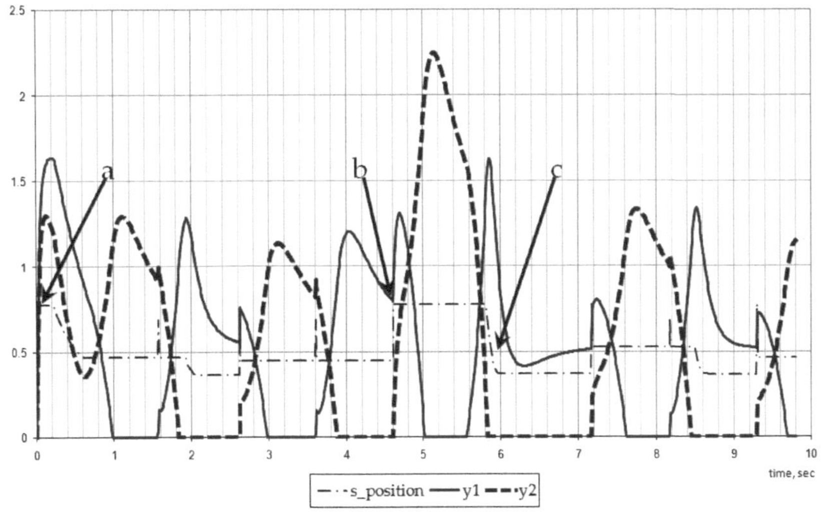

Fig. 9.29 Position controlled chewing: the startup (a), one cycle of food chewing (b) and the readjustment after removing the food (c)

9.5.3.2 Force Controlled

In the fully force controlled mode, the robot is programmed to respond to the force. The stronger the signals y_1 and y_2 are, the more force is applied on the jaw. Consequently, when running, the jaw doesn't close completely during an idle cycle. The sequence for this test series was the same as described above. The recording began after 10 s in a stable oscillation. After two cycles, the robot was fed with the same food sample for two cycles and then removed again. Fig. 9.30 shows the oscillator's input signal s, generated by the force processor.

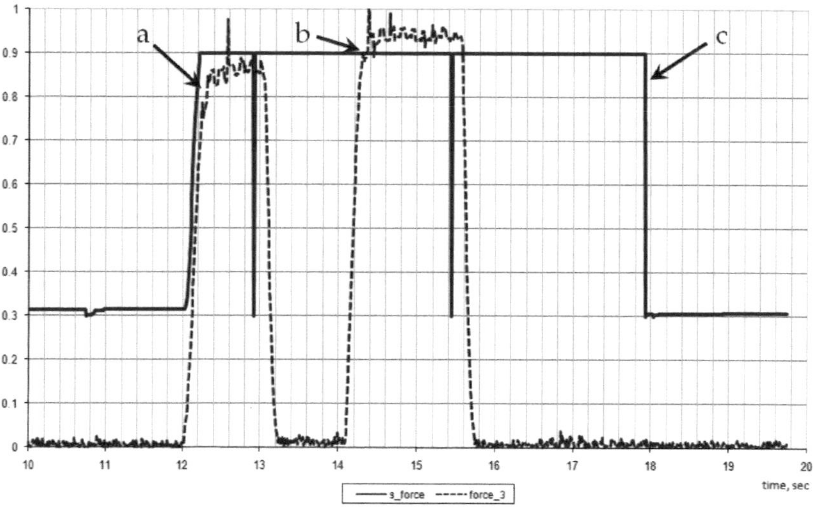

Fig. 9.30 Generation of the signal s using the **force processor**

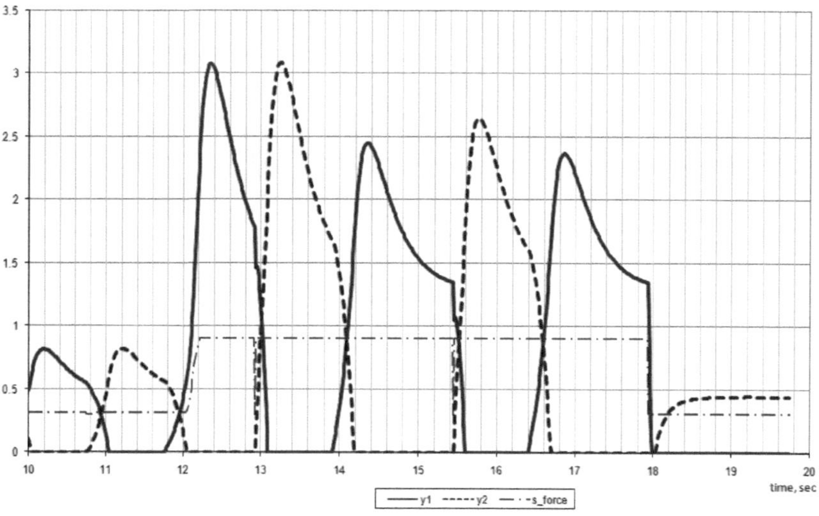

Fig. 9.31 Force controlled chewing: outputs (y_1 and y_2) and input (red) of the oscillator

After inserting the food sample into the mouth, as shown in Fig. 9.30, the force on the jaw rose during closing as the teeth come into contact with the sample (marked with a). Therefore, the oscillator's input signal s rose rapidly to increase the strength of the output signals. During the second cycle, the force on the jaw was increased again, because the pressure in the muscles became even higher (b). When reaching mark c in the graph, the input signal s was lowered again, as there was no longer any resistance in the mouth (c).

Fig. 9.31 shows the corresponding output signals y_1 and y_2. Again, the output signals adjusted immediately after the change was recognized by the oscillator. But compared to the oscillator's behavior in the position controlled mode, it changed to idle slowly. This is due to the fact that it took the force processor about a half cycle to perceive the change. Compared to the oscillator's behaviour in the position controlled mode, it changed to idle relatively slowly, but within one to two cycles of chewing [13, 14].

9.5.3.3 Hybrid Position and Force Controlled

For the next test series the oscillator's input signal was composed by the **Merger** subsystem, using 50% of the signal generated by the force processor and 50% of the signal generated by the position processor. All other settings were kept same as in the experiments outlined above.

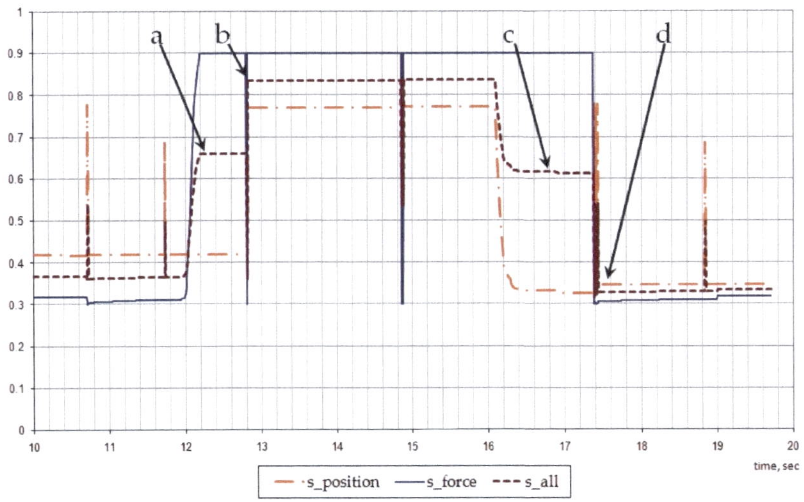

Fig. 9.32 Merged signal s, generated with 50% of the force signal and 50% of the position signal

Fig. 9.32 covers the time from 10s to 20 s after engaging the robot. The signal that was created using the position sensory input was drawn as the dash-dotted line, the signal generated by the force processor was solid and the composed signal was the dashed line. As in the previous experiment, the force processor detected the change as soon as the food was inserted (marked with a), whereas the position processor reacted about half a cycle later (b). After removing the food sample from the mouth, the force processor's generated signal was slower than the position processor's one (c & d). This resulted in a 'stepped' merged signal that was passed to the oscillator. That the system reacted faster and smoother on 'feeding' than in fully force controlled mode is one positive effect of combining both input signals.

Due to the inertia and nonlinear behaviours of the muscles, the jaw movement is disturbed at the times the food was applied and removed (Figs. 9.32 and 9.33). These results are in good agreement of the human chewing experiments that the closing muscles complete the adaptation of their force within one chewing cycle when the food is suddenly put in, or within one to two cycles when the food is removed from the mouth [13,14].

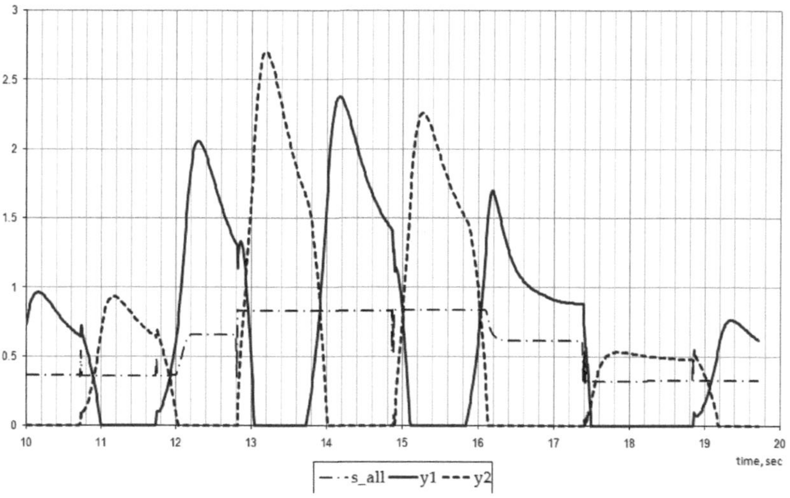

Fig. 9.33 Oscillator's output signal y_1 and y_2 in relation to its merged input

9.6 Summary

The Matsuoka oscillator was used to generate rhythmic and sensory patterns of muscular activities, and is intended as a motion generator and its sensory adjustment for human-like chewing robots. An assistive tool (via three GUIs) was developed to help analyze and design the oscillator. A full picture of the parameters' influence on various measurement of the oscillation was presented. The use of the oscillator was demonstrated in conjunction with a simplified neuromotor system of the jaw and simulations were performed to validate the capacity of the oscillator to generate additional voluntary muscular activities for various food resistances.

A hardware-in-the-loop simulation system was presented in which the chewing robot was actuated by two fluidic muscles, and commanded by the CPG based on the Matsuoka oscillator. The CPG's ability in generating and then adapting human-like chewing patterns of actuations was proven experimentally. The work done further confirms that, using this model, we would develop a chewing robot that would chew the way the human does. The work is also meaningful for its use to centrally controlling EMG-driven assistive or rehabilitative devices.

References

1. Anderson, K., et al.: The effects of bolus hardness on masticatory kinematics. J. Oral Rehab. 29, 689–696 (2002)
2. Foster, K., et al.: Effect of texture of plastic and elastic model foods on the parameters of mastication. J. Neurophysiol. 95, 3469–3479 (2006)
3. Daumas, B., et al.: Jaw mechanism modeling and simulation. Mech. Machine Theor. 40, 821–833 (2005)
4. Xu, W.L., et al.: A robotic model of human masticatory system for reproducing chewing behaviours. IEEE Robotics Automation Mag. 12, 90–98 (2005)
5. Takanobu, H., et al.: Integrated dental robot system for mouth opening and closing training. In: Proceedings of IEEE International Conference on Robotics & Automation, Washington DC, pp. 1428–1433 (2002)
6. Takanobu, H., Takanishi, A.: Dental robotics and human model. In: Proceedings of the 1st International IEEE EMBS Conference on Neural Engineering, Capri Island, pp. 671–674 (2003)
7. Xu, W.L., et al.: Design of a biologically inspired robot for foods chewing. IEEE Trans. Industrial Electr. 55, 832–841 (2008)
8. Xu, W.L., et al.: Kinematics and experiments of a life-sized chewing robot for characterising food texture. IEEE Trans. Industrial Electr. 55, 2121–2132 (2008)
9. Latash, M.L.: Neurophysiological basis of movement - human kinetics, Champaign, IL (1998)
10. Lund, P.: Mastication and its control by the brain stem. Crit. Rev. Oral Biol. Med. 2, 33–64 (1991)
11. Turker, K.S.: Reflex control of human jaw muscles. Crit. Rev. Oral Biol. Med. 13, 85–103 (2002)
12. van der Bilt, A., et al.: Oral physiology and mastication. Physiol. Behav. 89, 22–27 (2006)
13. Gonalez, R., et al.: Review: the use of electromyography on food tecture assessment. Food Sci. Tech. Int. 7, 461–471 (2001)
14. Peyron, M.A., et al.: Effects of increased hardness on jaw movement and muscle activity during chewing of visco-elastic model foods. Exp. Brain Res. 142, 41–51 (2002)
15. Ellias, S., Grossberg, S.: Pattern formation, contrast control, and oscillations in the short term memory of shunting on-center off-surround networks. Biol. Cybern. 20, 69–98 (1975)
16. Matsuoka, K.: Sustained oscillations generated by mutually inhibiting neurons with adaptation. Biol. Cybern. 52, 367–373 (1985)

17. Iwasaki, T., Zhang, M.: Sensory feedback mechanism underlying entrainment of central pattern generator to mechanical resonance. Biol. Cybern. 94, 245–261 (2006)
18. Kotoh, R., Mori, M.: Control method of biped locomotion giving asymptotic stability of trajectory. Autom. 20, 405–414 (1984)
19. Buchli, J., et al.: Engineering entrainment and adaptation in limit cycle systems, from biological inspiration to applications in robotics. Biol. Cybern. 95, 645–664 (2006)
20. Williamson, M.M.: Robot arm control exploiting natural dynamics. PhD Thesis. Massachusetts Institute of Technology (1999)
21. Berthouze, L., Lungarella, M.: Motor skill acquisition under environmental perturbations: on the necessity of alternate freezing and freeing of degrees of freedom. Adapt. Behav. 12, 47–64 (2004)
22. Matsuoka, K.: Sustained oscillations generated by mutually inhibiting neurons with adaptation. Biol. Cybern. 56, 345–353 (1987)
23. Koolstra, J.H.: van Eijden TMGJ Dynamics of the human masticatory muscles during a jaw open-close movement. J. Biomech. 30, 883–889 (1997)
24. Eijden, T.M.G.J., et al.: Architecture of the human jaw-closing and jaw-opening muscles. Anat. Rec. 248, 464–474 (1997)
25. Koolstra, J.H., van Eijden, T.M.G.J.: Combined finite-element and rigid-body analysis of human jaw joint dynamics. J. Biomech. 38, 2437–2439 (2005)
26. Low, K.H., et al.: On the development of a real time control system by using xPC Target: solution to robotic system control. In: Proceedings of IEEE, International Conference on Automation Science and Engineering, Canada, pp. 345–350 (2005)

Chapter 10
Knowledge System of Human Chewing Behaviours*

Abstract. Mastication is a complex process, influenced by numerous factors including those associated with an individual and the ingested food. Human chewing behaviour can be characterised by measuring mandibular movements and muscular activities during a masticatory sequence or by measuring the particle size distribution and rheological characteristics of the swallowed food mass. A formal description of the chewing behaviour is proposed in this chapter for use in constructively understanding the mastication process and assessing mastication performance. An object-oriented model is developed and described in Unified Modelling Language (UML). The chewing behaviour model is composed of three objects, one for the jaw's physiological apparatus, one for the properties defining the mastication process and foods being chewed, and a further one for the association of the properties. A complete representation of the chewing behaviour is achieved by linking the three object models via an additional class for chewing data that is collected experimentally. With the object model, the chewing behaviour is further instantiated by discovering knowledge hidden in the chewing database by data mining. A case study is presented to show the procedure of how the hidden knowledge can be discovered and the data mining results are interpreted in the context of food science.

10.1 The Need for Knowledge Systems

Mastication can be considered to be the first stage in food digestion. It is a complex process whereby food taken into the mouth is processed, often involving the reduction of particle size and the incorporation of saliva, into a form (bolus) suitable for swallowing. It is also brought approximately to body temperature before transfer to the stomach, where digestion, absorption and utilization begin [1].

* Reprinted with modification from Xu WL, Kuhnert L, Foster K, Bronlund J, Potgieter J and Diegel O (2007) Object-oriented knowledge representation and discovery of human chewing behaviours. Eng. Appl. Artificial Intell. 20:1000-1012, with permission from Elsevier.

W. Xu and J.E. Bronlund: Mastication Robots, SCI 290, pp. 271–287.
springerlink.com

Mastication is characterised by a rhythmic jaw opening and closing movement, whose main purpose is the formation of a swallow-able bolus [2]. The ability or ease with which an individual can form and swallow a bolus can largely affect food choice and therefore may impact on the nutritional status of the individual [3]. Numerous measurements can be used to characterise the masticatory process. Tracking jaw movement throughout a masticatory sequence can give information including mandibular displacement, velocities of opening and closing and masticatory frequency. Muscular activities can be measured by EMG and are thought to be associated with chewing forces. A food bolus expectorated just prior to swallowing can also be analysed for particle size and rheological charactersitics [4-8]. These parameters vary between subjects, due to differences in jaw geometry, teeth shape, the muscles of mastication, sensitivity to pain, and oral status. They also vary with changes in food texture such as elasticity, plasticity, hardness and adhesiveness [9-11]. The measurements are often gathered for specialised applications and are often incomplete for fully describing the chewing behaviour.

Because of the complex nature of the mastication process there has been considerable work in developing quantitative methods for evaluating the capability of a person to effectively chew foods. One such effort is to develop a chewing robot to simulate the chewing behaviour of a specific subject [12-14]. The robotic device would be used to quantitatively evaluate the dynamic changes to the texture of foods during chewing, which is vital information required in the development of new foods. It would also be used to objectively evaluate the differences between masticatory efficiency of edentulous and denture-wearing persons and how these differences are related to masticatory patterns.

A further research effort is a formal description of the chewing behaviour, which is the topic of this chapter. This is required to structure the mastication process and define its contributing attributes, thus guiding the design of chewing experiments in a systematic way. The formal description is also required to establish a knowledge based system of chewing with which the adaptive chewing of the robot would be attempted.

An object-oriented model [15] was used for this formal description as it provides not only a hierarchical and functional framework of the mastication process but also a scheme for the creation of instances containing the data associated with the mastication process. With the object model, semantic information (in the form of association rules), may also be embedded and advanced data manipulation methods can be used. Furthermore, the extension of an object-oriented model without the destruction of the existing structure is a major advantage. To visualize the object-oriented model, the Unified Modelling Language (UML) [16] was chosen.

10.2 Object-Oriented Knowledge Representation

10.2.1 Object-Oriented Model Describing Jaw Physiology

Fig. 10.1 depicts the UML object-oriented model for jaw physiology. Its first object is the "HumanSubject" class. It describes the human subject being involved in the chewing experiment. It has some basic atomic attributes ("id", "name", "age", "gender", "description"). The complex attributes "hasJaw" is of the type "Jaw" which creates an aggregation with the "Jaw" class and consequently shows that the "Jaw" is a part of the "HumanSubject". Additional basic attributes could be inserted to create a more detailed description of the human subject.

The "Jaw" class is one of the most important classes in this object-oriented model. It is connected to the classes: "Muscle", "Maxilla" and "Mandible" via an association and two aggregations. This indicates that the objects these classes define are a part of the jaw or that they are attached to the jaw. Therefore the "Jaw" class has three complex attributes which are "hasMuscle", "hasMaxilla" and "hasMandible".

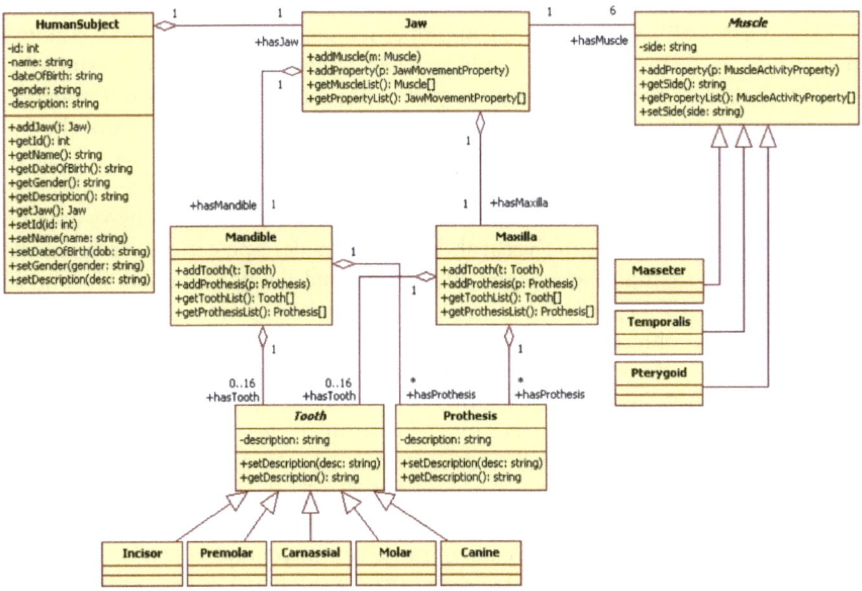

Fig. 10.1 Object-oriented model describing human jaw physiology

The "Muscle" class is defined as an abstract class as it does not need to be instantiated. It is a generalization of three inherited classes "Masseter", "Temporalis" and "Pterygoid" that are the muscle group classes used to instantiate the actual muscles of the human subject [17]. The atomic attribute "side" is used to clearly identify each one of the muscle instances. If additional muscle groups are to be added in the future (e.g. to increase the degree of detail) this is easily achieved by inserting new classes which are associated to the "Muscle" class via a generalisation association.

The classes "Maxilla" and "Mandible" which are part of the "Jaw" class host the instances of the classes "Tooth" and "Prothesis". These two classes serve a rather structural purpose as they do not have any other attributes apart from the complex attribute "hasTooth" and "hasProthesis" which create the aggregations between the above mentioned classes.

The "Tooth" and the "Muscle" class are structured analogically. Thus the "Tooth" class is the abstract generalisation of the classes inherited from it. All classes are completed by the attachment of methods used to manipulate their attribute values without directly accessing the attributes.

10.2.2 Object-Oriented Model for the Chewing Process

As the mastication process is often evaluated by jaw movement, muscular activity and food properties [4-7, 11], three classes of properties may be created accordingly, "Jaw Movement Properties", "Muscle Properties" and "Food Properties".

The class object "Jaw Movement Properties" describes how the jaw moves in 3D space. Basically all jaw movement properties can be extracted from the motion data which is recorded while a subject chews sample foods. The positions of markers, attached to the head and the jaw of the subject, are tracked throughout the whole chewing process. After subtracting the head movement from the jaw movement by applying translation and de-rotation methods [10], the 3D movement trajectories of the jaw and their time history can be obtained. The jaw movement properties often used in literature include the number of chews used before swallowing the food, the vertical, lateral and antero-posterior displacements, the frontal and sagittal plane angles of the jaw, and the opening and closing velocities [4, 5, 8, 10, 11].

To model this object in UML, the abstract information of the properties such as extrema, means, range, latency and frequency are employed. The class object "Muscle Properties" is used to describe how the muscles attached to the jaw act while chewing. The muscular activities, due to their close correlation to the biting/chewing force, may be used to describe the force applied to teeth and foods during chewing. The extraction of the muscular activity properties is comparable to the extraction of the jaw movement properties although the source of the data is different (the data is generated by the use of electromyography EMG) [6, 10, 11]. In EMG measurements surface electrodes attached to the skin above the masticatory muscle of a subject record the activities of each of the observed muscles while chewing sample foods. The data resulting from this procedure is used to create abstract properties analogical to the creation of the jaw movement properties (i.e., muscular activity extrema, range and latency).

The food properties, unlike the jaw movement properties and the muscle activity properties, do not describe the jaw itself or a part of the jaw. Instead they characterise a food sample which is chewed by the subject during the experiment. The complexity of this topic is considerably higher than both other property classes because a precise description of food demands the consideration of several different types of variables. This is, however, a well studied field of research in food science thereby providing a large amount of information on food properties and food property testing experiments. Thus, it is necessary to introduce sub-categories to structure the food properties due to the high quantity of available food properties. A first level of sub-categories that separate the food properties regarding the field of physics they belong to are, mechanical food properties, thermal food properties, chemical food properties, and electrical food properties.

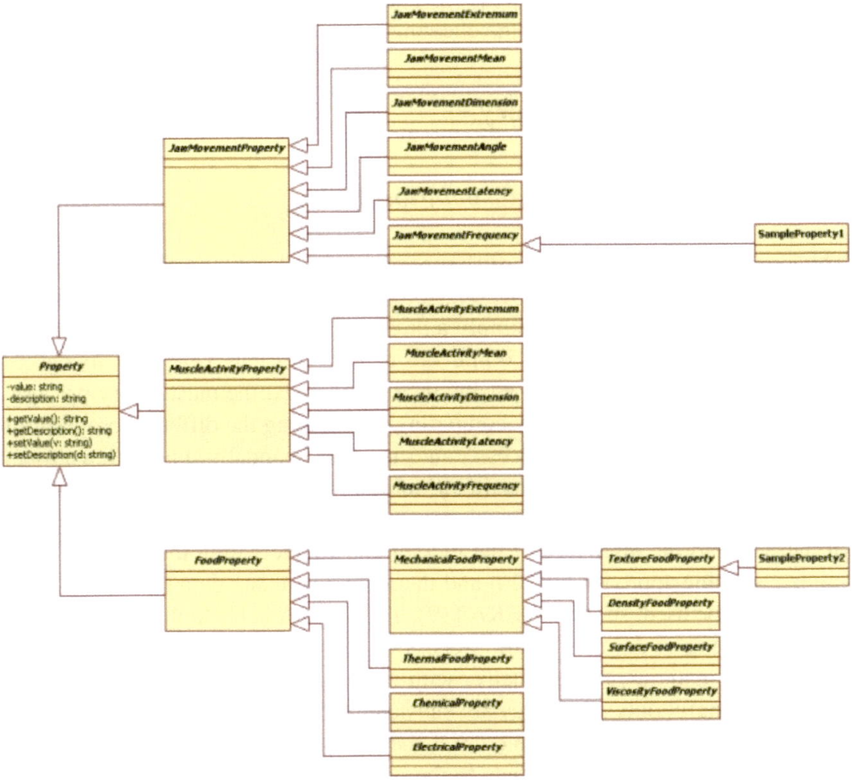

Fig. 10.2 Object-oriented model describing the chewing properties

As the mechanical and rheological food properties have got the highest significance to chewing, the other categories will not be discussed. Thus a second level of sub-categories divides the mechanical and rheological food properties into further sub-categories which are: texture properties, density properties, surface properties, viscosity properties.

The category-sub-category-concept is of highly hierarchical nature and therefore perfectly fits into object-oriented principles. As shown in Fig. 10.2 the top-level "Property" class, hosting general atomic attributes ("value", "description"), which are inherited to all originating classes, is the central class of this object-oriented model. The next hierarchical level is formed by three classes representing the main categories "JawMovementProperty", "MuscleActivityProperty" and "FoodProperty". These main categories are divided into sub-categories, as described earlier, by creating classes inheriting from the main category classes. Any number of hierarchical levels could be created following this procedure. All classes that are not part of the lowest hierarchical level are defined as abstract classes. They are not used to substantiate objects and rather serve a structuring purpose. In the end the lowest level of classes represents the properties actually being the scheme for instance creation. The objects instanced like this are used to store the actual data describing the chewing process.

10.3 Knowledge Discovery

10.3.1 Data Mining for Association Rules

The experimental testing of the different properties describing the chewing process is able to provide a large amount of data. Unfortunately, these datasets are rather trivial because they are simply a direct recording of what happened while a human subject was chewing. While the data is absolutely needed to describe chewing behaviour, it does not enable us to understand the masticatory process in detail as this is only possible by comparing and relating the different properties to, and with, each other. This process could be done manually, but with a growing number of datasets this is an unsuitable solution to the problem. This suggests the use of an automated approach which can cope with two problems: a large amount of datasets and the discovery of the hidden knowledge within the data. Data mining is a suitable approach [19, 20] and there are also many data mining tool suites available to be used, such as WEKA [19], TANAGRA [21], AlphaMiner [22].

Data mining is defined as the process of discovering patterns in data, which may be automatic or, more usually, semi-automatic. The patterns discovered become meaningful only when they are substantiated by appropriate interpretation in the context of the defined problem. Of the existing data mining algorithms, rule uncovering is the most appropriate to the mastication property associations in this research. The rule uncovering algorithm extracts every possible rule from a database. Due to the high number of discovered rules, two factors are often used to determine whether or not a rule is meaningful. They are *Coverage* or *Support,* the number of instances for which the rule's prediction is correct, and *Accuracy* or *Confidence,* the number of instances for which the rule's prediction is correct in relation to all instances the rule applies to [19].

10.3.2 Object-Oriented Model Describing Inter-property Association

To create the basis for a true understanding of the chewing behaviour, as well as to enable the reasoning and learning for applications like the adaptive chewing by robots [12], the properties need to be associated with each other. This is done by exploring the chewing data recorded during chewing experiments. To completely describe the chewing behaviour, a format that allows a formal description of the inter-property relations has to be created. The rule format was chosen as it is simple and effective.

A "rule of inference" is a scheme for constructing valid inferences. It establishes syntactic relationships between a set of formulas called premises (pre-conditions) and an assertion called a conclusion (post-condition). While these syntactic relations are used in the process of inference, new true assertions are arrived at from other already known ones [18].

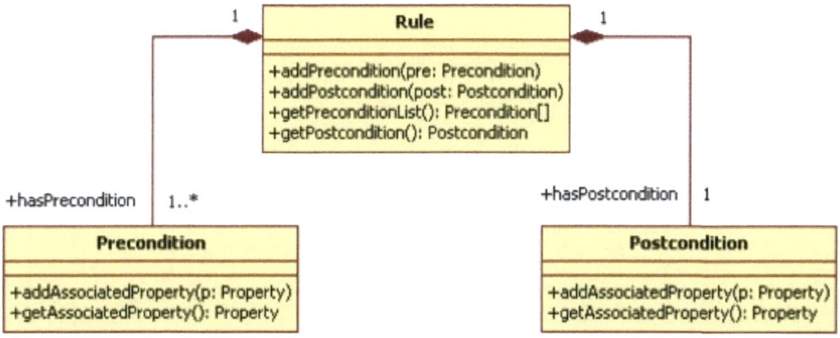

Fig. 10.3 Object-oriented model describing the inter-property association rules

This object-oriented model describing the property associations consists of three classes (Fig. 10.3). The "Rule" class is the central class as it is the container for a number of pre-conditions and one post-condition. The instances of the class "Precondition", attached to one specific rule object, are semantically meant to be connected with the logic "AND" operator to form the pre-condition-expression of the created rule of inference. The classes "Precondition" and "Postcondition" are linked to the property classes because they have a "value" attribute for defining the related property.

10.3.3 Object-Oriented Knowledge Representation

To complete the object-oriented model and to realise the communication between the three parts of the model, a new class, "ChewingSequenceDataSet" is added to the model, as shown in Fig. 10.4. It models the data belonging to the whole

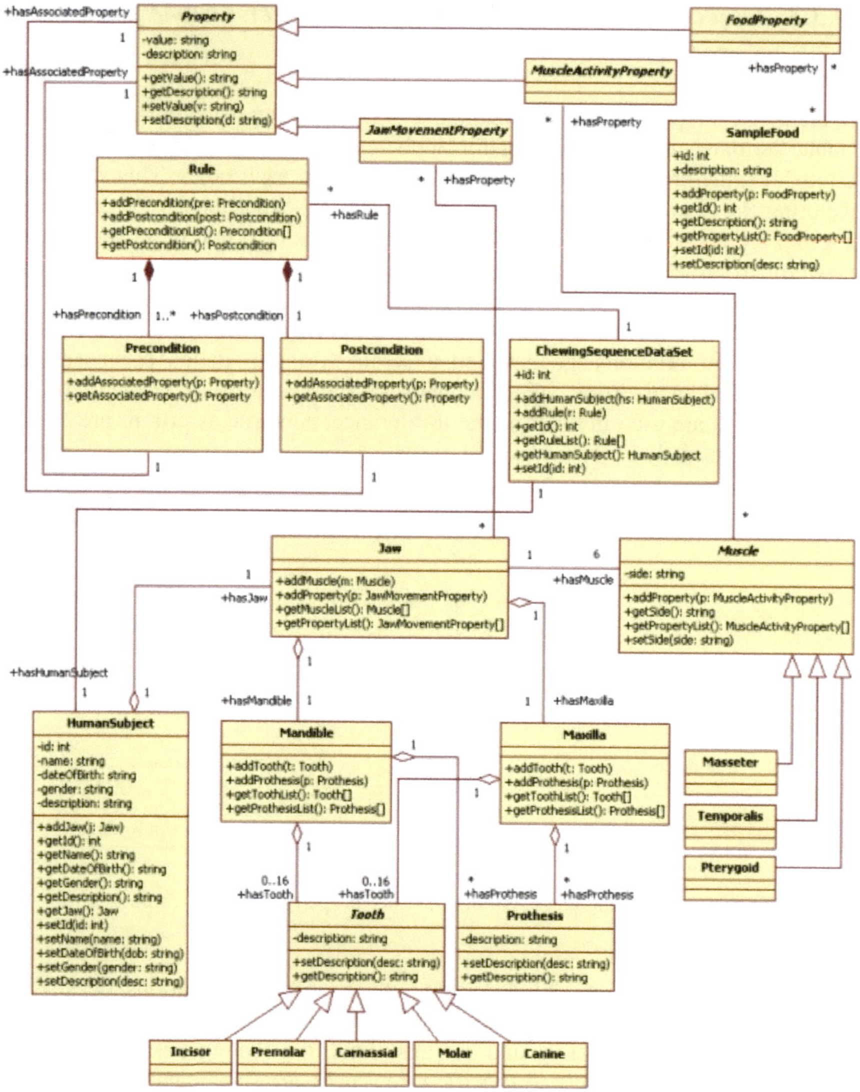

Fig. 10.4 Complete object-oriented model for chewing behaviour

chewing process including information about the human subject performing the chewing and the sample food being chewed. The rules describing the property relations are hidden in the "ChewingSequenceDataSet".

The "Jaw" class, the "Muscle" class and the "SampleFood" class are associated to their corresponding "Property" subclasses, (i.e. "JawMovementProperty", "MuscleActivityProperty" and "FoodProperty"). This adds the capability of attaching any number of property objects to the three classes mentioned above. These attached property objects are then used to store the chewing data.

10.3.4 Data Pre-processing

Huge amounts of datasets contain likely erroneous and correlated records that may have a highly negative effect on the outcome of the data mining process. Therefore, data must be pre-processed before the data mining can be performed. There are three basic data pre-processing techniques; attribute selection, attribute discretisation and data cleansing [17].

Attribute selection consists of removing irrelevant attributes from the datasets. Only the most relevant attributes to a particular problem remain. As most of the data mining algorithms are unable to deal with continuously-varying attributes, attribute discretisation needs to be performed. A pre-determined number of equal intervals are created to host the data. This is so-called equal interval binning. A similar but enhanced discretisation is the equal frequency binning or histogram equalisation, which is characterised by the fact that the same number of instances is inserted into each interval. For inaccurate and missing values (e.g. measurement errors, duplicate data and deliberate errors) and correct but irrelevant values (e.g. outlier) cleansing of the data must be done.

10.3.5 Implementation

The association rules discovered from data mining are intended to substantiate the chewing behaviour depicted in object-oriented model (Fig. 10.4). The object model interacts with the data mining process in the way shown in Fig. 10.5. Guided by the object model of the chewing behaviour, knowledge discovery begins with the gathering of chewing data via experiments, including food properties data, jaw movements and muscular activities data, followed by pre-processing. The data flow then splits up to be inserted into the object database and be used to create a flat database or a relational database. These two data management concepts allow the interaction with a wide variety of data mining tools via a standardised interface. The data mining tool is thus fed with data from a flat database file or a relational database and creates association rules that relate the mastication properties.

The chewing behaviour properties objects and the association rules just discovered complete the object database for the chewing knowledge system. This database can consecutively be used to create any application using the database's interface to access the needed data, which can be implemented by an object-oriented database management system (OODBMS) [23]. An OODBMS is the result of combining object oriented programming principles with database management principles. Object oriented programming concepts such as encapsulation, polymorphism and inheritance are enforced together with database management concepts such as the ACID properties (Atomicity, Consistency, Isolation and Durability), which lead to system integrity, support for an *ad hoc* query language and secondary storage management systems [23, 24]. Several features of OODBMS are comparable to relational database management systems (RDBMS). A class in an OODBMS is similar to a relation or table in RDBMS as well as a tuple in a RDBMS is similar to an instance of a class in an OODBMS.

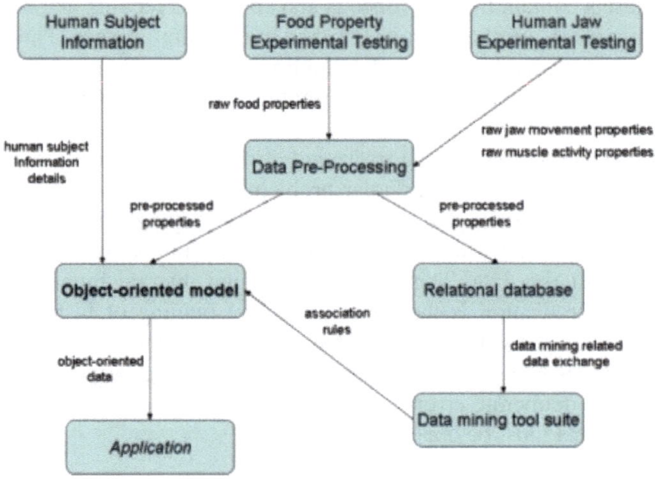

Fig. 10.5 Data flow between object model and data mining process

10.4 Case Study

This case study illustrates a sample application of the concepts developed above. The chewing data used was recorded previously while investigating the effect of texture on masticatory parameters during the chewing of plastic and elastic foods [11]. The WEKA (Waikato Environment for Knowledge Analysis) data mining suite was used to discover a priori unknown knowledge hidden in the chewing database.

10.4.1 Attributes of the Chewing Properties

The attributes making up the chewing datasets are given in Table 10.1. The jaw movement properties involve displacement, velocity, time and frequency. The muscle activity properties are the EMG readings of the "Temporalis" muscle and the "Masseter" muscle on the working and on the non-working side, in terms of the total muscular work for the entire mastication process and the total muscular work during a chewing cycle. The sample foods are characterized by two properties in terms of their group (elastic or plastic) and their hardness.

Table 10.1 Chewing properties for Case study

Jaw movement properties	Muscle activity properties	Food properties
NumberOfBursts (NbBursts) TimeForCompleteMastication (TMAS), s	TotalMuscularWorkForTheSequence (TRTOT), mV/s TotalMuscularWorkPerCycle (TMOY), mV/s	Group of food (Group), *elastic or plastic* Hardness of product (Product), *in terms of compression tests*
AverageCycleFrequency (Freq), 1/s SequenceFrequency = number of cycles/time of complete mastication (sFreq), 1/s MaximalHorizontalAmplitude (LatMax), mm MaximalVerticalAmplitude (AO), mm OpeningVelocity (OV), mm/s ClosingVelocity (CV), mm/s		

10.4.2 Pre-processing of Raw Data for Data Mining

Fig. 10.6 is a screenshot of how the original data looks, a flat database file in Microsoft Excel. It can easily be seen that some of the data records are incomplete and each attribute has its own value and range. So, the raw database needs to be processed before it is operated on by data mining. Fig. 10.7 shows the partial pre-processed database after the incomplete records are removed and each attribute is normalised in the range of 0 to 10.

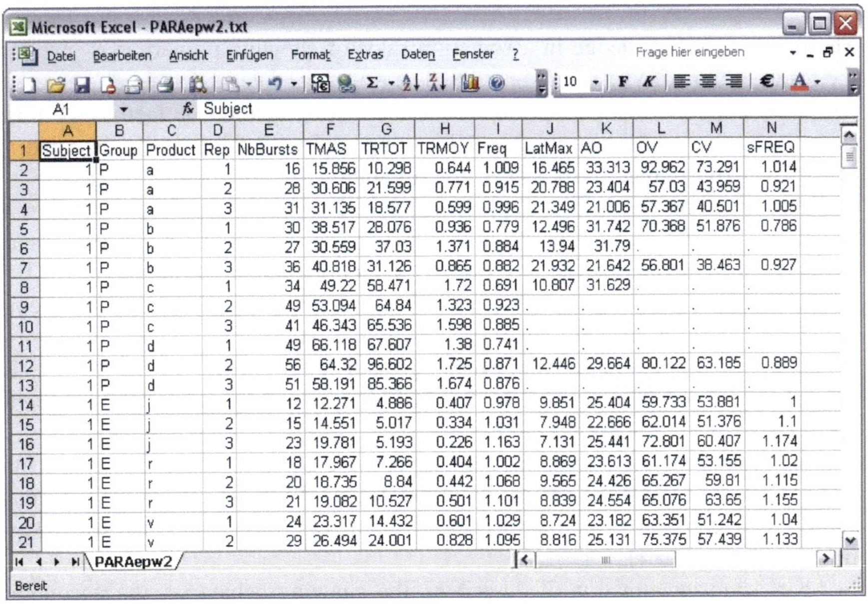

	A	B	C	D	E	F	G	H	I	J	K	L	M	N
1	Subject	Group	Product	Rep	NbBursts	TMAS	TRTOT	TRMOY	Freq	LatMax	AO	OV	CV	sFREQ
2	1	P	a	1	16	15.856	10.298	0.644	1.009	16.465	33.313	92.962	73.291	1.014
3	1	P	a	2	28	30.606	21.599	0.771	0.915	20.788	23.404	57.03	43.959	0.921
4	1	P	a	3	31	31.135	18.577	0.599	0.996	21.349	21.006	57.367	40.501	1.005
5	1	P	b	1	30	38.517	28.076	0.936	0.779	12.496	31.742	70.368	51.876	0.786
6	1	P	b	2	27	30.559	37.03	1.371	0.884	13.94	31.79	.	.	.
7	1	P	b	3	36	40.818	31.126	0.865	0.882	21.932	21.642	56.801	38.463	0.927
8	1	P	c	1	34	49.22	58.471	1.72	0.691	10.807	31.629	.	.	.
9	1	P	c	2	49	53.094	64.84	1.323	0.923	.	.	.	.	.
10	1	P	c	3	41	46.343	65.536	1.598	0.885	.	.	.	.	.
11	1	P	d	1	49	66.118	67.607	1.38	0.741	.	.	.	.	.
12	1	P	d	2	56	64.32	96.602	1.725	0.871	12.446	29.664	80.122	63.185	0.889
13	1	P	d	3	51	58.191	85.366	1.674	0.876	.	.	.	.	.
14	1	E	j	1	12	12.271	4.886	0.407	0.978	9.851	25.404	59.733	53.881	1
15	1	E	j	2	15	14.551	5.017	0.334	1.031	7.948	22.666	62.014	51.376	1.1
16	1	E	j	3	23	19.781	5.193	0.226	1.163	7.131	25.441	72.801	60.407	1.174
17	1	E	r	1	18	17.967	7.266	0.404	1.002	8.869	23.613	61.174	53.155	1.02
18	1	E	r	2	20	18.735	8.84	0.442	1.068	9.565	24.426	65.267	59.81	1.115
19	1	E	r	3	21	19.082	10.527	0.501	1.101	8.839	24.554	65.076	63.65	1.155
20	1	E	v	1	24	23.317	14.432	0.601	1.029	8.724	23.182	63.351	51.242	1.04
21	1	E	v	2	29	26.494	24.001	0.828	1.095	8.816	25.131	75.375	57.439	1.133

Fig. 10.6 Partial raw database

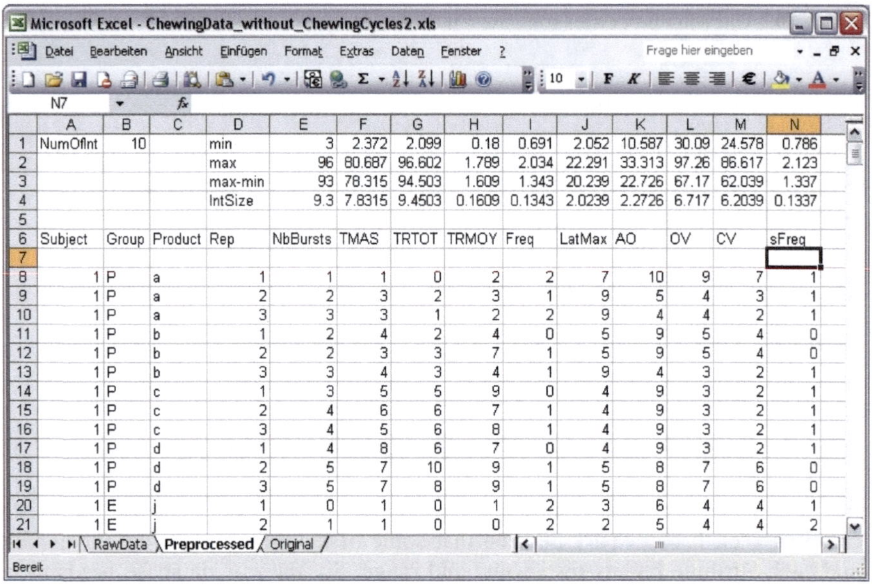

	A	B	C	D	E	F	G	H	I	J	K	L	M	N
1	NumOfInt	10		min	3	2.372	2.099	0.18	0.691	2.052	10.587	30.09	24.578	0.786
2				max	96	80.687	96.602	1.789	2.034	22.291	33.313	97.26	86.617	2.123
3				max-min	93	78.315	94.503	1.609	1.343	20.239	22.726	67.17	62.039	1.337
4				IntSize	9.3	7.8315	9.4503	0.1609	0.1343	2.0239	2.2726	6.717	6.2039	0.1337
5														
6	Subject	Group	Product	Rep	NbBursts	TMAS	TRTOT	TRMOY	Freq	LatMax	AO	OV	CV	sFreq
7														
8	1	P	a	1	1	1	0	2	2	7	10	9	7	1
9	1	P	a	2	2	3	2	3	1	9	5	4	3	1
10	1	P	a	3	3	3	1	2	2	9	4	4	2	1
11	1	P	b	1	2	4	2	4	0	5	9	5	4	0
12	1	P	b	2	2	3	3	7	1	5	9	5	4	0
13	1	P	b	3	3	4	3	4	1	9	4	3	2	1
14	1	P	c	1	3	5	5	9	0	4	9	3	2	1
15	1	P	c	2	4	6	6	7	1	4	9	3	2	1
16	1	P	c	3	4	5	6	8	1	4	9	3	2	1
17	1	P	d	1	4	8	6	7	0	4	9	3	2	1
18	1	P	d	2	5	7	10	9	1	5	8	7	6	0
19	1	P	d	3	5	7	8	9	1	5	8	7	6	0
20	1	E	j	1	0	1	0	1	2	3	6	4	4	1
21	1	E	j	2	1	1	0	0	2	2	5	4	4	2

Fig. 10.7 Partial database after pre-processing

10.4.3 Data Mining Results and Their Interpretation

The discovered knowledge in association rules of chewing properties is given in the result buffer of WEKA in Fig. 10.8. The full set of rules discovered is given in the Appendix. The rules are presented in descending confidence factor and those rules with lower confidence factor values (e.g., rules 42-57) may be regarded as less relevant but subjected to further investigation.

Take rule number 18, for example, which is depicted in an object as shown in Fig. 10.9. It states that IF the "NumberOfBursts" is "1" AND the "TotalMuscularWorkForTheSequence" is "0" THEN "TimeForCompleteMastication" is "1". This rule does not draw a definite conclusion of the type or product of food being associated with certain jaw movement features or level of the muscular activity. When it is inferred in a forward chaining manner [18] with rule number 31, IF "TimeForCompleteMastication" is "1" THEN "GroupOfFood" is "E", a definite conclusion can be drawn explicitly.

Consequently, some of the discovered rules, e.g., rules 1-10, are straightforward to interpret while the others, e.g., rules 18, 20, 43-45, are subjected to further careful investigation or inference before being embedded into the knowledge framework (Fig. 10.4).

Apart from the function of describing the chewing behaviour, the uncovered rules also enable a deeper understanding of how mastication is processed and what it depends on. A very simple interpretation of the aforementioned data mining

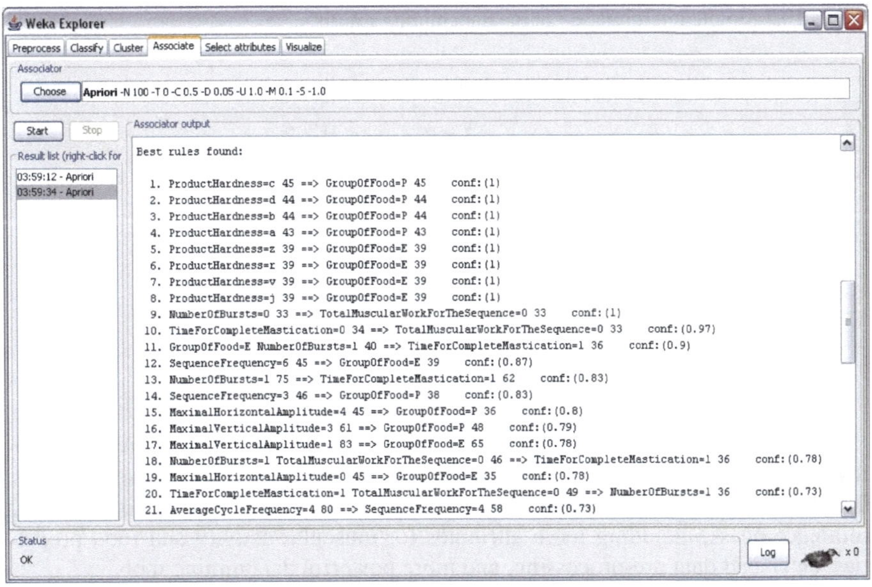

Fig. 10.8 Partial result buffer from WEKA

results is given in Table 10.2. In this case study, the foods are characterised in two groups "elastic" and "plastic" using the jaw movement and muscle activity properties. In this way it might be possible to determine whether a certain sample food being chewed is evaluated as either elastic or plastic by measuring jaw movement and muscular activity.

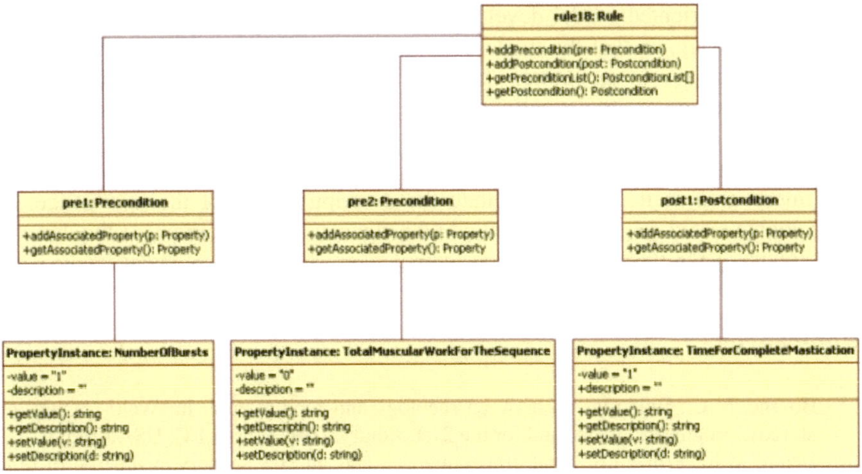

Fig. 10.9 Rule instance example with associated properties

Table 10.2 Simple interpretation of the data mining results

	Group of food	
	Elastic	Plastic
Sequence frequency	6	3–5
Maximal vertical amplitude	1–2	3
Maximal horizontal amplitude	0–2	4
Closing velocity	3–4	3–5
Total muscular work per cycle	1–2	—
Total muscular work for the sequence	0–1	—
Time for complete mastication	1–2	3
Number of bursts	1–3	—

Rules 18 and 31 have a confidence factor of 0.78 and 0.64, respectively, meaning 78% and 64% all instances can be predicted correctly by using those two rules. To achieve a higher confidence level, a few improvements can be made in the data mining process, including more attributes for muscular activity and food properties, enhanced data pre-processing, and more powerful data mining tools.

10.5 Summary

The chewing behaviour was described in an object-oriented knowledge representation involving three objects that define the mastication system and process. The chewing behaviour was substantiated by association rules in a chewing database discovered through data mining. The rules represent the relationships between the foods being chewed, the jaw movements and the muscular activities. A case study was presented to show how data mining was applied to discover the hidden knowledge and the results were interpreted.

The object-oriented model developed, describes the human chewing behaviour in a modular way, so that the model is extendable and evolvable to easily accommodate new objects in the future. This representation is useful in characterising the mastication process systematically, which is needed across fields such as biomedical science, food science and dentistry. It will also promote a chewing robot to chew foods in human way. The discovery of the association rules via data mining is still improvable as it does not include the descriptive part of the knowledge. A more sophisticated case is to be studied with rigorous interpretations of the discovered knowledge in the context of food science, dentistry and biomedical science.

References

1. Bourne, M.C.: Relationship between rheology and food texture. In: Welti-Chanes, et al. (eds.) Engineering and food for the 21st century. CRC Press LLC, USA (2002)
2. Heath, M.R., Prinz, J.F.: Oral processing of foods and the sensory evaluation of texture. In: Rosenthal, A.J. (ed.), pp. 18–26. Aspen Publishers, Inc., Gaithersburg (1999)

3. Martin, M.E.: Geriatric nutrition. In: Chernoff, R. (ed.) Oral health in the elderly. Jones and Barlett Publishers, Sudbury (1999)
4. Palla, S., et al.: Jaw tracking and temporomandibular joint animation. In: McNeil, C. (ed.) Science and practice of occlusion, pp. 365–378. Quintessence Publishing Co, Inc. (1997)
5. Anderson, K., et al.: The effects of bolus hardness on masticatory kinematics. J. Oral Rehab. 29, 689–696 (2002)
6. Agrawal, K.R., et al.: Food properties that influence neuromuscular activity during human mastication. J. Dent. Res. 77, 1931–1938 (1998)
7. Williams, S.H., et al.: Mechanical properties of foods used in experimental studies of primate masticatory function. Am. J. Primatology 67, 329–346 (2005)
8. Peyron, M.A., et al.: Effects of increased hardness on jaw movement and muscle activity during chewing of visco-elastic model foods. Exp. Brain Res. 142, 41–51 (2002)
9. Tortopidis, D., et al.: The variability of bite force measurement between sessions in different positions within the dental arch. J. Oral Rehab. 25, 681–686 (1998)
10. Gerstnert, G.E., Parekh, V.V.: Evidence of sex-specific differences in masticatory jaw movement patterns. J. Dent. Res. 76, 796–806 (1997)
11. Foster, K., et al.: Effect of texture of plastic and elastic foods on masticatory parameters. J. Neurophysiol. 95, 3469–3479 (2006)
12. Xu, W.L., et al.: A robotic model of human masticatory system for reproducing chewing behaviours. IEEE Robotics & Automation Mag. 12, 90–98 (2005)
13. Takanobu, H., et al.: Design of a mastication robot mechanism using a human skull model. In: Proceedings of IEEE/RSJ International Conference on Intelligent Robots and Systems, Yokohama, Japan, pp. 203–208 (1993)
14. Hayashi, H., et al.: A physiological control of chewing-like jaw movement for robotized jaw simulator JSN/ 2A. In: Proceedings of the 22nd Annual EMBS International Conference, Chicago, pp. 730–731 (2000)
15. Winston, P.H.: Artificial intelligence, 3rd edn. Addison-Wesley, Reading (1992)
16. Object Management Group, UML (2006), `http://www.uml.org/` (accessed February 20, 2006)
17. Hannam, A.C.: Jaw muscle structure and function. In: McNeil, C. (ed.) Science and practice of occlusion, pp. 41–49. Quintessence Publishing Co. Inc. (1997)
18. Russell, S., Norvig, P.: Artificial intelligence. Prentice Hall International Editions, New Jersey (1995)
19. Witten, I.H., Frank, E.: Data Mining: practical machine learning tools and techniques, 2nd edn. Morgan Kaufmann Publishers, San Francisco (2005)
20. Roiger, R.J., Geatz, M.W.: Data mining, a tutorial-based primer. Addison-Wesley, Reading (2003)
21. Tanagra, A free data mining software for research and education (2006), `http://eric.univ-lyon2.fr/~ricco/tanagra/` (accessed February 20, 2006)
22. AlphaMiner, An open source data mining platform (2006), `http://www.eti.hku.hk/alphaminer/` (accessed February 20, 2006)
23. Obasanjo, D.: An exploration of object oriented database management systems (2006), `http://www.25hoursaday.com/WhyArentYouUsingAnOODBMS.html` (accessed February 20, 2006)
24. McFarland, G., et al.: Object-oriented database management systems revisited. A report prepared for Air Force Research Laboratory, NY (2006), `http://www.dacs.dtic.mil/techs/oodbms2/` (accessed February 20, 2006)

Appendix

Complete data mining result buffer

=== Run information ===

Scheme: weka.associations.Apriori -N 100 -T 0 -C 0.5 -D 0.05 -U 1.0 -M 0.1 -S -1.0
Relation: ChewingData
Instances: 332
Attributes: 12
GroupOfFood
ProductHardness
NumberOfBursts
TimeForCompleteMastication
TotalMuscularWorkForTheSequence
TotalMuscularWorkPerCycle
AverageCycleFrequency
MaximalHorizontalAmplitude
MaximalVerticalAmplitude
OpeningVelocity
ClosingVelocity
SequenceFrequency
=== Associator model (full training set) ===

Apriori
=======

Minimum support: 0.1
Minimum metric <confidence>: 0.5
Number of cycles performed: 18

Generated sets of large itemsets:

Size of set of large itemsets L(1): 54

Size of set of large itemsets L(2): 61

Size of set of large itemsets L(3): 2

Best rules found:
1. ProductHardness=c 45 ==> GroupOfFood=P 45 conf:(1)
2. ProductHardness=d 44 ==> GroupOfFood=P 44 conf:(1)
3. ProductHardness=b 44 ==> GroupOfFood=P 44 conf:(1)
4. ProductHardness=a 43 ==> GroupOfFood=P 43 conf:(1)
5. ProductHardness=z 39 ==> GroupOfFood=E 39 conf:(1)
6. ProductHardness=r 39 ==> GroupOfFood=E 39 conf:(1)
7. ProductHardness=v 39 ==> GroupOfFood=E 39 conf:(1)
8. ProductHardness=j 39 ==> GroupOfFood=E 39 conf:(1)
9. NumberOfBursts=0 33 ==> TotalMuscularWorkForTheSequence=0 33 conf:(1)
10. TimeForCompleteMastication=0 34 ==> TotalMuscularWorkForTheSequence=0 33 conf:(0.97)
11. GroupOfFood=E NumberOfBursts=1 40 ==> TimeForCompleteMastication=1 36 conf:(0.9)
12. SequenceFrequency=6 45 ==> GroupOfFood=E 39 conf:(0.87)

13. NumberOfBursts=1 75 ==> TimeForCompleteMastication=1 62 conf:(0.83)
14. SequenceFrequency=3 46 ==> GroupOfFood=P 38 conf:(0.83)
15. MaximalHorizontalAmplitude=4 45 ==> GroupOfFood=P 36 conf:(0.8)
16. MaximalVerticalAmplitude=3 61 ==> GroupOfFood=P 48 conf:(0.79)
17. MaximalVerticalAmplitude=1 83 ==> GroupOfFood=E 65 conf:(0.78)
18. NumberOfBursts=1 TotalMuscularWorkForTheSequence=0 46 ==> TimeForCompleteMastication=1 36 conf:(0.78)
19. MaximalHorizontalAmplitude=0 45 ==> GroupOfFood=E 35 conf:(0.78)
20. TimeForCompleteMastication=1 TotalMuscularWorkForTheSequence=0 49 ==> NumberOfBursts=1 36 conf:(0.73)
21. AverageCycleFrequency=4 80 ==> SequenceFrequency=4 58 conf:(0.73)
22. ClosingVelocity=4 69 ==> GroupOfFood=E 50 conf:(0.72)
23. ClosingVelocity=3 57 ==> GroupOfFood=E 41 conf:(0.72)
24. AverageCycleFrequency=6 46 ==> GroupOfFood=E 33 conf:(0.72)
25. TotalMuscularWorkPerCycle=1 68 ==> GroupOfFood=E 48 conf:(0.71)
26. AverageCycleFrequency=3 51 ==> GroupOfFood=P 35 conf:(0.69)
27. AverageCycleFrequency=5 62 ==> SequenceFrequency=5 41 conf:(0.66)
28. SequenceFrequency=5 62 ==> AverageCycleFrequency=5 41 conf:(0.66)
29. TotalMuscularWorkForTheSequence=0 104 ==> GroupOfFood=E 68 conf:(0.65)
30. TimeForCompleteMastication=1 97 ==> NumberOfBursts=1 62 conf:(0.64)
31. TimeForCompleteMastication=1 97 ==> GroupOfFood=E 62 conf:(0.64)
32. TotalMuscularWorkPerCycle=1 68 ==> TotalMuscularWorkForTheSequence=0 43 conf:(0.63)
33. ClosingVelocity=5 57 ==> GroupOfFood=P 36 conf:(0.63)
34. SequenceFrequency=4 92 ==> AverageCycleFrequency=4 58 conf:(0.63)
35. SequenceFrequency=5 62 ==> GroupOfFood=P 39 conf:(0.63)
36. TimeForCompleteMastication=3 56 ==> GroupOfFood=P 35 conf:(0.63)
37. TotalMuscularWorkForTheSequence=1 84 ==> GroupOfFood=E 52 conf:(0.62)
38. MaximalVerticalAmplitude=2 60 ==> GroupOfFood=E 37 conf:(0.62)
39. NumberOfBursts=1 75 ==> TotalMuscularWorkForTheSequence=0 46 conf:(0.61)
40. MaximalHorizontalAmplitude=2 80 ==> GroupOfFood=E 48 conf:(0.6)
41. AverageCycleFrequency=5 62 ==> GroupOfFood=P 37 conf:(0.6)
42. OpeningVelocity=3 80 ==> GroupOfFood=P 47 conf:(0.59)
43. NumberOfBursts=1 TimeForCompleteMastication=1 62 ==> TotalMuscularWorkForTheSequence=0 36 conf:(0.58)
44. GroupOfFood=E TimeForCompleteMastication=1 62 ==> NumberOfBursts=1 36 conf:(0.58)
45. NumberOfBursts=1 TimeForCompleteMastication=1 62 ==> GroupOfFood=E 36 conf:(0.58)
46. TimeForCompleteMastication=2 81 ==> GroupOfFood=E 47 conf:(0.58)
47. OpeningVelocity=4 76 ==> GroupOfFood=P 44 conf:(0.58)
48. AverageCycleFrequency=4 80 ==> GroupOfFood=P 46 conf:(0.57)
49. TotalMuscularWorkPerCycle=2 64 ==> GroupOfFood=E 36 conf:(0.56)
50. NumberOfBursts=2 64 ==> GroupOfFood=E 36 conf:(0.56)
51. MaximalHorizontalAmplitude=1 70 ==> GroupOfFood=E 39 conf:(0.56)
52. NumberOfBursts=3 73 ==> TimeForCompleteMastication=2 40 conf:(0.55)
53. TotalMuscularWorkForTheSequence=1 84 ==> TimeForCompleteMastication=1 46 conf:(0.55)
54. NumberOfBursts=1 75 ==> GroupOfFood=E 40 conf:(0.53)
55. SequenceFrequency=4 92 ==> GroupOfFood=P 48 conf:(0.52)
56. NumberOfBursts=3 73 ==> GroupOfFood=E 37 conf:(0.51)
57. TimeForCompleteMastication=1 97 ==> TotalMuscularWorkForTheSequence=0 49 conf:(0.51)

Index

GPSR Compliance

The European Union's (EU) General Product Safety Regulation (GPSR) is a set of rules that requires consumer products to be safe and our obligations to ensure this.

If you have any concerns about our products, you can contact us on ProductSafety@springernature.com

In case Publisher is established outside the EU, the EU authorized representative is:

Springer Nature Customer Service Center GmbH
Europaplatz 3
69115 Heidelberg, Germany

Batch number: 10755329

Printed by Printforce, the Netherlands